Radar Interferometry

Remote Sensing and Digital Image Processing

VOLUME 2

Series Editor:

Freek van der Meer, *International Institute for Aerospace Survey and Earth Sciences, ITC, Division of Geological Survey, Enschede, The Netherlands and Department of Applied Earth Sciences, Delft University of Technology, The Netherlands*

Editorial Advisory Board:

RADAR INTERFEROMETRY

Data Interpretation and Error Analysis

by

RAMON F. HANSSEN
Delft University of Technology, The Netherlands

KLUWER ACADEMIC PUBLISHERS
DORDRECHT / BOSTON / LONDON

A C.I.P. Catalogue record for this book is available from the Library of Congress.

ISBN 0-7923-6945-9

Published by Kluwer Academic Publishers,
P.O. Box 17, 3300 AA Dordrecht, The Netherlands.

Sold and distributed in North, Central and South America
by Kluwer Academic Publishers,
101 Philip Drive, Norwell, MA 02061, U.S.A.

In all other countries, sold and distributed
by Kluwer Academic Publishers,
P.O. Box 322, 3300 AH Dordrecht, The Netherlands.

Printed on acid-free paper

Printed in the Netherlands.

Contents

Preface

This book is the product of five and a half years of research dedicated to the understanding of radar interferometry, a relatively new space-geodetic technique for measuring the earth's topography and its deformation. The main reason for undertaking this work, early 1995, was the fact that this technique proved to be extremely useful for wide-scale, fine-resolution deformation measurements. Especially the interferometric products from the ERS-1 satellite provided beautiful first results—several interferometric images appeared as highlights on the cover of journals such as Nature and Science. Accuracies of a few millimeters in the radar line of sight were claimed in semi-continuous image data acquired globally, irrespective of cloud cover or solar illumination. Unfortunately, because of the relative lack of supportive observations at these resolutions and accuracies, validation of the precision and reliability of the results remained an issue of concern.

From a geodetic point of view, several survey techniques are commonly available to measure a specific geophysical phenomenon. To make an optimal choice between these techniques it is important to have a uniform and quantitative approach for describing the errors and how these errors propagate to the estimated parameters. In this context, the research described in this book was initiated. It describes issues involved with different types of errors, induced by the sensor, the data processing, satellite positioning accuracy, atmospheric propagation, and scattering characteristics. Nevertheless, as the first item in the subtitle "Data Interpretation and Error Analysis" suggests, data interpretation is not always straightforward. Especially when the interferometric data consist of a superposition of topography, surface deformation, and atmospheric signal, it is important to recognize the characteristics of these signals to make a correct interpretation of the data. In this work, I hope to contribute to improved error analysis and data interpretation for radar interferometry.

This book owes significantly to the people I had the pleasure to work with during the past several years. First of all, I would like to thank Roland Klees for making it all possible, for supporting me to work abroad for such a long time, and for his supervision. My room mates Bert "Mr. Doris" Kampes and Stefania Usai provided a nice working environment and enough food for lengthy discussions.

I learned a lot from the MSc-students, whom I had the pleasure to advise during the last couple of years. Jaron Samson, Yvonne Dierikx-Platschorre, Ronald Stolk,

Claartje van Koppen, and Rens Swart, your work has contributed significantly to the results described in this book. Appendix A is based on ideas to combine GPS data with SAR interferograms for the correction or estimation of atmospheric error. The GPS data processing and analysis in this appendix was performed by Ronald Stolk and André van der Hoeven, supported by Hans van der Marel and Boudewijn Ambrosius.

The meteorological interpretation of the radar interferograms was only possible due to the close collaboration with several meteorologists. At the Royal Netherlands Meteorological Institute (KNMI) Sylvia Barlag, Frans Debie, Arnout Feijt, and many others participated in these analyses. Arnout Feijt developed the Meteosat water vapor channel parameterization using the GPS time series for one of the case studies of chapter 6, which enabled the Meteosat-InSAR validation. I am highly indebted to Tammy Weckwerth at the National Center for Atmospheric Research in Boulder, Colorado, who devoted much of her time to the interferogram interpretation leading to our Science paper. Our discussions in Boulder, as well as our email-battles improved my understanding of meteorology significantly. Stick Ware, thanks for making the link! I was introduced into the wonders of SAR amplitude imagery by Susanne Lehner, Ad Stoffelen, and Ilona Weinreich. The conformity between the observed wind patterns and the water vapor distribution provided a consistent support for the interferogram interpretation.

During the years we performed several common research projects with the Physics and Electronics Laboratory of TNO. Marco van der Kooij (now at Atlantis Scientific) and Erik van Halsema introduced me to interferometry and the Groningen land subsidence project. In a later stage, Jos Groot and Roel van Bree had a significant influence on my understanding of airborne interferometry (PHARUS) and in the Tianjin land subsidence project with EARS. At the Survey Department of Rijkswaterstaat, Erik de Min, Jur Vogelzang, and Yvonne Dierikx actively pursued the "practical relevance" issue, which kept me on track from time to time.

During my stay at the Institut für Navigation at Stuttgart University, Karl-Heinz Thiel, Jürgen Schmidt, and dr. Wu enabled me to obtain valuable experience in interferometric data processing. My roots in the understanding of radar lie at DLR-DFD in Oberpfaffenhofen. I am grateful to Richard Bamler, Michael Eineder, Nico Adam, and many others for giving me the opportunity to work with them, which was pleasing in many ways. At Stanford University I owe much to Howard Zebker for his support and hospitality. Discussions with him, Paul Segall, Sjonni Jónsson, and Weber Hoen were always enlightening. Now at the University of Hawaii, Falk Amelung was my tutor in geophysics. Falk, aloha for the Cerro Prieto Friday afternoon analysis. I am indebted to Ewa Glowacka at CICESE, Ensenada, and CFE, Mexico for their support in the Cerro Prieto study. David Sandwell is gratefully acknowledged for inviting me to visit Scripps Institution of Oceanography, which was a great experience.

Many people contributed to the GISARE field experiment, including the meteorological, the leveling, the GPS, and the SAR processing part. Installing corner reflectors in frozen ground at $-15°C$ requires special skills, we had a great team. Although the GPS-SAR experiment during the solar eclipse in 1999 failed to produce an observable ionospheric signal in the interferograms, it provided good experience thanks to all people contributing.

Special thanks go to Herman Russchenberg for his droplets, Michel Decré for fig. 2.1A, Riccardo Riva and Bert Vermeersen for the Izmit earthquake analysis, and Remko Scharroo for his precise satellite orbits.

The manuscript of this book greatly benefited from the meticulous reading of my defense committee: Richard Bamler, Philipp Hartl, David Sandwell, Tammy Weck-werth, Jacob Fokkema, Peter Teunissen, and Roland Klees, thank you for all your critical comments and valuable improvements.

Freek van der Meer, series editor, and Petra van Steenbergen at Kluwer Academic Publishers are gratefully acknowledged for making the monograph publication of this work possible in the *Remote Sensing and Digital Imaging Processing* series.

This book could not have been completed without the research funds of the Cornelis Lely Stichting, the Survey Department of Rijkswaterstaat, the European Space Agency, the Fulbright program, the Universiteitsfonds, and Shell International Production and Exploration B.V. All ERS SAR data were kindly provided by ESA. Meteorological data were made available by the Royal Netherlands Meteorological Institute (KNMI). Maps and visualizations have been implemented in GMT (Wessel and Smith, 1998) and in Matlab™.

Finally, I would like to thank Ardis for her understanding and support and for all the time I stole from her. I'm excited about the future with the three of us.

Ramon Hanssen
Delft, December 2000

Summary

Within a decade, imaging radar interferometry has matured to a widely used geodetic technique for measuring the topography and deformation of the earth. In particular the analysis and interpretation of the interferometric data requires a thorough understanding of the principles of the technique, the (potential) error sources, and the error propagation. This book reviews the basic concepts of radar, imaging radar, and radar interferometry, and revisits the processing procedure for obtaining interferometric products such as a digital elevation model or a deformation map. It describes spaceborne repeat-pass radar interferometry using a linear or Gauss-Markoff model formulation, which relates the interferometric observations to the unknown geophysical parameters. The stochastic part of the model describes the dispersion of the observations in terms of variances and covariances. Especially the influence of spatially correlated errors as induced by the satellite orbits and by atmospheric path delay are discussed. Mathematical models are presented that describe the spatial variability in the interferometric phase due to turbulent mixing of atmospheric refractivity and due to vertical atmospheric stratification. Using 52 SAR acquisitions, a systematic inventory of the characteristics of atmospheric signal in the radar interferograms is performed, using complementary meteorological data for the interpretation. Scaling characteristics are observed, which can be conveniently used to describe the power spectrum and covariance function of the atmospheric signal. The final variance-covariance matrix for the radar interferometric data is presented, including these spatially varying error sources. A number of case studies on deformation monitoring, such as land subsidence, earthquake deformation, and artificial reflector movement serve as examples of the application of interferometry and its error sources. The feasibility of the technique for practical geodetic applications is evaluated in relation to the geophysical phenomena of interest, yielding rules-of-thumb for its utilization. Finally, a novel application of interferometry for atmospheric studies, termed Interferometric Radar Meteorology, is presented and discussed. Maps of the vertically integrated water vapor distribution during the radar acquisitions can be obtained with a fine spatial resolution and a high accuracy. Several demonstration studies of this meteorological application are presented.

Nomenclature

List of acronyms

1D	One-Dimensional
2D	Two-Dimensional
ALD	Azimuth Look Direction
ALOS	Advanced Land Observing Satellite
AMI	Active Microwave Instrument
ASI	Agenzia Spaziale Italiana (Italian Space Agency)
AVHRR	Advanced Very High Resolution Radiometer
BTTB	Block-Toeplitz Toeplitz-Block
CFE	Comisión Federal de Electricidad, Mexico
CLARA'96	Clouds and Radiation experiment 1996
CMOD4	C-band Model 4 (empirically derived model function to relate normalized radar cross-section with wind speed and direction)
CNES	Centre National D'Etudes Spatiales
CPGF	Cerro Prieto Geothermal Field
CSA	Canadian Space Agency
CTRS	Conventional Terrestrial Reference System
DEM	Digital Elevation Model
DEOS	Delft Institute for Earth-Oriented Space Research
DFD	Deutsches Fernerkundungsdatenzentrum (German Remote Sensing Data Center)
DIAL	Differential Absorption Lidar
DLR	Deutsches Zentrum für Luft- und Raumfahrt e.V. (German Aerospace Center)
DPWV	Differential Precipitable Water Vapor
DUT	Delft University of Technology
ERS	European Remote Sensing Satellite
ESA	European Space Agency
fBm	Fractional Brownian Motion
FFA	Far Field Approximation
FFT	Fast Fourier Transform
FM	Frequency Modulated
GISARE	Groningen Interferometric SAR Experiment
GNSS	Global Navigation Satellite Systems
GPS	Global Positioning System
IEEE	Institute of Electrical and Electronics Engineers
IERS	International Earth Rotation Service
IGS	International GPS Service for Geodynamics
InSAR	Interferometric SAR / SAR interferometry
IPW	Integrated Precipitable Water
I/Q	In-Phase (real), Quadrature (imaginary)
IRM	Interferometric Radar Meteorology

ITRF	IERS Terrestrial Reference Frame
JERS	Japanese Earth-Resources Satellite
JPL	Jet Propulsion Laboratory
LOS	Line of Sight
LT	Local Time
MCF	Minimal Cost Flow
MIT	Massachusetts Institute of Technology
MSC	Mesoscale Shallow Convection
NAM	Nederlandse Aardolie Maatschappij B.V.
NAP	Normaal Amsterdams Peil (Dutch vertical reference datum)
NASA	National Aeronautics and Space Administration
NASDA	National Aeronautics and Space Development Agency (Japan)
NDVI	Normalized Differential Vegetation Index
NOAA	National Oceanic and Atmospheric Administration
PAF	Processing and Archiving Center
PDF	Probability Density function
PLW	Precipitable Liquid Water
PRARE	Precise Range and Rate Equipment
PRF	Pulse Repetition Frequency
PWV	Precipitable Water Vapor
radar	Radio detection and ranging
RAR	Real Aperture Radar
RCS	Radar Cross Section
RH	Relative Humidity
RMS	Root Mean Square
RRF	Range Reference Function
SAR	Synthetic Aperture Radar
s/c	Spacecraft
SIR	Shuttle Imaging Radar, Spaceborne Imaging Radar
SLAR	Side Looking Airborne Radar
SLC	Single-Look Complex
SLR	Satellite Laser Ranging / (Side Looking Radar)
SNR	Signal-to-Noise Ratio
SPOT	System Probatoire d'Observation de la Terre
SRTM	Shuttle Radar Topography Mission
SWST	Sampling Window Start Time
TEC	Total Electron Content
TECU	Total Electron Content Unit
TID	Traveling Ionospheric Disturbance
UTC	Universal Coordinated Time
VLBI	Very Long Baseline Interferometry
VTEC	Vertical Total Electron Content
WGS84	World Geodetic Survey 1984
WV	Water Vapor
X-SAR	X-band Synthetic Aperture Radar

List of symbols

^{0}	Initialization value, usually determined for reference surface
$\mathbf{A}$	Design matrix
$\mathbf{A}_k$	Part of the design matrix corresponding to observation k
B	Baseline
$B_{\parallel}$	Parallel baseline
$B_{\perp}$	Perpendicular baseline
$B_{\perp}^0$	Perpendicular baseline for range to resolution cell on the reference surface
$B_{\perp,\mathrm{crit}}$	Critical (perpendicular) baseline
B_h	Horizontal baseline

B_v	Vertical baseline
B_A	Azimuth time bandwidth
B_R	Range bandwidth
B_{Dop}	Doppler bandwidth
c	Speed of light $= 299792458$ m/s
C	Structure function coefficient
$\mathbf{C_{\hat{x}}}$	Variance-covariance matrix of the estimated parameters
$\mathbf{C_{\varphi}}$	Variance-covariance matrix
$\mathbf{C_s}$	Variance-covariance matrix of atmospheric delay
D_a	Antenna width (m)
$D\{.\}$	Dispersion operator, second moment, or variance-covariance matrix
D_i	Fractal dimension for a i-dimensional signal
D_k	Slant deformation for observation k
D_{χ}	Structure function of parameter χ
$E\{.\}$	Expectation operator, ensemble average, first moment, or mean
e	Partial pressure of water vapor (hPa)
f	Frequency
f_0	Carrier (center) frequency
f_{D}	Doppler frequency
f_{DC}	Doppler centroid frequency
f_P	Pulse repetition frequency (PRF) (cycles/m)
f_{ϕ}	Fringe frequency
h_p	Altitude of point p
h_{sp}	Height of the sub-ionospheric point (m)
$h_{2\pi}$	Height ambiguity (m)
H	Altitude; Haussdorff measure
H_k	Topographic height for observation k
I	Integrated Precipitable Water
k	Wavenumber
L	Effective number of looks
L_a	Antenna length (m)
$\mathbf{L1,L2}$	GPS carrier phase frequencies (L1=1.57542 GHz, L2=1.22760 GHz)
$\mathbf{L3}$	GPS ionospheric-free linear combination
m	Number of observations
n	Number of parameters (unknowns)
n_e	Electron content (m^{-3})
$n_{\shortparallel}$	Parallel error baseline
$n_{\perp}$	Perpendicular error baseline
N	Refractivity, Geoid height
pdf	Probability density function
P	Total atmospheric pressure (hPa)
P_{χ}	Power spectrum of parameter χ
$\mathbf{Q_{\varphi}}$	Real positive-semidefinite cofactor matrix
R	(Slant) range (m)
R_1	Slant range between reference (master) satellite and resolution cell (m)
R_2	Slant range between secundary (slave) satellite and resolution cell (m)
s_c	Chirp rate or Frequency slope (s^{-2})
$S_k^{t_i}$	Slant atmospheric delay for pixel k at t_i (m)
t	Time (s)
T	Temperature (K)
v	Velocity (m/s)
v	Hermitian product $y_1 y_2^*$
$v_{s/c}$	Velocity of spacecraft (m/s)
w_k	Integer ambiguity number
W	Liquid water content (g m^{-3})
$W\{.\}$	Wrapping operator $= \text{mod}\{\phi + \pi, 2\pi\} - \pi$
W_a	Radar footprint width in azimuth (3dB)
$\mathbf{x}$	Vector of parameters
$\hat{\mathbf{x}}$	Best linear unbiased estimator of the unknown parameters $\mathbf{x}$

$\vec{X}_{s/c}$	State vector of spacecraft
y_i	Complex observation (phasor) i
$Z(u,v)$	Amplitude spectrum of interferogram (u and v are wavenumbers)
$Z'(u,v)$	Amplitude spectrum of filtered interferogram
α	Baseline orientation angle; Spectral exponent
β	Error baseline orientation angle; Spectral index
β_r	Angular beamwidth in range direction (rad)
β_a	Angular beamwidth in azimuth direction (rad)
γ	Coherence (complex)
γ_{geom}	Coherence reduction due to geometric (baseline) decorrelation
γ_{DC}	Coherence reduction due to Doppler centroid decorrelation
$\gamma_{\mathrm{temporal}}$	Coherence reduction due to temporal decorrelation
γ_{vol}	Coherence reduction due to volume decorrelation
$\gamma_{\mathrm{processing}}$	Coherence reduction due to processing induced decorrelation
$\gamma_{\mathrm{thermal}}$	Coherence reduction due to thermal noise
Γ_χ	Semi-variogram of parameter χ
$\partial\theta$	Change in look angle
Δ_a	Focused azimuth resolution
Δ_r	Slant range resolution
ΔR	Slant range difference
$\Delta R_{\mathrm{defo},p}$	Slant range difference due to deformation
ε	Vector of observation errors
ζ	Local terrain slope w.r.t. reference surface (positive towards the radar): $\zeta = \theta_{\mathrm{inc}} - \theta_{\mathrm{loc}}$
θ	Look angle
θ_{inc}	Incidence angle, defined with respect to the global vertical at the scene
θ_{loc}	Local incidence angle, defined with respect to the local vertical at the scene: $\theta_{\mathrm{loc}} = \theta_{\mathrm{inc}} - \zeta$
ϑ	Reference phase (rad)
κ	$4\pi/\lambda$
λ	Radar wavelength
Λ	Geographic longitude in WGS84
μ	Mean
ρ	Local earth radius (m); Horizontal distance between two resolution cells.
ρ_χ	Autocorrelation function of parameter χ
σ^2	a priori variance factor
σ^0	"Sigma nought" or Radar Cross Section (RCS). Average power reflectivity or scattering coefficient per unit surface area
τ	Pulse length
τ_c	Effective pulse length
ϕ	Absolute interferometric phase, including reference phase
ϕ^{w}	Wrapped interferometric phase, including reference phase
Φ	Geographic latitude in WGS84
ϕ_r	Off-center beam angle in range direction (rad)
ϕ_a	Off-center beam angle in azimuth direction (rad)
ϕ_s	Squint angle (rad)
ϕ_N	Phase noise (rad)
φ	Absolute interferometric phase, reference phase subtracted
φ^{w}	Wrapped interferometric phase, reference phase subtracted
φ_k	Stochastic phase observation
$\underline{\varphi}$	Real stochastic vector of observations
ψ	Phase (rad)
ψ_{scat}	Scattering contribution to the phase value (rad)
Ω_k	Ratio between $B_\perp$ of deformation pair and $B_\perp$ of topographic pair
$\mathbb{N}$	Set of natural numbers
$\mathbb{R}$	Set of real numbers
$\mathbb{Z}$	Set of integer numbers
$\mathbb{C}$	Set of complex numbers
$\mathscr{F}$	Fourier transform

Chapter 1

Introduction

1.1 Motivation

The geodetic interpretation of measurements obtained by spaceborne repeat-pass radar interferometry is limited by unmodeled error sources. The research reported in this study is an attempt to situate the technique in a standard geodetic framework, analyze the characteristics of the errors, and improve data interpretation.

1.2 Background

Advances in geodesy during the last couple of decades can be characterized by a number of developments. First, there is a broadening of the spatial scales over which measurements can be performed. The use of artificial satellites boosted the definition and use of global reference systems and enabled the mapping of large areas on earth. Optical remote sensing from spaceborne platforms triggered fast developments in geodesy, especially using stereographic techniques, whereas point positioning entered a new era using global navigation satellite systems (GNSS).

Recognition of the earth's dynamics characterizes a second development. General acceptance of global plate tectonics does not only make every measurement a function of time, it also undermines the definition of reference systems (McCarthy, 1996). On more local scales there is considerable interest in monitoring deformations induced by earthquakes, volcanoes, glacier dynamics, post-glacial rebound, and anthropogenic deformation due to the exploration of mineral resources, geothermal energy, and water. Better understanding of these processes requires regular repeat measurements, high accuracy, and a fine spatial resolution. These requirements are motivated by considerations of public safety and hazard monitoring, sea-level rise and global warming, and controversial issues such as the underground storage of nuclear material. In this context, geodetic observations serve as crucial source of information for geophysical interpretation. Regarding deformations caused by human exploration legal and financial aspects play an important role (Taverne, 1993).

The accuracy of geodetic observations marks a third development. As a result of the increased number of alternative techniques the quality, in terms of the precision and reliability, needs to be balanced with the specific goals of the measurement campaign. Important elements are financial constraints, accessibility of the area, the duration of the measurement campaign, monumentation, and repeatability.

As a provider of geometric and gravimetric information, geodesy contributes to geophysics, geodynamics, and oceanography. Moreover, in a more indirect way space-geodetic observations can contribute to atmospheric physics. The influence of atmospheric refraction on the propagation velocity of electromagnetic waves is regarded as one of the most important error sources for distance measurement, one of the most elementary geodetic observables. Although space-geodetic techniques such as very long baseline interferometry (VLBI), radar altimetry, and global navigation systems such as GPS, were not designed for atmospheric studies, they contributed considerably to the global understanding of the atmosphere. Currently, GPS observations are used on a routine basis by meteorologists, e.g., for the observation of water vapor distributions.

Radar interferometry is perfectly embedded in this line of geodetic developments, combining the characteristics of large-scale imaging and quantitative observations of distances and angles. In terms of spatial scales, spaceborne radar interferometers are able to collect observations distributed over the entire earth. Quantitative ranging observations are currently collected over a swath width in the order of 100–500 km, with a resolution of about 20 m. As opposed to optical sensors, the active radar observations can be collected at nighttime, over areas with limited contrast, and through cloud cover. Using a *single-pass* interferometric configuration, in which two physical antenna's are used to collect the radar echoes, accuracies for elevation observations are in the order of 6 m for spaceborne sensors (Bamler and Hartl, 1998; Werner et al., 1993; Adam et al., 1999; Jordan et al., 1996), which is comparable to the optical techniques (Baltsavias, 1993; SPOT image company, 1995).

Nevertheless, the most spectacular application of radar interferometry is the observation of dynamic processes, using a *repeat-pass* configuration, in which two radar images are acquired by a single antenna, revisiting the area after a specific time interval. Coherent deformation of the earth's surface within the radar swath can be observed as a fraction of the radar wavelength, resulting in sub-cm ranging accuracies in the radar line-of-sight. The most stringent requirement for this application is that the archive of radar data contains acquisitions prior to the deformation event.

The background of this study was formed by the recognition of these capabilities for radar interferometry and the desire to consider the technique—which was conceived in an electrical engineering environment—from a more geodetic point of view. Although convincing demonstrations of the technique had been published, it was also known, or at least expected, that there were limitations affecting the applicability and accuracy. For example, for applications of spaceborne interferometry for deformation studies in the Netherlands, previous studies identified main problems due to the changing characteristics of the surface (*temporal decorrelation*) and atmospheric signal delay (van der Kooij et al., 1995). This study discusses these and other error sources and the way the errors propagate to the final parameters of interest. A conceptual approach is presented to regard the data using standard geodetic methodology.

The analysis of the atmospheric delay in the radar data has been one of the focus points of this study. Especially the results of ESA's ERS "tandem mission," which provided extremely coherent repeat-pass interferometric data, enabled an extensive data set of "atmospheric" interferometric products. The close collaboration with meteorologists lead to a remarkable result: apart from regarding atmospheric delay as a source of error for deformation studies, it could be analyzed and interpreted as a source of information for atmospheric studies. Another result from this new application of interferometry is its use in improving stochastic models for the atmospheric influence on other space-geodetic techniques such as GPS, radar altimetry, or VLBI, especially regarding small-scale spatial disturbances.

1.3 Problem formulation, research objectives and limitations

The achievements of InSAR in terms of deformation monitoring and topographic mapping have clearly demonstrated the unsurpassed capabilities of the technique. Nevertheless, this does not necessarily prove its feasibility for any potential application. Early applications of InSAR may be characterized as "technology driven," implying that its main goal was the demonstration of the capabilities. Only in recent years, attention is shifting to "application driven" developments. The decision whether or not to use InSAR for a specific application is not always straightforward. There are a number of problems that influence this decision.

SNR The signal-to-noise ratio[1] of the parameters under consideration is not always known. Many demonstration studies have focused on areas where the deformation or topography signal was dominant and error sources where marginal. Without denying their fundamental importance, these studies may often be characterized as rather opportunistic and ad-hoc. As application-oriented research shifts to problems with a much more critical SNR, the demand on identifying and modeling the error sources increases. The main error sources that influence the final results are (i) the contribution of atmospheric signal delay, (ii) the interferometric decorrelation due to temporal and geometric scattering characteristics, (iii) the unknown phase ambiguity number, and (iv) errors in supplementary information such as the reference satellite orbits and elevation models. Models of the error sources and their propagation into the parameters of interest have been limited to single point statistics, largely ignoring the quantitative description of covariance in the data, or even to a qualitative interpretation of possible artifacts only. A full stochastic model, expressed in terms of a variance-covariance matrix of the observations has not yet been defined.

Data availability The regular acquisition of spaceborne SAR data is influenced by power consumption considerations, downlink or data storage capabilities, and possible conflicting user requirements on the satellite's multi-instrument payload. Furthermore, orbit maintenance requirements for topographic map-

---

[1]Here the term SNR is used in an abstract sense, implying the relation between the parameters of interest and all disturbing factors, whether deterministic or stochastic.

ping are fundamentally different from those for deformation mapping. Other factors include the compatibility of polarization between the interferometric pairs, the used incidence angles, and the radar frequency. These restrictions limit the amount of useful interferometric data sets.

Ground truth Due to the unknown phase ambiguity number and the limited knowledge of the satellite's position, radar interferograms are essentially relative measurements. In order to relate these measurements to a reference datum, a priori information is required. Control points or absolute deformation measurements from other geodetic techniques are not always available, which influences the quality of the final results.

Surplus value The decision to use InSAR, instead of any other geodetic technique, for a specific application will always be weighted based on characteristics such as the spatial extent of the signal under study, the accessibility of the terrain, desired accuracy and reliability, repeatability, processing speed, uniqueness of the interpretation, and cost-effectiveness. This means that all these issues need to be known before the decision to apply a specific technique is made.

In order to improve the availability and value of interferometric products, these problems need to be considered.

Research objectives

The four categories that influence the decision to use InSAR for a specific geophysical application (SNR, data availability, ground truth, and surplus value) comprise the main limitations of the technique. It is beyond the scope of this study to comment on instrument or spacecraft design and operation considerations that determine data availability. Furthermore, when the radar interferometric observations and derived parameters are clearly defined as being relative, the availability and use of ground truth can be ignored. This study will focus on the first and the last item: the signal-to-noise ratio and the surplus value. The general problem statement addressed in this study is;

> How can the interpretation and analysis of repeat-pass spaceborne radar interferometric data be improved in a systematic way via a model-based quantification of the error sources?

From this formulation of the problem, four specific research questions are derived that give further direction to the treatment in this study.

1 How can the formation and the quality of a radar interferogram and its derived interferometric products be parameterized?

The formation of a radar interferogram is a procedure that involves the raw radar data acquisition and down-link from the spacecraft to the ground station on earth, the evaluation of the data quality, and the data processing using a series of specific operations. To enable a critical evaluation of the quality of each link in this chain, it is evident that their physical and mathematical properties need to be described.

The final result is a two-dimensional image representing some geophysical signal. A parameterization of the quality of this signal is not always straightforward, as errors with different characteristics will affect the result. Some errors will result in a bias of all interferometric phase values, some will have varying characteristics depending on the position in the image and some affect a single resolution cell only, while others increase the variance of all resolution cells. Especially resolving the superposed signals of topography, deformation, and atmosphere poses a significant problem in this respect.

A conceptual approach would be to express a radar interferogram in a mathematical model, consisting of a functional and a stochastic part. The functional model would relate the basic phase observations to the parameters of interest, while the stochastic model would describe the quality of the observations. As a next step, it could then be investigated how the observations in several interferograms can be combined in a single adjustment model.

2 Which error sources limit the quality of the interferogram and how does their influence propagate in the interferometric processing?

In order to make quantitative statements on the quality of an interferometric product, it is necessary to analyze the total chain from data acquisition via data processing to interpretation, and identify the error sources in every step. Errors introduced during the data acquisition will propagate into the final results, whereas specific processing errors will only contribute to the final part of the data processing. Due to the differences in the nature of the error sources, their effects can be very different as well. Methods that are optimal to parameterize the statistics of every individual step need to be developed.

3 To what extent do atmospheric refractivity variations affect the interferograms, is it possible to determine this influence using additional measurements, and how should it be modeled stochastically?

The effect of refractivity variations in the atmosphere is a known problem to space-geodetic observations such as SLR, VLBI, radar altimetry, and GPS. Nevertheless, its influence on radar interferometry has been ignored for several years, largely due to the assumptions of the all-weather capability of radar, which has been one of the main arguments for using radar instead of optical remote sensing from the beginning. The problem is now recognized as one of the main limitations of repeat-pass radar interferometry. The specific characteristics of different atmospheric situations can be analyzed to identify which constituents form the main bottleneck. Both a deterministic and a stochastic approach can be chosen to address this problem. In the deterministic approach, the anomalous signal is derived from additional meteorological or radio-geodetic measurements, and subtracted from the interferometric products. In the stochastic approach the unknown signal is modeled using a parameterization of its magnitudes and its spatial or temporal behavior.

4 Is it possible to analyze the atmospheric signal as a meteorological source of information?

The problem of eliminating or modeling the atmospheric signal from interferograms could possibly be inverted. Knowing the amount of deformation and the topographic component in an interferogram implies that the resulting product (the differential interferogram) reflects atmospheric delay variability only. From the ERS data set alone, thousands of interferometric combinations are available for atmospheric studies, with as the main advantages the very high spatial resolution and phase accuracy. Performing a meteorological interpretation of these interferograms could result in improving the value of radar interferometry for a new discipline.

Research limitations

To narrow the scope, this study will focus on the error sources in monostatic space-borne repeat-pass interferometry with main emphasis on differential interferometric applications. Error analysis and interpretation of signal induced by the troposphere is a key topic, combined with problems related to decorrelation and data processing. It will not elaborate on errors induced by "phase unwrapping," apart from discussing the essential topics related to it. It will be assumed that the phase ambiguities can be resolved without errors. Regarding SAR data processing, only the parameters that directly influence the interferometric processing and the quality and interpretation of the interferogram will be discussed.

1.4   Research methodology

In terms of methodology, different approaches are used to address the four main research questions listed above. To parameterize the formation and quality of the interferogram, the data acquisition and SAR focusing is regarded as a closed system, from which the complex point statistics are given in terms of the probability density function (PDF). Interferometric processing is described using Euclidean geometry and standard signal processing methods such as linear transfer functions and spectral representations. A formulation of the relation between the interferometric phase observables and different geophysical parameters is proposed using a linearized Gauss-Markoff model to allow for future data adjustment and filtering.

Error propagation studies are based on Euclidean geometry for satellite orbit errors, system simulations using ERS data characteristics and linear transfer functions for the data processing. Various practical experiments have been performed, e.g., for corner reflector movement and atmospheric signal delay. In these cases, cross-correlations with triangulation, leveling, GPS, and various meteorological observations are performed.

A statistical interpretation of the effects of atmospheric refractivity variations is performed using a series of interferograms created from 52 SAR acquisitions. Modeling of the results is based on turbulence theory and power law behavior, which is consequently translated to structure functions and covariance functions to derive the variance-covariance matrix for atmospheric signal. Since the atmospheric signal cannot be considered stationary for limited temporal and spatial intervals modeling includes these non-stationary effects. Physical sensitivity analyses are performed using forward modeling and ray-tracing experiments.

1.5 Outline

The current chapter is an attempt to embed radar interferometry as a new geodetic tool amidst the latest developments in geodesy. The main objectives addressed in this book are formulated using a main problem statement, divided in four specific research questions. Key methodologies are listed.

Chapter 2 is a review of radar fundamentals, extended to synthetic aperture radar and radar interferometry. It serves as a general reference for data processing issues, the use of supplementary data, and the physical aspects of signal propagation and scattering. This chapter is not intended to completely cover the field, but to provide proper reference needed for the subsequent chapters.

The chapters 3 and 4 present the description of radar interferometry in a standard geodetic setting, using a Gauss-Markoff formulation. In such a mathematical model, the relation between the observations and the unknown (geophysical) parameters is described by the functional model, the main topic of chapter 3. This model is then decomposed for three application areas: topographic mapping, deformation mapping, and atmospheric mapping.

Chapter 4 comprises the stochastic part of the Gauss-Markoff model and covers error and error propagation studies for the subsequent parts of data acquisition and processing, as well as a unified model combining these different contributions. The chapter starts with a short revisit of the necessary theoretical concepts, followed by a single-point statistics part (section 4.2–4.5) and a multiple-point statistics part (section 4.6–4.8). The latter part covers orbit errors and atmospheric errors.

Chapters 5 and 6 present the results of some case studies. Chapter 5 focuses on deformation monitoring and presents the results of experiments involving corner reflectors, identification of coherent scatterers, subsidence monitoring, and earthquake deformation monitoring.

Chapter 6 is self-contained and focuses on issues related to the use of radar interferometry for studies of mesoscale atmospheric dynamics. This is a novel application of the technique, based on the quantitative analysis of lateral signal delay gradients caused by atmospheric refractivity. Nevertheless, apart from these meteorological applications, the geodetic analysis of the atmospheric signal can provide new insights in the variability of refractivity, which influences all types of space-geodetic observations, such as GPS and VLBI. Especially the stochastic modeling of atmospheric delay can benefit from these observations, as shown both in chapter 4 and 6.

Conclusions and recommendations for future research are given in chapter 7.

Hints for reading

This book is intended to be of some value for geodesists, solid earth geophysicists, meteorologists, and radar scientists. For the first group of readers, it should bridge the gap between the electro-technical background of radar, SAR, and radar interferometry and the basic "language" of geodesy. This involves (i) the translation to a geodetic set of observation equations, (ii) the practical geodetic applications of the

technique, focused on accuracy and reliability, and (iii) the complementary value of
the technique amidst conventional geodetic techniques. For geodesists, chapters 3,
4, and 5 may be of direct interest. Chapter 2 discusses the necessary background,
whereas chapter 6 might be interesting to geodesists involved in space-geodetic techniques where atmospheric propagation plays an important role.

For the second group, solid earth geophysicists, the applications and limitations of
the technique are important. After reading mainly chapters 4, 5, and the conclusions
of this book, they should be able to decide whether radar interferometry is a feasible
technique for a specific geophysical problem.

Meteorologists, the third group, may have technical as well as scientific motives to
read this work. Technical intentions are mainly based on the fact that the future will
make more and more data available, on a routine basis, in which the propagation
of radio signals from satellites can be used to infer properties of the atmosphere.
Chapters 3 and 4 discuss the basics between the atmospheric parameters and the
propagation of radio waves. The scientific interest of meteorologists in radar interferometric data could arise due to the unique characteristics of the data. The radar
data are not obscured by cloud cover, can be acquired during day and night, and
deliver quantitative information on the entire atmospheric column, in contrast to
many other meteorological imaging data. Another unique feature of the data is the
very high spatial resolution and the high accuracy of the integrated refractivity measurements. In chapter 6, several case studies serve as examples for the atmospheric
value of radar interferometric data.

The last group, radar scientists, may refer to the references listed in chapter 2 to find
more technical aspects of radar data processing and interferometry. The evaluation
of the data accuracies and error sources, see chapters 3 and 4, especially regarding
the analysis of atmospheric signal, might be useful regarding future satellite missions. The design of future missions is influenced by the sensitivity of the radar for
disturbances and the surveying characteristics. For that matter, the required orbit
accuracies are of importance, as well as the sensitivity for tropospheric and ionospheric heterogeneities, the spatial range and magnitudes of these heterogeneities,
and the error propagation. This information is presented mainly in chapters 4 and 6.

Where numerical examples are discussed, radar parameters of the ERS SAR are
used, unless stated otherwise.

Chapter 2
Radar system theory and interferometric processing

This chapter reviews the basic concepts of radar and interferometry. It consists of a short history of radar, SAR, and interferometry, followed by an instrument description, listing the main components of a typical SAR instrument. A section on image formation describes the main processing issues in the formation of a complex SAR image based on raw satellite radar data. A major part of this chapter focuses on the interferometric processing cookbook, discussing the main issues and problems. Differential interferometry is covered in the last section of this chapter.

key words: *Radar, Signal processing, Interferometry*

2.1 Radar history and developments

"InSAR" is a nested acronym: Radio detection and ranging (radar), Synthetic Aperture Radar (SAR), and interferometric SAR (InSAR). A short review of the development of the subsequent techniques follows the acronyms in chronological order.

2.1.1 Radar

Radio detection and ranging (radar) refers to a technique as well as an instrument. The radar instrument emits electromagnetic pulses in the radio and microwave regime and detects the reflections of these pulses from objects in its line of sight. The radar technique uses the two-way travel time of the pulse to determine the range to the detected object and its backscatter intensity to infer physical quantities such as size or surface roughness. A monostatic radar uses only one antenna, both for transmitting and receiving, whereas in a so-called bistatic radar, the transmitting and receiving antennas are physically separated (Skolnik, 1962)

Two landmark discoveries in the development of radar were Maxwell's equations of electromagnetism in 1873, in "Treatise on Electricity and Magnetism," and Hertz's experiments in 1886. Hertz generated and detected the first known radio waves, and discovered that these waves were subject to reflection or *scattering* (Buderi, 1996). Although first radar systems were developed as early as 1903 for ship tracking and collision avoidance (the "telemobiloskop" of Hulsmeyer (1904)), pre-WW II military considerations boosted the development of radar in the 1930s and 40s (Curlander and McDonough, 1991).

The cavity magnetron was an important British invention which enabled the construction of small radars with short pulses, high transmitter power, and cm-scale wavelengths. Significant improvements of radar technology, triggered by the availability of the cavity magnetron, were achieved at MIT's Radiation Laboratory during the 40s, although similar developments occurred simultaneously at several locations around the world (Buderi, 1996). Basically all current forms of radar can be traced back to those early developments.

Although the technology mainly gained momentum through military applications in those early years, civil and scientific applications arose quickly afterward. In particular radio astronomy profited from the advances in technology. Earth-based radars proved ideal for studying celestial bodies such as the Moon, Venus, Mars, and the Sun. In January 1946, the first radar echo from the Moon was received (Buderi, 1996). Using the radio-astronomy antenna's in Arecibo and Goldstone, radar observations of the planets Venus and Mars were acquired in 1961 and 1963 respectively (Goldstein and Gillmore, 1963; Goldstein and Carpenter, 1963), retrieving four parameters of planetary bodies: velocity, time-delay, and intensity for the normal and crossed polarization (Goldstein, 1964, 1969). Important results from these measurements were the retrograde rotation of Venus and the improvement of the astronomical unit.

The first tests to bring a radar into space started with JPL efforts in 1962, and lunar radar observations from Apollo 17 proved successful in 1972. Nadir-looking radar altimeter measurements from earth orbiting satellites have been collected since Skylab (1973) and GEOS-3 (1975). Venus was first visited by a radar on Pioneer 1, launched 1978. It carried a nadir-looking radar altimeter which enabled a global mapping with an accuracy of about 300 m, albeit with a spatial resolution of about 50 km. The soviet Venera 15 and 16 improved that resolution to about 2 km (Curlander and McDonough, 1991; Henderson and Lewis, 1998).

2.1.2 Synthetic Aperture Radar

A specific class of radar systems are the *imaging* radars, such as Side-Looking (Airborne) Radar (SLR or SLAR) and later Synthetic Aperture Radar (SAR). The side-looking geometry of a radar mounted on an aircraft or satellite provided range sensitivity, while avoiding ambiguous reflections. The first SLAR's were incoherent radars—the phase information of the emitted and received waveforms was not retained. The resolution in the flight direction was obtained by using a physically long antenna, hence the name Real Aperture Radar (RAR). The practical restrictions on the antenna length resulted in a very coarse resolution in the flight direction, degrading with higher flight altitudes, even though high-frequency (K-band: 0.75–2.40 cm) radars were used, see eq. (2.3.4) on page 26.

Using a fixed antenna, illuminating a strip or swath parallel to the sensor's groundtrack, resulted in the concept of *strip mapping*. Although modern phased-array antennas are able to perform even more sophisticated data collection strategies, e.g., *ScanSAR*, *squint mode SAR*, and *spotlight SAR*, the strip map mode is still the most applied mode on current satellites.

The concept of using the frequency (phase) information in the radar signal's along-track spectrum to discriminate two scatterers within the antenna beam is generally contributed to Carl Wiley in 1951 (Wiley, 1954; Curlander and McDonough, 1991). Developments at the universities of Illinois and Michigan in the late 50s and early 60s culminated in the concept of the synthetic aperture radar. The key factor for these advances is the *coherent* radar, in which phase as well as amplitude are received and preserved for later processing. Moreover, the phase behavior needs to be stable within the period of sending and receiving the signal. As a result, an artificially long antenna can be created synthetically using a moving antenna, combining the information of many received pulse returns within the synthetic antenna length. This methodology leads to a dramatic increase in azimuth resolution of about three orders of magnitude.

Satellite remote sensing started with optical-mechanical scanners such as the Landsat satellites, working in the visible and infrared parts of the electromagnetic spectrum. Although these first spaceborne remote sensing images provided an unsurpassed new perspective of the planet, some strong limitations became evident soon. Almost continuous cloud cover, especially over tropical regions, lacking solar illumination of the polar regions for half the year, and the new physical properties to be derived from scattering in other parts of the spectrum triggered the development of satellite SAR missions.

The experiences of mainly airborne SAR tests in the 60s and 70s culminated in an L-band Synthetic Aperture Radar system onboard Seasat, a satellite launched in June 1978 primarily for ocean studies. Although a short-circuit ended this mission prematurely, after 100 days of operation, the SAR imagery was spectacular, and enabled SAR systems to be included in many satellites to come (Elachi et al., 1982). Main highlights of Seasat were the wealth of geologic information and ocean topography information that could be retrieved. The Seasat SAR imagery could be processed optically as well as digitally (Henderson and Lewis, 1998).

Based on the success of Seasat, several Space Shuttle missions carried a synthetic aperture radar, starting from 1981. The first instrument was the Shuttle Imaging Radar (SIR) laboratory. SIR-A, an L-band instrument with an incidence angle of 50°, operated for 2.5 days in 1981 and provided valuable engineering experience. SIR-B, an improved version of SIR-A, orbited the earth in 1984 and was able to steer its antenna mechanically to enable different look angles. Whereas all data from Seasat and SIR-A were recorded analogously on tapes, SIR-B was already equipped with full digital recording.

Cosmos-1870 was the first (S-band) SAR satellite of the former Soviet Union, launched in 1987 and orbiting at a height of 270 km. It operated for two years. ALMAZ-1, the second satellite was launched in 1991, and operated for 1.5 years.

The first satellite SAR mission to another planet, Venus, was launched May 1989, and named Magellan (Buderi, 1996). Magellan mapped 98% of Venus with a resolution of about 150 m from September 1990 until September 1992 (Henderson and Lewis, 1998) (Curlander and McDonough, 1991), during three consecutive imaging cycles. The combination of cycle I and III provided imaging from the same side, with a 5°–25° difference in look angle. This geometry provided sufficient parallax to obtain stereo-SAR elevation models with an accuracy of 150–200 m (Henderson and Lewis, 1998). Due to the large orbit separation, interferometry was only possible at a few locations were the orbits cross (Gabriel and Goldstein, 1988).

The European Space Agency (ESA) launched its first satellite equipped with a SAR in July 1991: ERS-1 (European Remote Sensing Satellite). Although planned for a 5-years life time, the satellite operated until March 2000. Designed primarily for monitoring polar oceans and ice, it operated under a look angle of 20.3°, resulting in strong topographic distortion. Nevertheless, the systematic data acquisitions, the orbit control, and the data distribution policy resulted in impressive advances in science, and major improvements in radar interferometry. An almost exact copy of ERS-1, ERS-2, was launched in April 1995, which enabled a so-called "tandem-mode" operation of the two satellites, a period of nine months in which ERS-2 followed ERS-1 in the same orbit, with a temporal spacing of 30 minutes. This resulted in groundtracks which repeated with an exact 24 hour interval. After 1996, ERS-1 acted as a backup satellite in hibernation mode, and was only activated for special occasions, mostly related to SAR interferometry. Such occasions were the eruption of volcanoes such as Vatnojöküll on Iceland, and severe earthquakes such as the Izmit earthquake in 1999 (Barbieri et al., 1999). Both ERS satellites operated from the same near-polar orbits at an altitude of 785 km. An image mode is used to image the earth's surface, while a wave/wind mode is used over oceans (Attema, 1991). The satellite data are used for many environmental purposes (oil slick monitoring, wind and wave field observations, transport of sediments, ice cover and movement) over the oceans. Over land, the data reveal useful information on land use, agriculture, deforestation, earthquake and volcanic deformation, and general geophysics.

Japan started its spaceborne SAR program launching the Japanese Earth-Resources satellite (JERS) in 1992. JERS was designed for solid earth remote sensing, and carried an L-band (23.5 cm) radar with a look angle of 35°, more favorable regarding topographic distortion. JERS operated until October 1998. The L-band acquisitions proved to be well-suited for repeat-pass interferometric applications, due to their reduced sensitivity for temporal changes in the scattering mechanisms at the earth's surface. Unfortunately, orbit control and maintenance was not as advanced as for the two ERS satellites.

Radar observations from the Space Shuttle continued in 1994, with SIR-C/X-SAR, a combined instrument developed by JPL, DLR, and ASI. The instrument orbited the earth in Spring and Autumn, to detect seasonal differences in the images. The mission operated for the first time in three frequency ranges: L-band, C-band, and X-band, with 23.5, 6, and 3 cm wavelengths respectively. The combination of the different frequencies extended and improved the interpretation of the data significantly (Lanari et al., 1996; Melsheimer et al., 1996; Coltelli et al., 1996). Interesting discoveries during the 1994 flights were the forgotten temples of Angkor, Cambodia, and the remains of an earlier version of the Chinese wall, close to the known one, while not recognizable at the surface.

Started in 1995, the Canadian Space Agency (CSA) operates Radarsat, a SAR satellite designed to perform especially Arctic observations of ice coverage, e.g., for shipping route planning, apart from many other applications (Ahmed et al., 1990). Radarsat has a ScanSAR mode and can acquire wide swaths. Although capable of interferometric measurements, orbit control and maintenance is limited, which restricts interferometric capabilities.

The experience with the repeat pass missions for topographic mapping, especially the problem of temporal decorrelation and atmospheric disturbances, culminated in the Shuttle Radar Topography Mission (SRTM) (Jordan et al., 1996). This Space Shuttle mission was performed between 11 and 23 February 2000 and used a single-pass configuration with a fixed 60 m boom to carry the two radar antennas. It mapped all land masses between 60°N and 58°S using C-band, and tiles of this area with a higher accuracy using X-band (Bamler et al., 1996b; Werner, 1998).

2.1.3 Interferometry

Compared with conventional geodetic techniques, one capability long remained out of reach for radar: the measurement of angles. Similar to a single human eye, which is essentially "blind" for the difference in distance to objects, it is impossible for a radar or SAR to distinguish two objects at the same range—but different angles—to the instrument. Nature readily provides the simple solution for the problem; the use of two sensors. It worked with two eyes, why not use two radars?

This idea, and the use of the phase information, cleared the way for interferometry. Using two SAR images, acquired either by two different antennas or using repeated acquisitions, it is possible to obtain distance as well as angular measurements. The use of the phase measurements (multiplicative interferometry) enabled the observation of relative distances as a fraction of the radar wavelength, and the difference in the sensor locations enabled the observation of angular differences, necessary for topographic mapping.

Basics of interferometry

Christiaan Huygens (1629-1695) developed the idea of the wave front in *Traité de la Lumière* (Huygens, 1690). He observed that two intersecting light beams did not bounce off each other as would be expected if they were composed of particles, and that an expanding sphere of light behaves as if each point on the wave front were a new source of radiation of the same frequency and phase. Interferometry of light and other electromagnetic signals is based on the wave front concept.

Two distinct types of interferometry can be distinguished: *additive* interferometry and *multiplicative* interferometry, see fig. 2.1. The signal resulting from the former is obtained by the incoherent summing of the amplitudes of two input signals. This results in so-called "amplitude" fringes, as in fig. 2.1A. One problem with this type of interferometer is that it is not possible to determine whether the interferometric signal is increasing or decreasing. Another problem is that the accuracy is only a

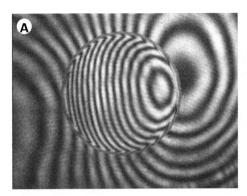 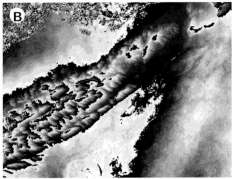

Fig. 2.1. Two types of interferometry. **(A)** Additive interferometry (amplitude fringes) of fluid thickness observed by laser illumination (courtesy M. Decré, Philips Research Laboratories) **(B)** multiplicative interferometry (phase fringes) for topographic mapping.

fraction of the amplitude cycle, which is often not very accurate. The application of additive (incoherent) interferometry is usually a consequence of limitations in detector technology, for example due the high frequencies of light. An example of an additive interferometer is the famous Michelson-Morley interferometer (1886).[1]

In multiplicative interferometry (fig. 2.1B) one is also able to measure the interferometric phase by coherent cross-multiplication of the two input signals. The phase data imply a better accuracy, as the fraction of the phase cycle is much easier to determine accurately. In 2D applications, a 2π phase-cycle is often referred to as a *fringe*.

Radar (SAR) Interferometry

Radio interferometry developed after the Second World War. In 1946, Ryle and Vonberg constructed a radio analogue of the Michelson-Morley interferometer and soon located a number of new cosmic radio sources (Tubbs, 1997). In the field of planetary mapping, the range-Doppler configuration enabled the mapping of radar reflectivity from iso-range lines and iso-Doppler lines (Evans and Hagfors, 1968). What remained unknown was the ambiguity between reflections from the northern and southern hemispheres of the planet. Rogers and Ingalls (1969) were the first to use interferometry to resolve this ambiguity, using two antennas. For the moon, the antenna beam width was small enough to resolve the north-south ambiguity, which enabled Zisk (1972b) to use interferometry for measuring elevation differences (Zisk, 1972a; Shapiro et al., 1972). These applications of interferometry are still used today (Margot et al., 1999a,b). For topographic mapping of the lunar poles, Margot et al. (1999a) reported a horizontal resolution of 150 m and a height resolution of 50 m,

---

[1]In 1868, the French physicist Fizeau (1819-1896) suggested using an interferometric method to measure stellar diameters by placing a mask with two holes in front of a telescope's aperture. He argued that the fringes would vanish at a separation related to the size of the star. Fizeau's ideas were pursued unsuccessfully by Stephan and, using a different concept, successfully by Michelson.

using two antennas at a distance of 20 km.

First experiments with airborne radar (SAR) interferometry for topographic mapping were performed by the U.S. military and a patent for obtaining elevation from phase difference images was filed in 1971 (Richman, 1982; Henderson and Lewis, 1998). Graham (1974) first published results with this method, using two antennas and coherent additive interferometry with optical processing techniques. Dual (single-pass) antenna airborne interferometry matured in the 80s, using digital processing techniques, and using coherent multiplicative interferometry (Goldstein et al., 1985). The first results, by Zebker and Goldstein (1986), produced *interferograms*, in which for every resolution element the amplitudes of the two images are multiplied, and the phases are differenced. This resulted in a topographic map of an area of 10 × 11 km, with an accuracy between 10 and 30 m. In 1987, Goldstein and Zebker showed that using two antennas mounted in the flight direction of the aircraft, *along-track interferometry* results in a sensitivity to motion of scatterers, relative to the stationary background. The application of interferometry with satellite data, using the *repeat-pass* method, where the satellite revisits an acquisition area after a certain period, was first demonstrated in 1987 by Li and Goldstein, (Li and Goldstein, 1987, 1990; Prati et al., 1990), using historic Seasat data, and by Gabriel and Goldstein (1988) and Goldstein et al. (1988) using the shuttle's SIR-B data. The consecutive application of repeat-pass interferometry using aircraft data was demonstrated by Gray and Farris-Manning (1993).

The estimation of topography had been the main focus for the early applications of radar interferometry, yielding elevation accuracies comparable with optical methods, albeit valuable due to the all-weather capability of radar. Nevertheless, the experiments with the repeat-pass configuration clearly demonstrated an even more spectacular application—deformation monitoring. The relative line-of-sight movement of scatterers with respect to a reference location in the image could be measured as a fraction of the wavelength, yielding cm to mm accuracies for L-band, C-band, and X-band radars (Gabriel et al., 1989). A problem in this application is that—for an effective baseline larger than zero—the deformation signal is always mixed with topographic signal. A suitable solution to this problem was *differential* interferometry, where the topographic signal obtained from a so-called *topographic interferogram* (or a reference elevation model) was scaled to the baseline conditions of the *deformation interferogram* and subtracted from it, yielding a *differential interferogram* (Gabriel et al., 1989). The first demonstration of differential radar interferometry for mapping the displacement field of the Landers earthquake was reported by Massonnet et al. (1993), who used a reference elevation model to remove the topographic phase signal. Zebker et al. (1994b) reported the so-called three-pass method, studying the same earthquake. This method uses two SAR acquisitions with a short temporal separation for retrieving the topographic phase signal, and the combination of one of these two with a third acquisition. The latter interferogram, which spanned the earthquake (the coseismic interferogram), was consequently corrected for topographic contribution by the first interferogram. Massonnet et al. (1994b) showed that even over time spans of one year, arid and undisturbed areas can still be used to combine two SAR images interferometricly, although they argued that atmospheric delay signal might influence their results.

The only solid earth objective of ERS-1 was the detection and management of land use. As interferometric applications of the SAR were not included in the design of the satellite, the many studies on topographic and deformation mapping were received with much interest by the scientific community, see (Hartl, 1991; Prati et al., 1993; Hartl et al., 1993; Hartl and Thiel, 1993; Prati and Rocca, 1993; Prati et al., 1993; Prati and Rocca, 1993; Small et al., 1993; Werner et al., 1993; Prati and Rocca, 1994a). The following section gives an overview of satellite SAR missions and their interferometric applications.

2.1.4   Missions and applications

Table 2.1 lists the most important SAR missions, showing their main characteristics in terms of life time, orbit, and SAR performance. The satellites indicated by a star are capable of interferometric applications. The dates of operation give an estimate of the historical archive of SAR data, and the orbital repeat-period indicates the sensitivity to temporal decorrelation and deformation rates. The altitude can be used as an indication for orbit stability, cycle period, and the influence of drag effects.

The SAR frequency and bandwidth are used to determine the radar wavelength, and hence its sensitivity to surface displacement, topographic height, temporal decorrelation, and range resolution, using equations discussed in sections 2.3 and 2.4. Using the incidence angle, height, frequency, and bandwidth parameters, the critical baseline $B_{\perp,\mathrm{crit}}$ can be calculated, which expresses the maximum horizontal separation of the two satellite orbits in order to perform interferometry, see eq. (4.4.11) on page 102. Finally, the swath width fixes the feasible spatial scales of the phenomena of interest.

The geodetic applications of spaceborne repeat-pass SAR interferometry can be categorized in roughly four disciplines (i) topographic mapping with a relative accuracy of 10–50 m, (ii) deformation mapping with mm–cm accuracy, (iii) thematic mapping based on change detection, and (iv) atmospheric delay mapping with mm–cm accuracy in terms of the excess path length.

Topographic mapping

Applications of spaceborne radar interferometry for creating digital elevation models (DEMs) have matured from early experiments, see Zebker and Goldstein (1986), to a standard geodetic tool in one decade. Today, commercially available software can be used by non-expert users, and companies provide services in creating geocoded DEMs from the available SAR data. Especially data from the ERS-1/2 "tandem mission" are currently used for these purposes (van der Kooij, 1999a). Due to the regular acquisitions of the ERS satellites, aimed at full coverage of the world, mosaicing of many interferometric elevation models enables the mapping of large areas, up to continental scales. Main applications of the elevation models include telecommunication, hydrological mapping and flooding predictions, cartography, and geophysics. The availability of topographic maps (status 1997) is shown in fig. 2.2,

Table 2.1. List of satellite SAR missions with their most important design parameters.

Mission	Year	ΔT (day)	H_{sat} (km)	f_0 (GHz)	B_R (MHz)	θ_{inc} (deg)	Swath (km)
Seasat⋆	1978	3	800	1.275	19.00	20(23)26	100
SIR-A	1981	–	235	1.278	19.00	(50)	50
SIR-B⋆	1984	–	235	1.282	12.00	(15)-(64)	10-60
Cosmos 1870	1987-89	var	250	3.000	*	(30)-(60)	20-45
ALMAZ	1991-92	–	300	3.000	uncoded	(30)-(60)	20-45
Magellan⋆	1989–92	var	290-2000	2.385	2.26	(17)-(45)	20
Lacrosse-1,2,3	1988/91/97	var	275	3.000	*	steerable	variable
ERS-1⋆	1991-92	3	790	5.300	15.55	21(23)26	100
	1992-93	35	790	5.300	15.55	21(23)26	100
	1993-94	3	790	5.300	15.55	21(23)26	100
	1994-95	168	790	5.300	15.55	21(23)26	100
	1995-2000	35	790	5.300	15.55	21(23)26	100
ERS-2⋆	1995-	35	790	5.300	15.55	21(23)26	100
JERS-1⋆	1992-98	44	568	1.275	15.00	26(39)41	85
SIR-C/X-SAR	Apr 1994	–	225	1.240	20.00	(15)-(55)	10-70
	Apr 1994	–	225	5.285	20.00		10-70
	Apr 1994	–	225	9.600	10–20	(15)-(45)	15-45
SIR-C/X-SAR⋆	Oct 1994	1	225	1.240	20 †	(55)	21-42
	Oct 1994	1	225	5.285	20 †	(55)	21-42
	Oct 1994	1	225	9.600	10-20	(15)-(45)	15-45
Radarsat⋆	1995-	24	792	5.300	11-30	(20)-(49)	10-500
SRTM⋆	2000	0	233	5.300	9.50	(52)	225
	2000	0	233	9.600	9.50	(52)	50
ENVISAT⋆	2001-	35	800	5.300	14.00	(20)-(50)	100-500
ALOS⋆	2002-	45	700	1.270	28/14	(8)-(60)	40-350
Radarsat-2⋆	2003-	24	798	5.300	12–100	(20)-(60)	20-500

ΔT, repeat period; H_{sat}, satellite altitude; f_0, carrier frequency; B_R, range bandwidth; θ_{inc}, incidence angle (values between parentheses denote nominal mid-incidence angles). ⋆Missions capable of interferometry. †The radar bandwidth was changed to 40 MHz for the one-day time interval. *Unknown. –Not applicable. SIR-A/B/C, X-SAR, and SRTM are instruments flown onboard the Space Shuttle. Lacrosse is a classified military Spotlight SAR mission (Richelson, 1991). Radarsat has a ScanSAR mode. The SIR-C data as used for interferometric applications are listed here. Compiled from (Gabriel and Goldstein, 1988; Curlander and McDonough, 1991; Rosen et al., 1996; Mouginis-Mark, 1995b; Evans et al., 1993; Henderson and Lewis, 1998; Raney, 1998; Radarsat, 1999).

stressing the lack of detailed topographic maps in many parts of the world.

Radar interferometric measurements of topography may seem similar to optical (stereographic) methods at first glance. Both methods need two "images" to infer height differences. The main technical difference, however, is that optical techniques are based on parallax, which is directly based on angle-measurements, whereas radar is a "ranging" device, which measures distances. Parallax measurements require the identification of homologous features in the terrain, which poses problems in, e.g., snow areas in polar regions or sand areas in deserts. Moreover, optical techniques need illumination by the sun and cannot penetrate through cloud cover. For example, in Europe only 10% of all acquired optical data are useful for topographic mapping due to cloud cover, whereas high-latitude regions are not illuminated by the sun for several months per year.

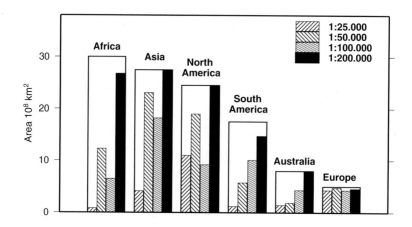

Fig. 2.2. Availability of topographic maps at different scales, status 1997. The y-axis indicates the continent size. Data according to Centre National D'Etudes Spatiales (CNES), Paris/Toulouse. Currently the results of the SRTM mission filled many of the gaps.

The main problems with topographic mapping using repeat-pass interferometry are layover, foreshortening, shadow, surface decorrelation (especially due to vegetation and the decrease of coherence over the time interval between the acquisitions), and the atmospheric signal in the data. The effective baseline length influences the latter two error sources. For example, a rather long baseline increases noise in the data due to geometric decorrelation, whereas it reduces the effect of atmospheric disturbances in the derived DEM dramatically, and vice versa. Note that single-pass interferometry is not influenced by these error sources, since the data are acquired at the same time, and atmospheric signal will cancel in the interferometric combination. The SRTM mission, using single-pass interferometry, is therefore specially designed to measure topography, yielding elevation data that are better and more reliable than repeat-pass data, albeit over a limited part of the world (Jordan et al., 1996; Bamler et al., 1996b).

Deformation monitoring

Since the observation of ground motion over agricultural fields by Gabriel et al. (1989), many applications of (differential) InSAR for deformation monitoring have been developed. As the scaling effect of the baseline, which influences the sensitivity for topographic mapping, is not present in deformation mapping, accuracies can reach the sub-cm level.

Applications extend to many forms of surface deformation. Table 2.2 shows that the applications can be organized in several categories. Seismic applications include the investigation of earthquakes, faults, and tectonics. The use of two acquistions to form an interferogram results in the capability of measuring *preseismic*, *coseismic*, and *postseismic* deformation. These terms are defined as deformation preceding a seismic event (two images before), spanning a seismic event (one image before and one after), or following a seismic event (two images after), respectively. These possibilities, using archived satellite data, enable geophysical and hazard studies with an accuracy and resolution unprecedented by any other geodetic technique. Coseismic deformation on scales of cm-m are observed by, e.g., Massonnet et al. (1993); Zebker et al. (1994b); Peltzer and Rosen (1995), Massonnet and Feigl (1995a,b), and Peltzer et al. (1999). Especially the combination of the radar data with GPS observations extends the analysis and interpretation of earthquakes considerably (Segall and Davis, 1997). Postseismic deformation has been observed by, e.g., Massonnet et al. (1994b, 1996b) and Peltzer et al. (1996).

In terms of volcano deformation monitoring, the same three categoeries can be distinguished. In case of volcanic eruption, *pre-*, *co-*, and *post-eruptive* measurements can be performed. Of course, the availability of an archive of pre-eruptive SAR observations for many volcanoes is important in the development of hazard forecasting capabilities. Although the detection and monitoring of deforming volcanoes is of major importance, it might be equally important to systematically monitor "sleeping" volcanoes that could pose a potential threat (Amelung et al., 1999).

Land subsidence, for example caused by mining activities, is a third category concerning deformation measurements. Many areas in the world are affected by subsidence (or uplift) due to the extraction of water, gas, oil, salt, or other mineral resources. Moreover, construction works such as tunnels often cause subsidence resulting in damage to the infrastructure. The feasibility of radar interferometry for these applications depends on the subsidence rates, i.e., deformation gradient, in combination with the influence of decorrelation due to land use or vegetation and of atmospheric signal in the interferograms. Although the temporal decorrelation poses a significant limitation for obtaining continuous deformation maps, it is observed that urban areas or specific natural or anthropogenic features remain coherent for extended time intervals, (Usai, 1997; Usai and Hanssen, 1997; Hanssen and Usai, 1997; Usai and Klees, 1998; Ferretti et al., 1998; Usai and Klees, 1999a,b). Recent advances using stacks of coregistered SAR images show that many permanent (coherent) scatterers can be identified, especially over urbanized areas (Ferretti et al., 1999b, 2000). Since conventional coherence estimators use estimation windows that are often too wide to identify a single scatterer, the systematic analysis of many images is inevitable for a robust identification procedure.

Glacier and ice motion in the order of meters can be conveniently studied using interferometry as well, see e.g. Goldstein et al. (1993). Due to the inaccessibility of the terrain, e.g., Greenland or the Antarctic, the interferometric measurements provide invaluable information on glacier dynamics, which can be used as input for more advanced problems such as global warming and sea-level rise. Depending of the speed of the ice sheet and glacier velocity, a 35 day repeat interval may be too long for the SAR images to stay within the grid distance needed for coregistration conditions. In such cases, additional SAR techniques such as speckle tracking (Gray et al., 1999) may be neccesary in combination with the interferometric combination of small patches in the SAR data. Many studies in this field have been performed, see table 2.2.

Table 2.2. Major geophysical events studied using radar interferometry, and main references.

Location	References
EARTHQUAKES	
Kobe, Japan	Ozawa et al. (1997)
Landers earthquake, California	Massonnet et al. (1993); Massonnet and Adragna (1993); Zebker et al. (1994b); Peltzer et al. (1994); Massonnet et al. (1994b), Feigl et al. (1995); Peltzer et al. (1996); Massonnet et al. (1996b); Hernandez et al. (1997); Price and Sandwell (1998); Michel et al. (1999)
Hector earthquake, California	Sandwell et al. (2000)
Manyi earthquake, Tibet	Peltzer et al. (1999)
Izmit earthquake, Turkey	Barbieri et al. (1999); Hanssen et al. (2000a); Reilinger et al. (2000)
Eureka valley, California	Massonnet and Feigl (1995b); Peltzer and Rosen (1995)
Northridge earthquake, California	Massonnet et al. (1996a); Murakami et al. (1996), Kawai and Shimada (1994)
Kagoshima-kenhokuseibu earthquake, Japan	Fujiwara et al. (1998)
Nuweiba earthquake, Gulf of Elat (Aqaba)	Baer et al. (1999); Klinger et al. (2000)
Grevena earthquake, Greece	Meyer et al. (1996); Clarke et al. (1996)
Colfiorito, Umbria-Marche, Italy	Stramondo et al. (1999)
Creep San Andreas fault/Parkfield	Rosen et al. (1998); Bürgmann et al. (2000)
VOLCANOES	
Vatnajöküll, Iceland	Roth et al. (1997); Thiel et al. (1997)
Krafla spreading segment, Iceland	Sigmundsson et al. (1997)
Etna, Italy	Massonnet et al. (1995); Briole et al. (1997); Delacourt et al. (1997); Lanari et al. (1998); Williams and Wadge (1998)
Iwo Jima, Japan	Ohkura (1998)
Izu peninsula, Japan	Fuliwara et al. (1998)
Katmai, Alaska	Lu et al. (1997); Lu and Freymueller (1999)
Kilauea, Hawaii	Mouginis-Mark (1995a); Rosen et al. (1996); Zebker et al. (1996, 1997)
Soufriere Hills, Montserrat	Wadge et al. (1999)
Campi Flegrei, Italy	Usai et al. (1999); Avallone et al. (1999)
Yellowstone caldera, Wyoming	Wicks et al. (1998)
Piton de la Fournaise, Reunion	Sigmundsson et al. (1999)
Long Valley, California	Thatcher and Massonnet (1996)
Unzen, Japan	Fujii et al. (1994)
Galapagos	Mouginis-Mark (1995a); Jónsson et al. (1999); Amelung et al. (2000a)

Continued on page 21

Table 2.2. —*Continued*

ANTHROPOGENIC SUBSIDENCE/UPLIFT

Geothermal fields	Massonnet et al. (1997); Hanssen et al. (1998a); Jónsson et al. (1998); Carnec and Fabriol (1999); Fialko and Simons (2000)
Las Vegas, Nevada	Amelung et al. (1999)
Paris, France	Fruneau et al. (1998)
Napels, Italy	Tesauro et al. (2000)
Antelope Valley, California	Galloway et al. (1998)
Gardanne, France	Carnec et al. (1996)
Pomona, California	Ferretti et al. (2000)
Groningen, Netherlands	Hanssen and Usai (1997); van Bree et al. (2000)
Imperial valley, (swelling) California	Gabriel et al. (1989)

GLACIER/ICE MOTION (selected)

Antarctica/Patagonia	Goldstein et al. (1993); Hartl et al. (1994a,b); Wu (1996); Rott and Siegel (1997); Rott et al. (1998); Joughin et al. (1999)
Greenland	Joughin (1995); Kwok and Fahnestock (1996); Joughin et al. (1996); Rignot et al. (1997); Joughin et al. (1997); Mohr (1997); Mohr et al. (1998); Joughin et al. (1998); Hoen and Zebker (2000)

Other, more specific, applications of deformation monitoring include the amount of mass extraction from a quarry, see Hartl and Thiel (1993) or the movement of man-made corner reflectors, see Hartl and Xia (1993), Hartl et al. (1993), and Prati et al. (1993).

Thematic mapping

Changes in the scattering characteristics of the ground surface between the two radar acquisitions can result in a change in the coherence—the degree of similarity between the two images—in the interferogram. Whenever the scattering mechanisms remain unchanged, the degree of coherence is highest, and only influenced by the acquisition geometry, data processing, and thermal noise.

The analysis of the decrease in coherence associated with different types of land use has been reported by Zebker and Villasenor (1992); Askne and Hagberg (1993); Hartl and Xia (1993); Wegmüler et al. (1995); Yocky and Johnson (1998). Specific applications include: forestry, (Wegmüller et al., 1995; Wegmüller and Werner, 1995; Wegmüller and Werner, 1997), flood monitoring, (Geudtner et al., 1996), lava streams, (Rosen et al., 1996; Zebker et al., 1996; Lu and Freymueller, 1999), ice penetration, (Hoen and Zebker, 2000), land slides, (Rott et al., 1999), and fires, soil moisture changes, or vegetation growth.

A problem in the analysis of the interferometric coherence measurement is the separation of the different sources of decorrelation. For example, it is necessary to separate geometric decorrelation due to baseline characteristics from temporal decorrelation in order to make a correct interpretation of the physical phenomena (Hoen and Zebker, 2000). Zebker and Villasenor (1992) and Gatelli et al. (1994) showed that for surface geometric decorrelation, a linear model can be used, relating the perpendicular baseline to the amount of decorrelation. Whenever volume decorrelation occurs, e.g., in forest or ice applications, this model needs to be extended, see Hoen and Zebker (2000).

Atmospheric delay mapping

The sensitivity of the radar signal delay for atmospheric refractivity variation is usually considered to be a nuisance in radar interferometry. However, the abundance of SAR acquisitions over areas where topography is known, and surface deformation is absent, makes the interpretation of atmospheric signal in the data an interesting goal for meteorology and the study of atmospheric dynamics. Nevertheless, due to the low repetition frequency of current SAR satellites, these applications will have a rather opportunistic character. Current studies therefore focus on demonstration and validation of the technique, referred to as Interferometric Radar Meteorology (IRM), rather than providing operational capabilities. Some demonstration and validation studies are discussed in chapter 6 (Hanssen et al., 1999, 2001, 2000b).

2.2 Sensor characteristics

Synthetic aperture radars are available in many different configurations, although the basic modules are usually the same. Most important, to enable SAR (and InSAR) processing, the radar needs to be coherent, at least within the time span between sending and receiving one pulse. In a coherent radar the phase of the transmitted signal is retained and used as a reference against which the returned signal can be compared. Here we will discuss essential hardware components of the instrument, using the Active Microwave Instrument (AMI) on board of the ERS satellites as an example.

The SAR is one of the two functions of the AMI, which consist of a 10×1 m area antenna, a SAR processor, a pulse generator, transmitter, and some other sub-systems. The second function of the AMI, the wind scatterometer, uses several common sub-systems. As a result, both functions cannot be operated in parallel. Due to the power consumption of the SAR in image mode, it can only be operated 12 minutes per 100 minute orbit, of which maximally 4 minutes in eclipse (Attema, 1991).

During its operation, the SAR processor initiates the command to generate a radar pulse with a duration of 37.1 μs. This pulse is generated and since it needs to be shaped into a "chirped pulse" with varying frequency, the signal is fed into a dispersive delay line. This device produces a linear frequency modulated (FM) chirp, over a range of 15.5 MHz. The rate at which consecutive chirps are generated is the pulse repetition frequency (PRF), which is programmable in the range 1640–1720 Hz. The typical value is 1680 Hz, resulting in 10 transmitted pulses before the first echo is received, see fig. 2.5. The chirp is passed to the transmitter and up-converter part, which shifts the signal to a carrier frequency of 5.3 GHz. The high-power amplifier amplifies the signal by approximately 45 dB, until it reaches the desired peak power of 4 kW for the phased-array antenna. It then passes via the waveguide to the antenna subsystem, consisting of a circular assembly and the SAR antenna. The circular or switch assembly assures that the high-power transmit signal is isolated from the sensitive low-noise amplifier for the echo signal.

When the echo signal is received by the SAR antenna, it is routed via the waveguide, through the circulator assembly, to the receiver. The received radar echo is then amplified in a low-noise amplifier and mixed with the local oscillator signal to provide a down-converted signal at the intermediate frequency of 123 MHz. Apart from the echo, also a sample of the transmitted waveform is down-coverted, which serves as the replica signal for on-ground SAR processing. After routing these signals through a pulse compressor they are sent to the analog-to-digital I/Q converters, producing the real (in-phase) and imaginary (quadrature) part of the signal. Although the SAR detector uses 6 bits for the replica and calibration pulses, it uses 5 bits for the image data, i.e. 5 bits (I), 5 bits (Q). With a sampling frequency of 18.96 MHz, this produces 10 bits of digital code every 52.74 ns, which are directly downlinked to receiving stations on earth. Usually, these values are stored in one byte for standard processing, i.e., 1 byte (I) and 1 byte (Q).

The Doppler effect due to the satellites velocity in an inertial reference system, in combination with the earth's rotation in the same system, introduces a frequency shift in the received data. Since the beam width is small, these frequency shifts are limited to a range of about 1500 Hz, well within the PRF of 1680 Hz. However, since the additional effect of the earth rotation varies with latitude, the whole bandwidth varies around the zero-Doppler location. This effect is reduced by applying yaw-steering, where the total antenna beam is electronically shifted forward and backward, in a range from $0°$ at the poles to about $±4°$ at the equator (Attema, 1991; Alaska SAR Facility, 1997). This results in a Doppler Centroid frequency that is less than 900 Hz from the zero-Doppler direction (Bamler and Schättler, 1993).

2.3   Image formation

The antenna of the ERS-SAR is aligned along the satellite's flight path to direct a narrow beam sideways and downwards ($20.3°$ to the right of nadir) onto the earth's surface to obtain strips of imagery of about 100 km in width, see fig. 2.3. Imagery is built up from the time delay and strength of the return signals, which depend primarily on the roughness and dielectric properties of the surface and its range to the satellite.

The "synthetic" aperture of SAR uses the redundancy in subsequent pulse returns that elucidate a single point scatterer. The inverse problem of reconstructing the scatterers response from a series of pulse return signals is referred to as SAR focusing or synthetic aperture processing. This methodology, first demonstrated by Graham (1974) based on the concepts of Wiley (1954), improves azimuth resolution from the 4.5 km beam width for a single pulse to approximately 5 m for the full synthetic aperture.

First focusing has been developed using optical processors, based on the concepts of holography (Hovanessian, 1980). Currently, all processors are electronic (digital).

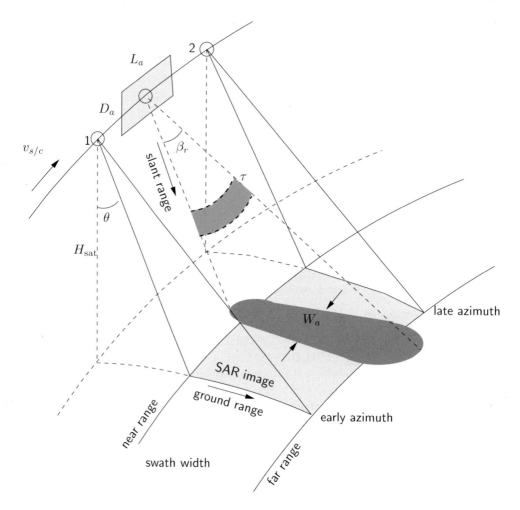

Fig. 2.3. Imaging geometry of a SAR acquisition. Due to the spacecraft velocity $v_{s/c}$ the pulses illuminate a swath parallel to the satellite track. The footprint of a single pulse is indicated by the dark shaded area. The total SAR image starts in azimuth at the "early azimuth" time, and lasts until the "late azimuth" time. In range direction, it covers an interval from "near range" to "far range," which corresponds to the light shaded area in ground range.

Several electronic algorithms for SAR focusing have been developed: range-Doppler (Bennett and Cumming, 1979; Wu et al., 1981; Curlander and McDonough, 1991), seismic migration (Prati et al., 1990), the PRISME architecture (Massonnet et al., 1994a), and chirp scaling (Runge and Bamler, 1992; Raney et al., 1994). See also references in Bamler (1992), Massonnet and Feigl (1998), and Otten (1998). The ERS SAR data used in this study were either preprocessed to SLC-format by one of the ESA PAFs, or processed using the SAR processor developed at JPL and Stanford

University, which applies a range-Doppler algorithm (Zebker et al., 1994b).

The azimuth resolution of a focused SAR image is closely related to the dimensions of the radar antenna. A small antenna with a broad beamwidth contains high-frequency information of a point scatterer's response, resulting in a broader azimuth bandwidth (Bamler and Hartl, 1998). The bandwidth is in the order of $2v_{s/c}/L_a$, resulting in an azimuth resolution of about $L_a/2$. Although it might seem that reducing the antenna size might increase resolution to infinity, the effective gain of the antenna is proportional to the square of its aperture. Therefore, a reduction in antenna size will be a trade-off with the maximum attainable signal-to-noise ratio (Henderson and Lewis, 1998).

In the following subsections, the process of image formation or focusing is briefly reviewed. As mentioned above, there are many SAR processing algorithms, and some of those regard SAR image focusing directly as a two-dimensional problem. Nevertheless, here we try to describe the process as consisting of smaller steps. We start with describing the antenna pattern, then describe the range pulse waveform, and finally the range compression. Then the Doppler centroid concept is introduced, followed by range migration correction and azimuth focusing. This section concludes with the imaging geometry of a single SAR image.

2.3.1 Antenna pattern

A rectangular antenna with length L_a and width D_a (see fig. 2.3) and a uniform current density produces a (normalized) *antenna pattern* as a function of the off-center beam angle ϕ, see Olmsted (1993); Curlander and McDonough (1991); Zebker (1996):

$$\text{sinc}^2(\frac{D_a}{\lambda}\phi_r)\text{sinc}^2(\frac{L_a}{\lambda}\phi_a), \qquad (2.3.1)$$

where ϕ_r and ϕ_a are the off-center angles in range and azimuth direction respectively. Evaluating eq. (2.3.1) in azimuth direction, for constant ϕ_r, this pattern is shown in fig. 2.4.

Usually, in radar systems the energy pattern is considered constant between the half power (3 dB) angles. In the sinc^2 pattern in range, the half power is reached at $\frac{D_a}{\lambda}\phi_r = -0.443$ and $\frac{D_a}{\lambda}\phi_r = +0.443$. The beam widths β_r and β_a are therefore

$$\beta_r = 0.886\frac{\lambda}{D_a}, \quad \text{and} \qquad (2.3.2a)$$

$$\beta_a = 0.886\frac{\lambda}{L_a}. \qquad (2.3.2b)$$

For ERS-1/2, using the characteristics from Table 4.2, page 101, the theoretical values for the beamwidths are

$$\beta_r = 2.870°, \quad \text{and} \quad \beta_a = 0.287°. \qquad (2.3.3)$$

In fact, for ERS-1/2 the beamwidth of the main lobe is slightly broadened in range direction in order to get the power distributed more evenly across the full swath.

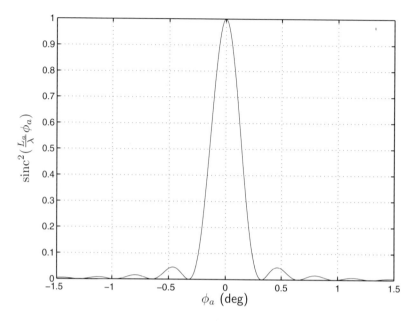

Fig. 2.4. The normalized one-way antenna pattern $\mathrm{sinc}^2(\frac{L_a}{\lambda}\phi_a)$ in azimuth direction (constant ϕ_r).

This results in practical beam widths of 5.4° and 0.228° in range and azimuth, respectively (Attema, 1991).

The radar footprint width W_a in azimuth direction at a distance R is now

$$W_a = \frac{\lambda}{L_a}R, \qquad (2.3.4)$$

i.e., approximately 4.8 km for ERS. For *focused spot scanning*, or real-aperture radar (RAR), this is the maximum attainable resolution (Fitch, 1988). In range, the footprint width or swath width is determined by the time interval of the pulse return registration.

2.3.2   Range modulation

The transmit/receive configuration of the ERS-SAR is sketched in fig. 2.5. The radar emits short, high-energy microwave pulses to earth and records the echoes from each pulse. A monostatic radar uses one single antenna to transmit as well as receive the pulses. The pulse rate is defined by the pulse repetition frequency (PRF=1680 Hz). The pulse waveform can be written as (Bamler and Schättler, 1993)

$$s(t) = g(t)\,e^{j2\pi f_0 t}, \qquad (2.3.5)$$

where $g(t)$ is the complex envelope and f_0 is the carrier frequency (5.3 GHz). The resolution in range Δ_r, i.e., the shortest range difference at which two scatterers can be distinguished, for a rectangular envelope depends on the pulse length τ:

$$\Delta_r = c\tau/2, \tag{2.3.6}$$

and therefore a shorter pulse envelope will result in a higher resolution in range direction. On the other hand, to obtain a high-SNR radar image, high peak powers are desirable, since the energy of the received pulse is 10^{-11} orders of magnitude smaller than the energy of the emitted pulse. These two considerations, short pulses vs. high peak powers, are conflicting since the amount of energy the instrument can emit in a finite time span is limited, and as a consequence there is a limit in reducing the length of the pulse. For ERS, with $\tau = 37.1$ μs, we find the minimal distance between two resolvable points in range direction to be $\Delta_r \approx 5.5$ km, or 14 km in ground range. Note that this is independent of the altitude of the satellite.

The problem of the limited pulse length is circumvented by phase-coding the envelope. In a linear frequency modulated (FM) chirp waveform the frequency $f(t)$ of the pulse increases linearly with slope s_c, i.e.,

$$f(t) = s_c t \quad -\tau/2 < t < \tau/2. \tag{2.3.7}$$

Deriving the phase signal from (2.3.7) by integration yields $\phi(t) = \pi s_c t^2$ or

$$g(t) = e^{j\pi s_c t^2}, \tag{2.3.8}$$

omitting the integration constant. The bandwidth ranging from $f(-\tau/2)$ to $f(\tau/2)$ is now defined as $B_R = s_c\tau$. It can be shown that using matched filtering on the received pulse, also referred to as a *pulse or chirp compression technique*, the effective range resolution improves to

$$\Delta_r = c/(2B_R) \approx 9.6 \quad \text{m}, \tag{2.3.9}$$

corresponding to $\sim$25 m in ground range (Curlander and McDonough, 1991). See Papoulis (1991) for the matched filter principle. The range enhancement factor can be found from the time-bandwidth product:

$$\tau B_R = s_c \tau^2, \tag{2.3.10}$$

which is 576.6 for an ERS SAR image. This resolution corresponds with an effective compressed pulse length $\tau_c = 64$ ns.

The radar echo can be described as the convolution of the transmitted waveform and the surface reflectivity and corresponds with one line in range direction. Considering the echo of a point scatterer with phase shift φ_s, the sensor receives a delayed replica of the transmitted waveform:

$$r(t) = g(t - \Delta t)\, e^{j2\pi f_0(t-\Delta t)}\, e^{j\varphi_s}, \tag{2.3.11}$$

where Δt is the round-trip time between transmitting and receiving. Amplitude scaling factors are omitted here. The linear operation of *quadrature demodulation*,

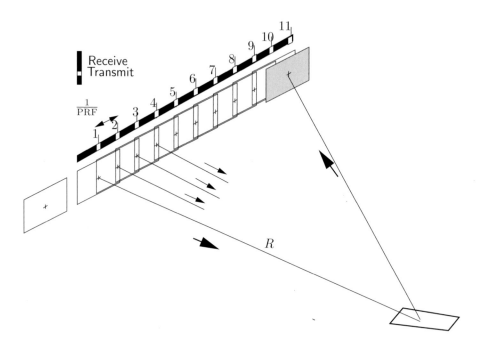

Fig. 2.5. SAR pulse transmitting and receiving mode. For the ERS configurations, a pulse emitted at position 1 is received between position 10 and 11. The time interval between the pulse transmission is the inverse PRF. Therefore, the spacecraft moves approximately 40 m between transmitting and receiving the pulse.

using the local oscillator, shifts the signal to a frequency band centered at zero-frequency, which results in a strict linearity in further signal processing (Curlander and McDonough, 1991, p.136). This yields

$$r(t) = g(t - \Delta t)\, e^{-j2\pi f_0 \Delta t}\, e^{j\varphi_s}. \tag{2.3.12}$$

On board of the satellite, the signal is digitized to complex samples and downlinked to a receiving station, along with a host of other engineering data (chirp replica, noise measurement, calibration pulse) (Alaska SAR Facility, 1997).

2.3.3   Range or chirp compression

The first step in the SAR processing involves the improvement of the resolution in range direction, by compressing the chirped wave form discussed in section 2.3.2. The matched filtering procedure requires the replica of the transmitted chirp. These replicas can (i) be computed, the so-called range reference function (RRF), or (ii) directly retrieved from the downlinked data, as discussed in section 2.2. In the raw ERS data, a replica of the transmitted chirp is available every 24 pulses. To compute the RRF, eq. (2.3.8) is evaluated, using $f_s\tau$ points, where f_s is the sampling frequency and τ is the pulse length. See fig. 2.6 for the shape of the reference

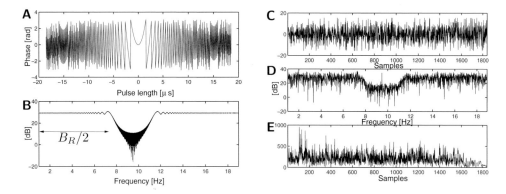

Fig. 2.6. Chirp compression. Panel **(A)** shows the phase of the chirp, passing the zero frequency in the middle. **(B)** The amplitude of the spectrum of the chirp in dB. The bandwidth of 15.5 MHz can be clearly distinguished. **(C)** One range line of raw ERS data, showing only the real values ranging between -15.5 and 15.5. The mean spectrum of 256 lines is shown in **(D)**, and can be compared with (B). The amplitude of the range compressed image is shown in **(E)**.

function.

The chirp or RRF is usually weighted with a Hamming filter to reduce sidelobe effects, zero-padded to the next higher power of 2, and transformed to the Fourier domain. The zero-padded raw SAR data of one pulse are Fourier transformed as well, and multiplied with the RRF. Transforming this product back to the time domain yields the range compressed signal (Curlander and McDonough, 1991; Zebker, 1996; Price, 1999).

2.3.4 Doppler centroid

The antenna beamwidth in azimuth, β_a, causes a point scatterer to be imaged within the beam for a number of consecutive pulses. During these pulses its relative velocity with respect to the radar changes, causing a Doppler-like effect. Strictly, the analogy with the Doppler effect is incorrect, as this is a continuous phenomenon, whereas SAR acquisitions can be regarded as a discrete stop-and-go process (Bamler and Schättler, 1993).

The antenna pattern of the ERS-SAR is aimed almost exactly perpendicular to the sum of the flight direction vector and the velocity vector due to the earth's rotation. Nevertheless there will always be a slight *squint angle* ϕ_s of the spacecraft, steering the antenna pattern away from the zero-Doppler direction, cf. fig. 2.7.

The Doppler frequency f_D for a target approaching with velocity v is $f_\mathrm{D} = 2v/\lambda$. In the SAR geometry in fig 2.7A and B this is

$$f_\mathrm{D} = \frac{2v_{s/c}}{\lambda} \sin \phi_s \sin \theta, \tag{2.3.13}$$

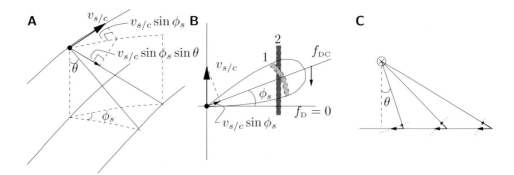

Fig. 2.7. The Doppler centroid and relevant parameters. **(A)** 3D view of the satellite, squint angle ϕ_s and look angle θ. The velocity vector $v_{s/c}$ can be decomposed to the direction towards the target. **(B)** shows a top view of the configuration, with the antenna pattern and Doppler centroid direction. Surface elements in area 1 are in the same range distance from the sensor, whereas area 2 depicts the path of a surface element through the antenna pattern. **(C)** is a view in the flight direction of the satellite, showing the variation in the velocity component over range.

where $v_{s/c}$ is the velocity vector of the spacecraft in an earth-fixed frame and θ is the look angle. Note that this equation determines the sign of the phase observations in the sequel, since a velocity towards the sensor, or a decrease in range ρ, results in a positive f_{D}. Therefore, $f_{\mathrm{D}} = 2v/\lambda = (2/\lambda)(-\partial\rho/\partial t)$. Since the frequency is obtained by differentiating the phase, with $f = (1/2\pi)(\partial\phi/\partial t)$, we find that a decrease in phase corresponds with an increase in range, i.e., $\phi = (-4\pi/\lambda)\rho$.

The center frequency of the passage of a point scatterer through the antenna beam is termed the Doppler centroid frequency, f_{DC}, whereas *zero-Doppler*, $f_{\mathrm{D}} = 0$, denotes the direction in which the Doppler frequency is equal to zero. This direction is perpendicular to the flight direction. Often the phase data in the focused SAR image are *deskewed*, implying that the phase values correspond with the zero-Doppler phase. Deskewed, or "zero-Doppler" processing means that the data are always observed effectively perpendicular to the flight track, which is convenient for keeping track of time, during geocoding, and coarse coregistration, although these steps can be applied during the post-processing as well.

The variation of the Doppler frequency during the passage of the scatterer through the beam is expressed by the *Doppler bandwidth*, B_{Dop}. For a narrow antenna beam in azimuth direction, the Doppler bandwidth is well approximated by (Raney, 1998)

$$B_{\mathrm{Dop}} = \frac{2\beta_a v_{s/c}}{\lambda}, \qquad (2.3.14)$$

where β_a is the azimuth beamwidth of the antenna and $v_{s/c}$ is the relative spacecraft velocity. The radar PRF must be sufficient to sample the Doppler spectrum unambiguously, so that PRF $> B_{\mathrm{Dop}}$. The Doppler bandwidth determines the azimuth

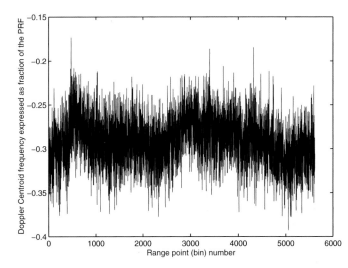

Fig. 2.8. Doppler centroid frequency estimates expressed as a fraction of the PRF for 400 lines of raw data over Java, Indonesia. The data are obtained by differencing the range bins in adjacent lines. The Doppler centroid frequency is estimated to be −0.29 times the PRF.

resolution of the focused SAR image, using

$$\Delta_a = v_{s/c}/B_{\mathrm{Dop}}. \qquad (2.3.15)$$

A typical Doppler bandwidth for ERS is 1377 Hz, depending on the definition of the beamwidth. Note that higher azimuth resolutions are possible using a shorter wavelength, a wider beamwidth, or a lower (faster) orbiting spacecraft.

In fig. 2.7B, a surface element in shaded area 2, after range migration, will have a slightly altered phase value for the next pulse. In that time, the imaged area will be almost the same, the only difference being the relative velocity towards the sensor. This way, we can use the phase difference between neighboring pulses, for the same range distance to estimate the Doppler centroid coarsely. Averaging a number of those differences will give a reasonable estimate, as demonstrated in fig. 2.8. More accurate algorithms are discussed in Madsen (1989). In the example of fig. 2.8, an average phase shift of −1.83 radians is observed from the data, which corresponds with −0.29 cycles. With a PRF of 1679.902 Hz, this corresponds with an estimated Doppler Centroid Frequency of −488.92 Hz. The relative velocity of the spacecraft is 7554.27 m/s, and using eq. (2.3.13) yields a squint angle $\phi_s = -0.27$ degrees. For focusing the SAR data, f_{DC} is important for SNR maximization and for suppressing azimuth ambiguities (Geudtner, 1995). For the focused SAR data, $f_{\mathrm{DC}} \neq 0$ is equivalent to a spectrum that is shifted in azimuth direction. Such shifts are important to consider for interferometric applications, e.g. during interpolation procedures.

Note that f_{DC} varies over range, as sketched in fig. 2.7C. This variation is caused by the variation of the incidence angle over range, cf. eq. (2.3.13), in combination with complicating factors such as earth rotation and topography. Scatterers that have a velocity component in the direction of the radar appear displaced in the focused image.

2.3.5   Range migration

Compare the passage of a scatterer through the beamwidth, area 2 in fig. 2.7B, with the position of scatterers in a range bin, area 1 in fig. 2.7B. It is obvious that the single scatterer appears shifted in different range bins in consecutive pulses, even if the data are acquired under zero-squint. This effect (range migration) makes SAR focusing an inherent 2D process. Although different concepts are available for solving this problem, a common solution is to shift the returns from range cells to a common range bin in order to apply 1D transforms. This procedure is evaluated in blocks of range-compressed data, Fourier transformed in azimuth. The range migration correction results in a non-integer shift of the data, requiring sinc-interpolation (Curlander and McDonough, 1991). During range migration correction it is necessary to decide between zero-Doppler (deskewed) processing or Doppler centroid (skewed) processing, as both strategies result in different migration corrections.

2.3.6   Azimuth focusing

Azimuth compression or focusing is a procedure analogous to range compression. The concept of matched filtering, the convolution of a reference function to the received signal can be applied using the chirp as reference function in range and the predicted Doppler-rate (or phase-history) of a scatterer as reference function in azimuth. The azimuth reference function is computed for each range bin and Fourier transformed. The product of the transformed filter with the Fourier transformed raw data is transformed back to the time domain, yielding a focused SAR image. The basic concepts of this methodology can be found, e.g., in Curlander and McDonough (1991).

In effect, focusing the SAR data yields a final azimuth resolution which is dependent of the azimuth time bandwidth B_A and the velocity of the sensor, see eq. (2.3.15).

2.3.7   Multilook processing

Multilook processing during image focusing can be applied to improve radiometric accuracy at the cost of image resolution. This is usually applied in the range-Doppler domain, by segmenting the Doppler spectrum in a number of subsets, or *looks*, and summing these (independent) looks, see Elachi (1988); Curlander and McDonough (1991). For interferometric applications single-look processing is commonly applied, using the full azimuth bandwidth. In the interferometric SAR processing, the term multilooking is also used for simply averaging a number of resolution cells to improve phase statistics, see section 2.5.6. Phase statistics are discussed in detail in chapter 4.

2.3.8   Imaging geometry

The reference frame of a focused SAR image is spanned by the range-azimuth coordinates. Since the radar is side-looking, terrain elevation will result in geometric distortions in the SAR image. In fact, even the variation in the projection of the reference surface (ellipsoid) in range direction causes geometric distortions, due to the varying incidence angle. Figure 2.9 shows the effects of foreshortening (A), layover

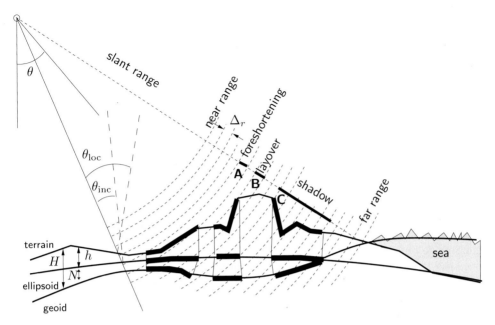

Fig. 2.9. Imaging geometry and reference systems. Radar echos are recorded between the concentric circle segments indicating near range and far range. A range resolution cell with size Δ_r, is indicated by the spacing between the concentric lines. Area **A** indicates foreshortening, **B** indicates layover, and **C** shadow. Ground-range coordinates can be referenced to the ellipsoid. Interferometric heights are referenced to the ellipsoid as well. To obtain orthometric heights H, the geoid height N needs to be accounted for. The look angle θ is defined w.r.t. the (geocentric) state vector, whereas the incidence angle θ_{inc} and the local incidence angle θ_{loc} are defined w.r.t. the local vertical to the ellipsoid and the local terrain, respectively.

(B), and shadow (C), caused by the oblique viewing geometry. The radar coordinate system can be transformed to a ground-range/azimuth coordinate system situated on the ellipsoid.

Resolution and posting

Resolution is defined as the minimal distance at which two distinct scatterers with the same brightness can be uniquely discerned as separate signals (Born et al., 1959). For any system, the fundamental resolution is equal to one over the system bandwidth, which corresponds with the width of the impulse response. These measures of time can be easily transformed into spatial measures. In range this is expressed in eq. (2.3.9), in azimuth in eq. (2.3.15). A resolution cell is defined as the illuminated area responsible for the radar reflection data mapped to a single pixel. Therefore, the resolution cell size is determined by the resolution in range and azimuth.

Pixel size, on the contrary, is a confusing measure, since it is often used in the same context as resolution. A pixel, however, is an infinitesimally small point in which

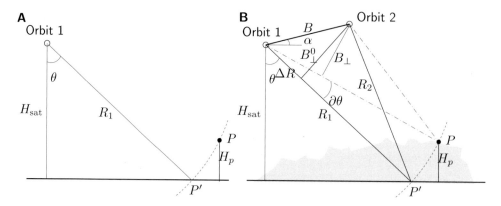

Fig. 2.10. Single-pass and interferometric configuration for a height H_p. **(A)** The difference between point P' on the reference surface $(H_{p'} = 0)$ and point P at the same range but at height H_p. Both points cannot be resolved in a single SAR image, since they are mapped in one resolution cell, and the change in look angle θ cannot be measured. **(B)** Interferometric configuration: two SAR sensors mounted on a platform are separated by a baseline B, and observe the complex response at range R_1. The change in look angle, $\partial\theta$, can be determined from the interferometric phase. Using the range R_1 and the height H_{sat} of the platform it is possible to determine the height H_p of point P.

the complex digitized signal value corresponding with a resolution cell is centered. A pixel, by definition, does not have a physical size, although the representation of a grid of pixel values as an image might suggest so. An acceptable alternative for pixel size is *posting*, the spatial distance between two pixel-nodes, or pixel spacing.

For systems that have the same bandwidth as sampling frequency, posting and resolution cell size have identical values. For most SARs, however, signal processing algorithms require oversampling, or correlation between adjacent pixel values. In the spectral domain this corresponds with a system bandwidth that is smaller than the Nyquist frequency. For ERS focused SAR images, an oversampling ratio of ∼1.223 is used, corresponding with 50% overlay between adjacent resolution cells.

2.4 SAR interferometry

Comparable to a single static radar, the range direction information in a synthetic aperture radar only conveys information on the distance to the sensor at a certain azimuth time. The distance measurement follows from the time observations of the pulse returns by the local oscillator, resulting in a relative accuracy of half the posting in range direction, typically 4–5 meters. As the phase information of a resolution cell has a uniform probability density function, see chapter 4, it does not contain any useful information.

Figure 2.10A shows that distance measurements are not capable of distinguishing two points P' and P at the same slant range, but displaced horizontally in ground-range. Although the backscatter intensity can give an indication for the presence of topography, this information cannot be made quantitative to a high accuracy. It is evident that this problem could be solved by the observation of angular differences between a point P' at a reference body and a point P at a height H_p above this reference body, with the same range R_1 to the sensor. This is equivalent with measuring cumulative angular differences between neighboring resolution cells. In fact, this is what SAR interferometry provides by observing both points from a slightly different geometry, see fig. 2.10B. The effective distance between the two sensors, measured perpendicular to the look direction, is referred to as the *perpendicular* or *effective baseline*, $B_\perp$. Because the instrument is not capable of directly measuring the small angular differences, this information needs to be derived from the distance measurements between both sensors and the resolution cell on earth, applying some simple trigonometry as indicated in fig. 2.10B. Thus, the basic problem of SAR interferometry is the determination of these distance differences.

The required accuracy for measuring the distance differences between sensors and resolution cell is in the mm-range. As the ranging information determined by the range resolution is three orders of magnitude worse, it is not applicable for this type of accurate observations. The phase observations of the received echo provide the solution for this problem. Disregarding atmospheric propagation delay for the moment, the phase observation for a single resolution cell can be regarded as the sum of two contributions: the phase proportional to the distance and the phase due to the scattering characteristics of the resolution cell. Although the scattering phase component is unpredictable, it is a deterministic quantity, i.e., if the phase measurement would be repeated under exactly the same conditions, it would yield the same result. Under these circumstances we state that the imaging is *coherent*. The degree of coherence is a direct measure for the similarity between the two observations. As a consequence, the phase *difference* between two sensors for a coherent system is only dependent on the difference in range, as the scattering phase contributions cancel.

In practice this method imposes high demands on the geometric configuration—parameterized by the *spatial baseline*—and the interferometric processing of the SAR data. Moreover, for repeat-pass interferometry the temporal separation of the two acquisitions—referred to as the *temporal baseline*—can result in changing scattering characteristics due to weathering, vegetation, or anthropogenic activity. All these parameters affect the degree of coherence up to total decorrelation.

In the following sections we will briefly discuss the influence of the parameters mentioned above. A more thorough mathematical treatment of these concepts can be found in chapters 3 and 4.

2.4.1 Phase observation, topographic height, and surface deformation

Using the interferometric configuration sketched in fig. 2.10B, we derive the physical and the geometrical relationships between the two phase observations to obtain topographic height and surface deformation estimates.

Both SAR images are composed of a regular grid with complex values, or *phasors*, y_1 and y_2, which can be decomposed in an amplitude and a phase component using

$$y_1 = |y_1| \exp(j\psi_1)$$
$$y_2 = |y_2| \exp(j\psi_2). \tag{2.4.1}$$

After aligning and resampling the y_2 grid to corresponding locations in the y_1 grid, complex multiplication yields the complex interferogram:

$$v = y_1 y_2^* = |y_1||y_2| \exp(j(\psi_1 - \psi_2)). \tag{2.4.2}$$

The observed phase values ψ_{1p} and ψ_{2p} in the two images for resolution cell P are

$$\psi_{1p} = -\frac{2\pi \, 2R_1}{\lambda} + \psi_{\text{scat},1p},$$
$$\psi_{2p} = -\frac{2\pi \, 2R_2}{\lambda} + \psi_{\text{scat},2p}, \tag{2.4.3}$$

where R_1 and R_2 are the geometric distances and $\psi_{\text{scat},1p}$ and $\psi_{\text{scat},2p}$ are the contributions of the scattering phases in both images. For now, we ignore phase contributions due to signal propagation delay. The origin of the minus sign has been discussed in section 2.3.4. In case the scattering characteristics are equal during both acquisition, i.e., $\psi_{\text{scat},1p} = \psi_{\text{scat},2p}$, the interferometric phase can be written as

$$\phi_p = \psi_{1p} - \psi_{2p} = -\frac{4\pi(R_1 - R_2)}{\lambda} = -\frac{4\pi \Delta R}{\lambda}, \tag{2.4.4}$$

and its derivative is

$$\partial \phi_p = -\frac{4\pi}{\lambda} \partial \Delta R. \tag{2.4.5}$$

Geometrically, the path length difference ΔR can be approximated as (cf. fig. 2.10B)

$$\Delta R = B \sin(\theta - \alpha). \tag{2.4.6}$$

This approximation is known as the *far-field* or *parallel-ray* approximation, see Zebker and Goldstein (1986). Due to the 2π phase ambiguity and orbit inaccuracies it is not possible to derive ΔR from the geometry. However, the relation between changes in ΔR and θ is easily found to be

$$\partial \Delta R = B \cos(\theta^0 - \alpha) \partial \theta, \tag{2.4.7}$$

where the initial value of θ^0 is obtained for the reference surface. Combining the physical phase observations in eq. (2.4.5) with the geometric configuration expressed in eq. (2.4.7) the relation between an interferometric phase change and the change in the look angle θ is found to be

$$\partial \phi = -\frac{4\pi}{\lambda} B \cos(\theta^0 - \alpha) \partial \theta. \tag{2.4.8}$$

The interferometric phase change can be defined as the difference between the measured (unwrapped) phase ϕ and the expected phase for the reference body ϑ derived from the orbit geometry, hence

$$\partial\phi = \phi - \vartheta. \tag{2.4.9}$$

The height of the satellite above the reference body is known, and can be expressed as

$$H_{\text{sat}} = R_1 \cos\theta, \tag{2.4.10}$$

and the derivative for a resolution cell P with range R_{1p} gives the relationship between a change in look angle θ due to a height difference ∂H_{sat}:

$$\partial H_{\text{sat}} = -H_p = -R_{1p} \sin\theta_p^0 \partial\theta. \tag{2.4.11}$$

Note that this definition is independent of the choice of the reference body, although fig. 2.10B is sketched for a flat earth. Furthermore it has to be noted, cf. fig. 2.10A, that H_p is the measured height for range resolution cell P, since P and P' are in the same resolution cell. Using eqs. (2.4.8) and (2.4.11) we derive the relationship between the height H_p above the reference body and the phase difference $\partial\phi_p$:

$$H_p = -\frac{\lambda R_{1p} \sin\theta_p^0}{4\pi B_{\perp,p}^0} \partial\phi_p, \tag{2.4.12}$$

with

$$B_{\perp,p}^0 = B\cos(\theta_p^0 - \alpha), \tag{2.4.13}$$

see fig. 2.10B. The initial value θ_p^0 is found for an arbitrary reference surface (e.g., a sphere or ellipsoid). A recursive scheme is used to find new values for θ_p at a specific height above this reference surface. Inserting $\partial\phi_p = 2\pi$ in eq. (2.4.12) yields the *height ambiguity*—the height difference corresponding with a 2π phase shift:

$$h_{2\pi} = |\frac{\lambda R_{1p} \sin\theta_p^0}{2B_{\perp,p}^0}|. \tag{2.4.14}$$

To conclude this evaluation, we combine the influence of topography, H_p, and surface displacement, D_p, on the interferometric phase differences, relative to the reference body. Using eqs. (2.4.5), (2.4.7) and (2.4.12), we find

$$\partial\phi_p = -\frac{4\pi}{\lambda}(D_p - \frac{B_{\perp,p}^0}{R_{1p} \sin\theta_p^0} H_p). \tag{2.4.15}$$

Since the measured interferometric phase is the sum of the reference phase ϑ and the deviations:

$$\phi_p = \vartheta_p + \partial\phi_p, \tag{2.4.16}$$

where the reference phase is defined by

$$\vartheta_p = \frac{4\pi}{\lambda} B \sin(\theta_p^0 - \alpha),$$ (2.4.17)

we find

$$\phi_p = \frac{4\pi}{\lambda}(B \sin(\theta_p^0 - \alpha) - D_p - \frac{B_{\perp,p}^0}{R_1 \sin\theta^0} H_p).$$ (2.4.18)

Using ERS parameters, this implies that for an effective baseline $B_\perp$ of 100 meters, a height difference H_p of 1 meter yields an interferometric phase difference of approximately 4.5 degrees, which is well below the noise level of some 40 degrees, and is therefore practically undetectable. However, in the differential case, a change D_p of 1 cm in the range direction, yields a phase difference of 127 degrees, which is easily detectable.

2.4.2   Differential interferometry

Differential interferometry aims at the measurement of ground deformation using repeat-pass interferometry. Since line-of-sight displacements enter directly into the interferogram, independent of the baseline, it can be measured as a fraction of the wavelength. Unfortunately, non-zero baselines will always cause some sensitivity w.r.t. topography in the interferogram. The length of the perpendicular baseline deviates from zero since (i) it is often not possible—or not desirable—to manoeuvre the satellite in a zero-baseline orbit, and (ii), the baseline varies with the look angle, yielding only one zero-baseline range-bin in the interferogram. Of course it depends on the range of topography in the scene whether the baseline is significant.

There are several ways to construct a differential interferogram. The *two-pass* method uses an external elevation model that is converted into radar coordinates, scaled using the baseline, and subtracted from the interferogram (Massonnet et al., 1993). Since for many areas in the world (crude) elevation models are available this can be a feasible approach. Of course, errors in the DEM will propagate into the deformation results, depending on the baseline characteristics.

A second method is the so-called *three-pass method* (Zebker et al., 1994b), see section 3.5.1. Here, an extra SAR acquisition is used and combined with a suitable partner acquisition to create the so-called *topographic pair*. This pair is assumed to have no deformation, a suitable baseline providing sensitivity to topography, and sufficient coherence. This pair is unwrapped, scaled to the baseline characteristics of the *deformation pair*, and subtracted from it, yielding the differential interferogram, the so-called *differential pair*.

In order to perform the three-pass method, using a topographic pair derived from SAR image A and B and a deformation pair from image A and C, the effect of the baseline difference between the two pairs should be taken into account. The baselines can be scaled after the reference phase is subtracted from both interferograms (Zebker et al., 1994b).

The three-pass method can only be applied when both the topographic pair and the deformation pair have a common image. This image is then used as a reference to align the other two. Due to the available baselines for a specific scene, it might be that the topographic pair and the deformation pair cannot be chosen with a common image. For example, the baselines for all candidate partners of the deformation pair are too large, reducing coherence in the topographic pair and hampering phase unwrapping. In that case, the *four-pass method* can be used, where the topographic pair and the deformation pair are independent, i.e. they share no common SAR acquisition, see section 3.5.2. As long as sufficient alignment of the two interferograms can be performed the methodology is comparable to the three-pass method. Practical differences with the three-pass method are manifested in the influence of atmospheric signal in the three or four scenes and in the alignment.

Regarding the atmospheric signal, the three-pass method has one identical atmospheric contribution for an arbitrary pixel in the common image. Since the topographic pair is scaled up or scaled down with scaling factor $B_{\perp,\mathrm{defo}}/B_{\perp,\mathrm{topo}}$ to match the baseline characteristics of the deformation pair, also the atmospheric signal in the topographic pair is scaled up or down. The influence of the scaling needs to be considered carefully in the interpretation of the results. In section 3.5.1, the functional model for differential interferometry is discussed in more detail.

2.4.3 Conditions for interferometry

SAR interferometry only works under coherent conditions, where the received reflections are correlated between the two SAR images. Evidently, this is the most important condition for interferometry. Loss of coherence, known as *decorrelation*, can be due to a number of driving mechanisms. The effect of some sources of decorrelation, e.g., as introduced by the alignment and interpolation of the images, can be reduced by using well-designed filtering procedures. Other sources of decorrelation are more significant and non-reversible. The two most important conditions are related to the phase gradient and the temporal variation in the physical distribution of the elementary scatterers.

The phase gradient condition can be conveniently described in the spectral domain. The temporal bandwidth of the SAR images in range corresponds with a spatial bandwidth due to the projection on the earth's surface. A phase gradient in range of n cycles/pixel corresponds with a spectral shift between the spectra of both acquisitions of nf_s Hz, where f_s is the sampling frequency. The spectral shift results in a decreased overlap between the corresponding parts of the spectrum (the signal) and an increasing non-overlapping part of the spectrum (the noise). Due to the limited bandwidth, a phase gradient larger than B_R/f_s cycles/pixel (approximately 0.822 for ERS) results in a zero overlap between the spectra, hence a complete loss of correlation. The occurrence of this situation is dependent on: the length of the perpendicular baseline, the steepness of the topographic slopes, and/or the gradient of the surface deformation.

If topography w.r.t. an ellipsoidal reference surface is neglected, loss of correlation due to the length of the perpendicular baseline occurs at approximately 1 km for

ERS, commonly referred to as the *critical baseline*, $B_{\perp,\text{crit}}$. A topographic slope towards the sensor decreases the critical baseline locally. The relation between the fringe frequency f_ϕ [cycles/m], the perpendicular baseline $B_\perp$, the incidence angle θ_{inc}, and the local terrain slope ζ (positive for slopes tilted towards the sensor) can be expressed by (Gatelli et al., 1994; Bamler and Hartl, 1998):

$$f_\phi = \frac{\partial\phi/2\pi}{\partial R} = -\frac{2B_\perp}{\lambda R \tan(\theta_{\text{inc}} - \zeta)}. \qquad (2.4.19)$$

Using the critical phase gradient of 0.882 cycles/pixel ($f_\phi = 0.882/7.91$ cycles/m), we find a critical baseline of $\sim$1100 m. Equivalently, a phase gradient due to surface deformation in range direction needs to be less than 0.822 cycles/pixel in the radar line of sight, corresponding to 0.882 $\lambda/2$ m per 7.9 m or 2.9 mm/m. This gradient of approximately 10^{-3} forms an important upper limit for detecting slopes or deformation gradients in the interferogram. It excludes, e.g., areas close to an earthquake faulting area, where deformation gradients above 10^{-3} may occur.

Figure 2.11, modified from Massonnet and Feigl (1998), gives an overview of the limitations of the ERS SAR system, regarding deformation studies, although the change in range could also be due to topographic slopes. The detectable combinations of a change in range and the width of the phenomenon are limited within the five bold lines, indicated in the figure. The upper deformation gradient of 10^{-3} is shown, as well as a lower deformation gradient of 10^{-7}, or 1 cm per 100 km. The lower 10^{-7} gradient is mainly due to orbit inaccuracies or long wavelength atmospheric gradients.

The limits to the left-hand and right-hand side of the white area are given by the physical limitations of the system, specifically the resolution and the swath width, respectively. Finally the thermal noise of the ERS-AMI system imposes an upper limit on the phase accuracy of about 12°, or 1/30th of a fringe, which is equivalent to 1 mm in range between any two pixels.

After identifying these conditions for the application of interferometry, we can evaluate whether specific geophysical phenomena comply with these conditions. For example, events such as the rupture near an earthquake fault, or strong volcanic deformations can easily exceed the 10^{-3} limit. On the lower end of the plot, events such as earth tides, loading, or postglacial rebound are hard to detect since they have too long spatial wavelengths with respect to the imaging system and the 10^{-7} gradient limit. Of course, swaths including several consecutive images might be used to circumvent some of these problems.

Some of the case studies reported in chapter 5 are indicated in fig. 2.11. For example, the vertical movement of a corner reflector is at the edge of the range of detectable deformations, with a width of only a few pixels and a change in range of less than one-half phase cycle. For subsidence phenomena, as in the Cerro Prieto Geothermal Field (CPGF) or the Groningen gas field, it is evident that these phenomena have a strong time contraint. With a maximum subsidence rate of $\sim$7 mm/yr and a spatial extent of $\sim$20 km, the deformation could become marginally detectable after 1 year, although a time span of 5 years is more suitable for this phenomenon. On the other

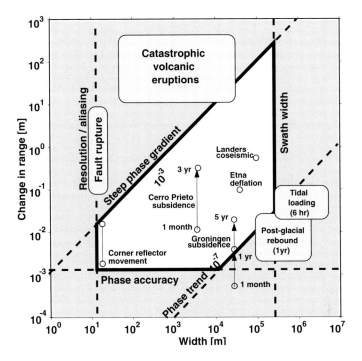

Fig. 2.11. Range of applications for deformation mapping, modified from Massonnet and Feigl (1998). The horizontal axis shows the characteristic spatial width of the imaging system and several geophysical phenomena. The vertical axis expresses the deformation gradient, which can be translated to a phase gradient for C-band repeat-pass SAR interferometry. The effects of temporal decorrelation, atmospheric error signal, and residual topographic phase errors are not included.

hand, subsidence rates at CPGF of ∼1 cm/month over an area of only a few km across are much easier to measure.

Temporal decorrelation, not indicated in fig. 2.11, is the second major limitation for the application of repeat-pass interferometry, and a key problem in large parts of the world when using large time intervals. For temporal decorrelation to occur, the incoherent sum of the variable scatterers within a pixel should be a significant fraction of the coherent sum of stable scatterers. Weathering, vegetation, or anthropogenic activities are common causes of temporal decorrelation.

A decorrelation effect can be directly related to the variance of the interferometric phase and propagates to the parameters that are inferred from this phase. This is discussed in depth in chapters 3 and 4.

Apart from error sources causing direct decorrelation, which is often easily observable as noise in the interferogram, there are often limitations in distinguishing between the different geophysical or induced sources of coherent phase variation. As stated previously, the observed interferometric phase is a coherent superposition of effects from topographic height, surface deformation between the acquisitions, the state of the atmosphere during both acquisitions, and possible errors in the orbit of the sensor. The interpretation of a SAR interferogram, aimed at only one of these applications, is therefore dependent on prior knowledge about the other contributions. For instance, in two-pass differential interferometry, an error in the a priori elevation model yields artifacts in the deformation maps.

Other limiting conditions can occur in the interferometric processing sequence. Especially the process of phase unwrapping, in which the phase ambiguities are resolved, can lead to local or global errors in the inferred height or deformation maps.

Other limitations in the interpretation of interferometry depend on the application of the products. For example, for DEM generation, micro-topography may have to be considered. Within each resolution cell the actual height may vary considerably, for example by buildings. It depends on the dominant scatterers within the resolution cell, i.e., a roof top, what the resulting phase will be and therefore, which height will be measured. In layover areas, scattering from various separated locations may contribute to the phase observation, cf. fig. 2.9, leading to an ambiguity that cannot be solved for. For areas that lack significant return signal (low backscatter areas, i.e., due to specular[2] reflection) the phasors are too short to provide a useful phase observation. Finally, strong scatterers may create significant sidelobes in the interferogram. These sidelobes can contaminate the phase observations in the neighboring resolution cells significantly, which can result in an erroneous interpretation of these phase values.

2.5 Interferometric processing overview

This section gives an overview of the sequence of most important decisions and processing steps to create interferograms and geocoded products from SAR data. The interferometric processing can start with focused complex SAR data, sometimes referred to as Single-Look Complex (SLC) data, which may be available as a product from the agency exploiting the satellite. Nevertheless, raw (unfocused) SAR data are often preferred over SLC data, since they are usually cheaper, can be delivered faster, and exclude the possibility of different focusing strategies at the various processing facilities. In the processing steps described in the sequel, it doesn't matter whether image focusing, as described in section 2.3, has been performed by a third party or if it is included in the interferometric processing chain. The only exception is the azimuth filtering, which is not necessary using raw data since the mean of the Doppler centroid frequencies of both images can be used in image focusing. For SLC data, it is not known beforehand for which interferometric combinations the data will be used. Therefore, the data are focused with respect to the zero-Doppler frequency. To suppress noise introduced by the non-overlapping parts of the image spectra, azimuth filtering may be required to remove these parts.

---

[2]Specular reflection is the redirection of an electromagnetic wave at an equal but opposite angle, as described by Snell's Law in optics. Hence, the scattered signal is directed away from the radar (Raney, 1998).

2.5.1   Image selection

Image selection is perhaps one of the most vital decisions in the application of radar interferometry, assuming that several SAR acquisitions are available. The criteria depend on the specific application of the study. Main decisions regard the type of sensor, the availability of the data, the temporal and spatial distribution of the baselines, and the characteristics of the terrain and atmosphere during the image acquisitions.

The sensor and platform characteristics include important parameters such as wavelength, bandwidth, SNR, orbit inclination, and repeat period. Spaceborne repeat-pass radar interferometry is mainly feasible from L-band ($\lambda = 23$ cm) to X-band ($\lambda = 3$ cm). Below L-band, ionospheric signal will deteriorate the observations and above X-band the instrument is too sensitive for the weather situation. The wavelength will also influence the fringe density due to topography or deformation. The sensitivity for topography or deformation is a combination between wavelength and SNR. The range-bandwidth of the sensor determines the length of the critical baseline and the range resolution. Platform characteristics such as orbit inclination and the repeat interval determine the coverage of the earth, the occurrence of polar gaps, and the revisit times between SAR acquisitions. Power considerations in combination with other instruments on the platform can limit the operating time of the sensor per orbit. Finally, the availability of precise tracking devices and the orbit maintenance procedures influence the accuracy of the interferometric baseline.

The next concern after deciding on the type of sensor and platform is the data availability. For satellites such as ERS, JERS, and Radarsat, convenient tools are available on-line to browse through the archive of acquired images. Due to the overlap between the SAR swaths of neighboring orbits, as well as the availability of ascending as well as descending orbits, a certain area on earth may be imaged from different viewing geometries. This enables the reduction of image distortion effects and resolving different components of deformation vectors. Note that it is not possible to create interferometric pairs from SAR images acquired from different orbital tracks. Due to the maximum baseline restrictions, the viewing directions of the two acquisitions should differ less than 4 minutes of arc for ERS. For scientific satellites such as the ones listed above, it is possible to request future acquisitions of a certain area, using a specific imaging mode.

Focusing on archived data, the distribution of temporal and spatial baselines between SAR acquisitions can be used to decide on the feasibility of interferometry for a specific application. In order to obtain a quick overview of the possibilities, a graphical representation of the spatial and temporal baselines can be convenient, see fig. 2.12. The spatial (perpendicular) baselines give an indication of the sensitivity to topographic height, the amount of decorrelation due to the phase gradients, and the effectiveness of the phase unwrapping. Depending on the roughness of the terrain, convenient perpendicular baselines for topographic mapping range between 100 and 500 m. The temporal baselines for topographic mapping should ideally be as short as possible, to minimize the effect of temporal decorrelation. For interferograms spanning surface deformation, a minimal perpendicular baseline is preferred to re-

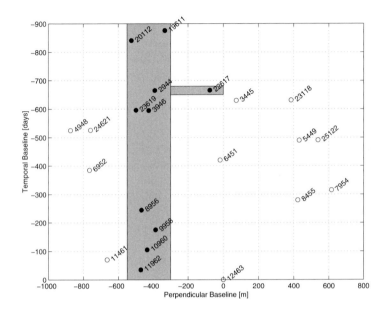

Fig. 2.12. Baseline plot, showing the distribution of SAR acquisitions in time and space. The x-axis shows the (relative) spatial separation of the images perpendicular to the look-direction, the y-axis the time interval w.r.t. the most recent acquisition. The interferometric combination of two SAR images results in a temporal and spatial (perpendicular) baseline.

duce phase signal due to topography and noise due to the phase gradients. The ideal temporal baseline depends on whether the deformation is expected to be continuous in time or instantaneous, as e.g. earthquake deformation. For continuous deformation processes, the expected deformation rate should be used to choose temporal baselines which ensure a sufficient signal-to-noise ratio between the deformation and all other effects in the interferogram (noise, atmospheric signal, residual topographic signal). For instantaneous processes, a short temporal baseline spanning the event is usually ideal.

Some characteristics of the terrain are important to consider during the image selection process. Relevant characteristics are roughness, elevation range, type and amount of vegetation, and the influence of anthropogenic activities. The first two terrain characteristics are evaluated in combination with possible baselines and sensors, and may lead to the combination of ascending and descending interferograms. The latter two characteristics give an qualitative indication of the amount of decorrelation that can be expected. For example, on coarse scales (> 100 km) the monthly Normalized Differential Vegetation Index (NDVI) as derived from NOAA AVHRR observations can be used to approximate the amount of decorrelation due to vegetation in tandem imagery, see van der Kooij (1999b). To avoid seasonal effects, such as snow or deciduous trees, it can be advantageous to select data from the same time of year. If surface deformation is the goal of the study, it is important to consider the type, scale, and amount of deformation to be expected.

The amount of atmospheric signal to be expected during the acquisitions can be a final consideration in selecting the data. Information about the state of the atmosphere can be derived from satellite imagery or ground-based synoptic meteorological data. Especially indications of convectivity in the troposphere, often related with vertical refractivity structures, can be an indication for significant atmospheric signal in the acquisitions. Likewise, choosing night-time instead of day-time acquisitions will reduce the likelihood of atmospheric signal.

The considerations listed above may even lead to an optimal selection of potential acquisitions and interferometric combinations. However, they may also lead to the conclusion that radar interferometry is not the most suitable geodetic tool to fulfill the goals of the study. Either way, proper consideration and evaluation of the options is of major importance.

2.5.2 Preprocessing

If data processing starts from raw data they need to be checked for inconsistencies, such as missing lines or sampling window start time (SWST) changes, before focusing the images. If the number of missing lines is limited, the good lines heading or trailing the inconsistencies can be copied to fill the gaps. This will only slightly deteriorate image focusing. SWST changes within the image can be accounted for by aligning both sides.

For interferometric processing, the spectra of the files should have maximum overlap in azimuth direction. For pre-processed SLC data, where the desired interferometric combinations are not known beforehand, an azimuth filtering procedure might be necessary, see section 2.5.5. When raw SAR data are used, and the interferometric combinations are known in advance, maximum spectral overlap in azimuth direction can be obtained by estimating the mean Doppler centroid frequency, f_{DC}, in each image. Since f_{DC} determines the center frequency for the azimuth window used by the SAR processor, there should be maximum overlap between the spectra of both images. This can be obtained by using the average value of the two f_{DC} estimates.

2.5.3 Coregistration

The sub-pixel registration of both focused SAR images is a strict requirement for interferometric processing. Depending on the start-stop times of a particular section in the orbits (corresponding with the operation of the SAR), there can be an along-track shift between the two SAR images, up to several thousands of lines. The across-track shift is equal to the length of the parallel baseline divided by the posting in range ($c/2f_s$), which can amount up to tens of pixels.

Since cross-correlation techniques for optimal alignment tend to be slow for very large search windows, the procedure is usually separated in two steps: *coarse* and *fine* coregistration. In the coarse coregistration, the offsets are approximated either by defining common points in the image by visual inspection or by using the satellite orbits and timing as a reference. The subsequent fine coregistration usually applies automatic correlation techniques to obtain sub-pixel alignment accuracy.

Coarse registration

The raw SAR data from ERS are tagged by a counter using a 3.4 ms interval. With a relative velocity of 7.1 km/s on the ground, this corresponds with approximately 25 m in azimuth direction. Using precise orbit information the position of a point at a certain range/azimuth position can be determined within some 25–50 m.

Such an approach usually applies two procedures. First, for an arbitrary point on the master's orbit and an arbitrary range the position of a pixel on the ellipsoid is determined. Next, given this position an iterative algorithm searches along the slave orbit until the correct Doppler position is found from which this pixel is observed. Both orbit positions (in the master and the slave orbit) correspond to an exact acquisition time with respect to their images start times (the acquisition times of the first azimuth line) and using the pulse repetition frequency the azimuth line numbers can be computed for both images. The along-track shift is the difference between these line numbers. In range direction the distances to the common pixel, divided by the range posting, yields the across-track shift. The procedure is comparable to the geocoding of the interferogram, described in more detail in section 2.5.13.

Fine registration

Once the relative shifts between the two SAR images are determined within tens of pixels in azimuth and a few pixels in range, the fine (sub-pixel) registration can be performed. Just and Bamler (1994) have shown (for distributed scatterers) that coregistration to an accuracy of 1/8th of a pixel yields an almost negligible (4%) decrease in coherence, see also section 4.4.6, as long as the data are acquired with relatively small squint angles. Coherent registration techniques apply the full complex (amplitude and phase) data to perform a complex cross correlation. These techniques can be very accurate, but are generally not very robust. For situations with large effective baselines, where many fringes appear in the estimation window due to the side-looking geometry and topography, or with poor phase fidelity these methods may fail. Incoherent techniques apply either only the (squared) amplitude of the signal or even only the phase.

The cross-correlation of the powers (squared amplitudes) of the two images is a commonly applied method for coregistration. The correlation peak indicates the offset vector between the images in range and azimuth direction. For sufficiently correlated images the offsets can be determined with an accuracy of 1/20th of a single-look pixel, corresponding with 20 cm in azimuth and 1 m in ground-range for ERS. Note that since the cross-correlation product has a doubled bandwidth compared with the two SLC images, aliasing will be introduced. To avoid this, both datasets need to be oversampled by at least a factor two before calculating the cross-correlation. Prati and Rocca (1990) oversample a set of windows of 100×100 pixels in both complex images by a factor of 8 using an FFT interpolation. The relative shift between the two images is then determined for every window.

Once the relative registration shifts are determined for a set of windows, ideally evenly distributed over the whole scene, the shifts can be regarded as a set of displacement vectors to restrict a two-dimensional polynomial to determine the displacement vectors for every single pixel in the slave image. The polynomial transformation might be up to a fifth degree, although usually offsets and stretch parameters in both directions are sufficient, comparable to a bi-linear model. See Brown (1992) for a general review paper on coregistration and Michel et al. (1999) for an applied overview. Coregistration for SAR interferometry is discussed in Gabriel and Goldstein (1988); Gabriel et al. (1989); Prati et al. (1989); Prati and Rocca (1990); Weydahl (1991); Lin et al. (1992); Hartl and Xia (1993); Zebker et al. (1994c); Geudtner et al. (1994); Geudtner (1995); Carrasco et al. (1995); Coulson (1995); Homer and Longstaff (1995); Schwäbisch and Geudtner (1995), and Samson (1996).

2.5.4 Resampling and interpolation

The interferometric combination of the two complex images requires evaluation of the complex values in one of the two at the pixel positions of the other. Although the implementation may be different, this resampling can be regarded as consisting of two subsequent steps: (i) reconstruction of the continuous signal from its sampled version by convolution with an interpolation kernel, and (ii) sampling of the reconstructed signal at the new sampling locations. The choice of the interpolator kernel (especially its length) requires a tradeoff between interpolation accuracy and computational effort. Using two subsequent one-dimensional interpolation algorithms, the amount of floating point operations is in the order of k^3 for an interpolation kernel with k points. Kernel sizes vary between 1 (nearest neighbor), 2 (linear), and 4, 8, or 16 for extended kernels. The corresponding interpolation times are in the order of k^3, $8k^3$, $64k^3$, $512k^3$, and $4096k^3$, respectively. In section 4.4.5, it is shown that 4-point or 6-point kernels are usually sufficiently accurate. Moreover, for larger baselines spectral shift filtering in range increases the oversampling ratio so that even 2-point kernels may be of sufficient quality. The same holds in azimuth direction, if the Doppler centroids are different, as may occur using ERS-1 and ERS-2 combinations. For high-quality interferogram processing, the complex data are oversampled by a factor of 2 before the complex multiplication of both datasets, to avoid aliasing effects. Then, linear (2-point) interpolation is sufficiently accurate, see fig. 4.11.

The interpolator kernel is centered at zero-frequency. Therefore, the suppression of the spectrum of the signal is most severe for the higher frequency. As the signal spectrum of SAR data is only centered in range direction and not in azimuth, a Doppler centroid frequency that deviates significantly from zero-Doppler will be distorted non-symmetrically. To avoid this, it is recommended to shift the signal temporary to the zero-Doppler frequency, perform the interpolation, and shift the interpolated signal back to the original Doppler centroid frequency.

2.5.5 A priori filtering

Phase noise in the interferogram can lead to problems in the phase unwrapping (high number of residues) or hamper data interpretation. Although the effects in the interferogram might be similar, the driving mechanisms behind the noise may

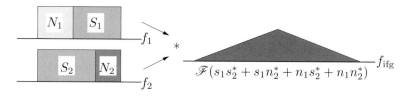

Fig. 2.13. The convolution of two spectra with a distinct signal (S_1, S_2) and noise (N_1, N_2) component, where $\mathscr{F}(s_1) = S_1$. In the convolution (the product of the original data), the noise components n_1 and n_2 appear smeared over the spectrum.

be different and require a different approach. To account for this difference, we distinguish between a priori filtering and a posteriori filtering. A priori filtering is applied to the original SAR data, before interferogram generation, whereas a posteriori filtering is applied after the interferogram has been formed—if applicable even to stacks of interferograms. A posteriori filtering is discussed in section 2.5.8. Complex multilooking is a filter operation commonly applied during interferogram formation.

The main characteristic of the error sources attacked by a priori filtering is the separability of signal and noise in the spectral domain. Consider two image spectra that contain a distinct signal part and noise part, as in fig 2.13. The complex multiplication of the two images, as applied during interferogram formation, corresponds with a convolution of the two spectra. In the convolution, the noise parts of the two original images will get smeared over all frequencies in the spectrum of the resulting interferogram. Since the noise components cannot be effectively separated at this stage, a posteriori filtering of the interferogram for this type of noise is not very effective.

Spectral shift compensation

Spectral shift compensation aims at increasing the SNR of the product by eliminating the noise components in the two datasets (Gatelli et al., 1994; Prati and Rocca, 1994b; Just and Bamler, 1994). Increasing the SNR improves coherence, hence better phase statistics and less problems during phase unwrapping. Slightly different approaches need to be applied in range and azimuth direction.

The origin of the spectral shift problem is the difference between the object spectrum and the data spectrum, as sketched in fig. 2.14A. The object spectrum is a function of the object properties, in this case the radar reflectivity of the terrain, and is independent of the imaging system. In contrast, the data spectrum reveals the characteristics of the imaging system—its bandwidth contains the measurable width of the spectrum, and its Nyquist frequency is determined by the system's sampling frequency. Ideally the object spectra should be mapped identically to the data spectra in order to apply multiplicative interferometry.

A spectral shift is caused by the difference in the local incidence angles towards the two sensors, see fig. 2.14B. If both sensors would be at exactly the same position and look in the same direction, no phase-differences would occur in the image, hence,

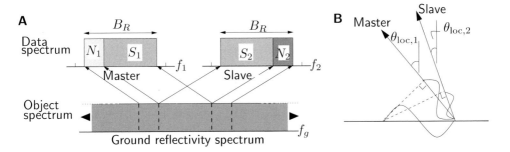

Fig. 2.14. Mapping of the wavenumber domain (object spectrum) of the surface reflectivity to the frequency domain in the SAR data spectrum. **(A)** A different mapping of the object spectrum onto the data spectra of the two images causes the same wavenumbers to appear shifted in the two SAR data spectra. **(B)** The wavenumber shift in range is caused by the different local incidence angles, θ_{loc}, resulting in different projections.

no fringes. Since the time-domain (the image) will be equal for both sensors, the spectral domain is equal as well. Consider now a different geometry, with sensor 2 at a slightly different position from sensor 1. Now phase differences—fringes—will occur in the image. Albeit the *location* of the data spectrum is indifferent (the data do not sense a difference in the local incidence angle), its *contents* have shifted to other frequencies, cf. fig. 2.14A. In other words, the ground-to-slantrange mapping causes a wavenumber shift—the wavenumbers of the object spectrum have shifted to other frequencies in the data spectrum.

The problem of the wavenumber shift is that both images have an overlapping spectral part, which contains information, and two non-overlapping parts, which can be considered as noise for interferometric purposes. Therefore, a bandpass filter needs to be applied to eliminate the two parts containing noise. To tune the bandpass filter it is necessary to determine the wavenumber shift as accurate as possible. The fringe frequency reflects the amount of spectral shift and hence the amount of decorrelation due to the non-overlapping parts of both spectra. It can be estimated locally from the data—as long as there is sufficient correlation between the two datasets—or from the orbit state vector information. In the latter case, the change in range between near range and far range is used to compute the average phase gradient (fringe frequency), corresponding to one spectral shift value for the entire image. However, for topography with a variety of slopes ζ, the filtering can be too coarse—resulting in an unnecessary loss of resolution—or not significant enough—resulting in an unnecessary loss of coherence. In the former case, the filtering can be performed per range line, or even in small patches to adapt its performance to local slopes. The patches are complex multiplied to form an interferogram strip which is Fourier transformed. The fringe frequency f_ϕ [cycles/m] follows from the peak in the amplitude spectrum. The fringe frequency is related to the perpendicular baseline $B_\perp$, the incidence angle θ_{inc} and the local slope of the terrain ζ in the zero-Doppler direction, cf. eq. (2.4.19)

(Gatelli et al., 1994; Bamler and Hartl, 1998):

$$f_\phi = \frac{\partial \phi}{2\pi \partial R} = -\frac{2B_\perp}{\lambda R_1 \tan(\theta_{\text{inc}} - \zeta)}, \tag{2.5.1}$$

where R_1 is the slant-range to the reference satellite. After the spectral weighting (Hamming in the case of ERS) is removed, the non-overlapping parts of the spectra of both datasets can be set to zero-values, and the Hamming window can be applied again to the newly formed spectra, while considering the new bandwidths $B_{R,1} = B_R - |f_\phi|$. The smaller bandwidth translates in a reduced resolution in range direction, see eq. (2.3.9). Therefore, range filtering can yield a significant increase in correlation between the two images at the expense of a reduced resolution (Gatelli et al., 1994).

Spectral shift in azimuth direction

The procedure in the azimuth direction depends on the input products for the interferometric processing. Working with raw radar data, the Doppler centroid frequencies $f_{\text{DC},1}$ and $f_{\text{DC},2}$ of the two datasets can be estimated, see section 2.3.4. By processing both datasets at the average (center) f_{DC} value, the relative spectral shift can be accounted for. Note, however, that this approach does not account for the sinc^2 weighting function, cf. fig. 2.4, on the antenna pattern (Schwäbisch, 1995a). SLC data have been processed at a fixed Doppler Centroid frequency, and as a result tuning the processing f_{DC} for a specific interferogram combination is not possible anymore. In this case, both spectra will appear shifted in the SLC data spectrum. Similarly as in range direction, the non-overlapping parts of the spectrum will result in noise in the interferogram.

The difference in f_{DC} between both images can be caused by (i) the different squint angles of the sensors, or (ii) the convergence of the orbits. Both possibilities have the same effect: a point on the terrain is viewed from two different positions. The different imaging geometry causes a different mapping of ground-range to slant-range, therefore an increased/decreased sensitivity for e.g. higher frequencies, which is translated to a shift in the spectral domain of the surface within the data spectrums bandwidth.

Bandpass azimuth filtering can be performed in order to remove the non-overlapping parts of the azimuth image spectra. At the same time the oversampling by a factor of two can be performed, so the output data are twice as large as the input data. These operations are performed in the frequency domain. Both images need to be filtered. The weighting function used by the SAR processor needs to be selected beforehand (often Hamming weighting)

Azimuth filtering may be necessary whenever the Doppler centroid frequencies of the two SAR images differ too much. This is a result of a difference in the squint angles of the radar antennas. An effective squint angle can be introduced by the earth's rotation. For example, the orientation of the antenna beam of Seasat was orthogonal to its orbital plane, causing $\pm 3°$ squint variation for each orbit due to the earth's rotation. The phased array antennas of the ERS systems, however, apply a technique called "yaw-steering," which largely corrects for earth rotation

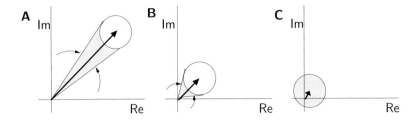

Fig. 2.15. Phasor with error ellipse in the complex plane. The errors in the real and imaginary direction are assumed to be equal. By comparing **(A)** and **(B)** it is clear how the amplitude of the phasor influences the phase variance, since the phase variation in situation (A) is significantly less than in situation (B). In **(C)** the extreme case is shown in which the phase observation is nearly worthless.

effects. Effectively, the antenna beam is maintained orthogonal to the spacecraft's groundtrack, which translates in a symmetry with respect to zero-Doppler within a tolerance of 5% (Henderson and Lewis, 1998). For this reason, effective azimuth squint angle variation for ERS is limited.

Doppler centroid frequency estimation for SLC images

When SLC images are used as input, the Doppler centroid frequency variation (mainly over range) can be estimated to obtain the input values for azimuth filtering. The frequency can be estimated over a strip of SLC imagery, small in azimuth but wide in range using a discrete Fourier transform. Areas with homogeneous intensity or water areas are particularly suitable. Although these estimates are not very accurate they are often good enough for locating the zero-padding location (for oversampling) and coregistration.

2.5.6 Interferogram formation

A complex interferogram is constructed by a pointwise complex multiplication of corresponding pixels in both datasets, see eqs. (2.4.1) and (2.4.2):

$$y_1 y_2^* = |y_1| \exp(j\psi_1) \, |y_2| \exp(-j\psi_2) = |y_1||y_2| \exp(j(\psi_1 - \psi_2)). \qquad (2.5.2)$$

Although especially the phase $\phi_p = \psi_{1p} - \psi_{2p}$ of pixel P is of importance for interferometry, the complex interferogram is usually retained for future computations. Apart from the fact that the amplitude of the phasor contains important information on the SNR of the phase observations, see fig. 2.15, the complex representation does not have the artificial 2π phase jumps. This is convenient, e.g., when orbital phase gradients need to be subtracted from the interferogram. Furthermore, a posteriori multilooking to improve phase variance cannot be applied with optimal accuracy when the amplitude information is discarded.

Oversampling

The complex multiplication of two aligned and interpolated SLC datasets corresponds with a convolution in the frequency domain. Consequently, the resulting interferogram will have a doubled bandwidth. Although the raw radar data have been truncated to a bandwidth smaller than the sampling frequency, both in range and in azimuth, the convolution of both datasets will yield spectral contributions above the Nyquist frequency, which results in aliasing effects, disturbing especially the short wavelength signal in the interferogram. To avoid aliasing effects in this part of the spectrum, the data are oversampled by a factor 2 (or possibly less whenever the bandwidth is already reduced due to azimuth/range filtering) before evaluating the complex multiplication. The resulting interferogram is usually low-pass filtered to obtain the same resolution as the SLC images. In this approach aliasing effects are eliminated.

Taking looks

To decrease the noise in the interferogram, Goldstein et al. (1988) propose a *multilook* approach, in which the complex interferogram data in a specified window are simply averaged. Such a boxcar convolution approach corresponds with a two-dimensional sinc multiplication in the spectral domain. In section 4.2.2 it is shown analytically that the PDF of the interferometric phase improves in the multilook case (Lee et al., 1994). Multilooking can be performed simultaneously with the complex multiplication, and is often applied to a range-azimuth ratio that yields approximately square pixels, such as 1 : 5, 2 : 10, etc. The procedure corresponds to boxcar filtering followed by subsampling. Note that the term *multilooking* strictly has another meaning in SAR processing, as discussed in section 2.3.7.

2.5.7 Reference phase

Using the satellite state vectors, the orbits of both satellites are being modeled by a polynomial of the first or second degree during the observation period of the two images. To determine the expected phase behavior for a reference body, the distances between the orbits and the reference body need to be evaluated and differenced for every resolution cell. Standard reference bodies are global ellipsoids such as WGS84 or locally best-fitting ellipsoids such as Bessel. The procedure can be summarized in four steps:

1. at a few positions (times) along orbit 1, the ranges to a few equally distributed points in the interferogram area are determined,

2. for these points, the position at the time of imaging along orbit 2 is retrieved together with the corresponding ranges,

3. the range differences are determined for every reference point, converted to phase differences, and

4. interpolated (or evaluated) for every resolution cell of the interferogram.

In the first step, the Doppler frequency used for the SAR focusing, the range time, and the definition of the reference body are used to find the location of point P, by solving a set of three equations (Schwäbisch, 1995a).

- The point is effectively viewed from an angle defined by the processing Doppler frequency (zero-Doppler for ESA SLC products), cf. eq. (2.3.13),

$$f_{\mathrm{D}} = \frac{2|\vec{v}| \sin \phi}{\lambda} = \frac{2\vec{R}\dot{\vec{R}}}{|\vec{R}|}, \tag{2.5.3}$$

where $\vec{v}$ is the velocity vector, ϕ the squint angle, and $\vec{R}$ and $\dot{\vec{R}}$ are the range vector and its derivative in time respectively.

- Point P is located on the reference body.

- The observed range corresponds with the difference between the state vector and the positioning vector of the reference point.

These three equations can be solved iteratively. The satellite state vectors are given in a geocentric earth-fixed coordinate system, such as the Conventional Terrestrial Reference System (CTRS), and the reference surface needs to be known in the same system.

In the second step the range from point P to the second orbit is evaluated, under the condition that this location satisfies the Doppler centroid mapping for the second image. The third and fourth step are straightforward.

Note that in some cases the satellite orbits are not known or not known with sufficient accuracy. In those cases, the baseline can be approximated from the data, especially the range offset vectors (a measure for the parallel baseline) and the fringe frequency (which is a function of the perpendicular baseline) (Zebker et al., 1994b).

2.5.8 A posteriori filtering

Apart from complex multilooking during interferogram formation, or spectral shift filtering prior to interferogram formation, other filters have been proposed to enhance the quality of the interferogram. Isotropic filters such as square boxcar filters tend to perform well in flat areas, but fail in areas where the phase gradient is high with respect to the kernel size. For example, if there are several phase cycles within such a filter kernel, the resulting (filtered) phase will have a value which does not necessarily reflect a "mean" value. Non-isotropic filters take into account that the fringe frequency often has a dominant direction, and adjust the filter kernel adaptively to the fringe frequency and slope.

Schwäbisch (1995b) uses a direction dependent filter that uses the local fringe frequency for the orientation. The frequency domain shape of the lowpass filter is a two-dimensional Gaussian, where the extent of the filter is highly non-isotropic to suppress frequencies perpendicular to the dominant fringe frequency direction.

Goldstein and Werner (1998) propose an adaptive filter, based on smoothing the spectrum of the interferogram $Z(u, v)$ in patches using the amplitude of the spectrum. The spectrum is weighted by multiplication with its own (smoothed) intensity to the power of an exponent α:

$$Z'(u, v) = Z(u, v) |Z(u, v)|^\alpha, \tag{2.5.4}$$

where $Z'(u, v)$ is the spectrum of the filtered interferogram. Hence, for $\alpha = 0$, the multiplication factor is one, and no filtering applies. However, for larger α-values, the filtering will be significant. The methodology ensures that the signal in the spectrum (assumed to have strong amplitudes) will be enhanced, whereas the noise (assumed to have a significantly lower amplitude) is relatively suppressed. This way, the filter is non-isotropic and changing to the characteristics of the terrain. For example, for water areas, where all frequencies are assumed to have comparable amplitudes, the filter will not enhance or suppress anything. On the other hand, a topographic slope will have a distinguishable amplitude in the spectrum, which will be enhanced by the filter. The filter is applied on overlapping patches in the interferogram. After inverse Fourier transformation, they are reassembled with a linear taper in x and y direction.

2.5.9 Phase unwrapping

For interferometric applications, the principal observation is the two-dimensional relative phase signal, which is the 2π-modulus of the (unknown) absolute phase signal. Although the forward problem—wrapping the absolute phase to the $[-\pi, \pi)$ interval—is straightforward, the inverse problem is one of the main difficulties in radar interferometry, essentially due to its inherent non-uniqueness and non-linearity, see Ghiglia and Pritt (1998) for general reference. Variable phase noise, as well as geometric problems such as foreshortening and layover are the main causes why many proposed techniques do not perform as desired. Without additional information, the assessment whether the results of phase-unwrapping are trustworthy will always be based on strong assumptions on the data behavior.

The proposed methodology of analyzing interferometric data as a geodetic technique, which will be described in chapters 3 and 4, is an attempt to approach the problem of phase unwrapping as an adjustment and filtering problem. This approach is based on the availability of more than one interferogram of the area under consideration. First promising results using such an approach have recently been presented by Ferretti et al. (1997). Here, for the sake of completeness, we briefly discuss the more conventional algorithms and methods.

Let us denote the observed (wrapped) phase ϕ^w as a function of the unknown true phase ϕ, with

$$\phi^w = W\{\phi\} = \mod\{\phi + \pi, 2\pi\} - \pi, \quad \text{with} \tag{2.5.5a}$$
$$\phi = -4\pi \Delta R / \lambda + \phi_N = 2\pi k + \phi_N, \tag{2.5.5b}$$

where W is the wrapping operator, $\Delta R = R_1 - R_2$ is the difference in range to the two satellite positions, $\phi_N \in [-\pi, \pi)$ expresses additive phase noise, and k

is the integer ambiguity number. This is largely following the notation of Bamler and Hartl (1998). Without any assumption, it is not possible to solve the integer ambiguity number k. Usually, based on a priori knowledge of the terrain, it is assumed that the phase gradient between adjacent pixels is limited to the $[-\pi, \pi)$ interval; the smoothness criterion. Hence, if the observed phase gradient is larger than $+\pi$, a cycle is subtracted, and if it is less than $-\pi$, a cycle is added. Now, the phase gradients of the wrapped phases are assumed to be equal to the gradients of the true phases. As long as the true phase gradients are small—less than one-half cycle—and unaffected by noise, this assumption is valid. The actual phase unwrapping is then performed by integrating the phase gradients, e.g., simply by a one-dimensional (flood-fill) summation of the phase gradients, starting from an arbitrary seed location. However, if one of the estimated gradients is erroneous, this error will propagate throughout all subsequent pixels and may affect the entire image.

The smoothness criterion implies that the true phase gradient field exhibits the characteristics of a conservative potential field, i.e., the curl of the field is equal to zero (Kellogg, 1929; Bamler and Hartl, 1998):

$$\nabla \times \nabla \phi = 0. \tag{2.5.6}$$

The gradient $\nabla \phi$ of the true phase is unknown, and has to be estimated based on the gradient $\nabla \phi^w$ of the wrapped phase. For a specific pixel with coordinates (i,j), the best estimate is usually denoted by

$$\hat{\nabla}\phi_{i,j}^w = E\left\{ \begin{bmatrix} W\{\phi_{i+1,j}^w - \phi_{i,j}^w\} \\ W\{\phi_{i,j+1}^w - \phi_{i,j}^w\} \end{bmatrix} \right\}. \tag{2.5.7}$$

Since noise in the phase data will yield errors, in general $\hat{\nabla}\phi_{i,j}^w$ does not necessarily equal $\nabla \phi_{i,j}$, resulting in global phase errors after integration.

The most basic evaluation of the smoothness criterion in a digital image is on 2×2 neighboring pixels, see fig. 2.16:

$$\begin{aligned} r = &\; W\{(\phi_{i+1,j}^w - \phi_{i,j}^w)\} + W\{(\phi_{i+1,j+1}^w - \phi_{i+1,j}^w)\} \\ &+ W\{(\phi_{i,j+1}^w - \phi_{i+1,j+1}^w)\} + W\{(\phi_{i,j}^w - \phi_{i,j+1}^w)\} \end{aligned} \tag{2.5.8}$$

where the value of r can be zero (no residue), $+1$-cycle (positive residue) or -1-cycle (negative residue) (Goldstein et al., 1988). The occurrence of residues in the wrapped phase image is an indication of noise or undersampling and inconsistent with the smoothness condition. It will yield global phase errors in the unwrapped interferogram. Residues can be discharged by "connecting" nearby positive and negative residues with each other, so that the total interferogram is unloaded. Using this approach, true phase gradients that exceed one-half cycle can be identified and accounted for.

Many new phase unwrapping algorithms have been proposed since SAR interferometry became an active field of research. Here we briefly discuss some of the more popular ones.

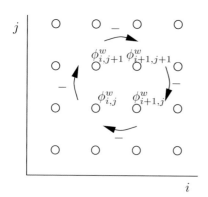

Fig. 2.16. Definition of a residue. If the circles reflect the centers of pixels, the summation over the phase differences (gradients) between four neighboring pixels determines whether the loop is neutral (no residue) or whether a positive or negative residue appears.

Residue-cut method In the residue-cut method, also known as the *branch-cut* or *(minimum spanning) tree* algorithm, all residues are connected (unloaded) with branch cuts, under the condition that the total length of all cuts is minimal (Goldstein et al., 1988), see also Arfken (1985) and Ching et al. (1992). The algorithm starts at one residue and searches within a window of 3×3 around it for another residue. If a residue of different sign is found, the branch cut is applied and the algorithm continues to the next residue. If the found residue is of equal sign, the search continues from this new position, and repeats until all plus and minus residues are unloaded, hereby spanning a "tree" (Chen and Zebker, 2000a). If there are no new residues within the search region, but the found ones are not unloaded yet, the search radius is doubled to prevent the algorithm from breaking off. While integrating the phase gradients, the cuts constitute an obstacle through which the integration cannot continue. As a result, isolated regions can occur in the interferogram, surrounded by cuts. An important characteristic of the residue-cut algorithm is the fact that the estimated phase gradients can only have discrete $k2\pi$ offsets with $k \in \mathbb{Z}$. In practice, many trees will form in noisy areas such as water surfaces, layover regions, and regions affected by temporal decorrelation.

Least-Squares Method (Ghiglia and Romero, 1994, 1996; Pritt, 1996; Hunt, 1979) These methods appear in a weighted and unweighted fashion, and are based on minimizing the sum of the quadratic errors between the unwrapped phase gradients and the estimated phase gradients in a global way. FFT techniques can be applied to solve the least-squares technique in an efficient way (Ghiglia and Romero, 1994). Considering the problem in a Green's-function formulation can also be regarded as a least-squares method (Fornaro et al., 1996a,b). Bamler et al. (1996a) proved that least-squares methods for phase unwrapping can introduce errors which propagate through the image, effectively by underestimating slopes (gradients) in the interferogram, see (Bamler et al., 1998).

Minimal cost flow methods (Costantini, 1996, 1998; Flynn, 1997; Chen and Zebker, 2000a) Minimal cost flow (MCF) or network flow methods consider the
phase unwrapping problem as a global minimization problem with integer variables. Comparing the difference between the estimated and the unknown phase
gradient of the unwrapped phase should yield $k2\pi$ differences, with $k \in \mathbb{Z}$.
MCF methods follow the residue-cut methods, adding integer cycles to cross
the cuts. Positive and negative residues are the surplus and demand nodes,
and the "flow" that equalizes the nodes can travel along paths determined by
the cost function. An optimal solution is reached if the predefined objective
function is minimized. A practical approach for the definition of costs for
topography or deformation mapping is given in Chen and Zebker (2000b).

Linear integer combinations of interferograms, as discussed in section 3.5.3, is a
methodology to reduce the fringe rate, which can facilitate phase unwrapping (Massonnet et al., 1996c)

2.5.10   Patch unwrapping

Phase unwrapping methods applying branch-cuts, as well as interferograms containing, e.g., islands, long waterways, or joint areas of low coherence such as moving
lava flows are likely to contain isolated patches with sufficient coherence for successful phase unwrapping, surrounded by incoherent areas or branch-cuts. The phase
gradient integration necessarily starts from an arbitrary seed, and can continue only
within the limits of the region surrounded by branch-cuts. After unwrapping as
many of these regions as possible, the connection of the separate patches is here
referred to as patch unwrapping.

Placing seeds—the starting points for phase gradient integration—in a dense grid
minimizes the possibilities that coherent patches are missed. Proper bookkeeping is
required to label the unwrapped patches and prevent unwrapping starting from a
seed in a patch that has been unwrapped already. The final output is an interferogram, with separately unwrapped patches and a map with patch numbers. The task
of patch unwrapping is theoretically as ill-posed as phase unwrapping, unless some
sort of prior information on the terrain or the deformation pattern is added. For
example, for a short perpendicular baseline the height ambiguity might be so large
that an integer jump between patches is very unlikely. Often manual intervention is
necessary to apply the patch unwrapping. In section 5.4 two possibilities for patch
unwrapping are presented and demonstrated.

2.5.11   Differential interferogram generation

Producing a differential interferogram requires the two interferograms (the deformation pair and the topographic pair) to be aligned to the same grid. In three-pass
differential interferometry this requirement is satisfied by using a common reference
image, onto which the other two are coregistered and resampled. The four-pass
technique requires a repeated coregistration and interpolation. First the two interferograms are formed separately, followed by the coregistration of one of the pairs

onto the other. After all images are in the same grid, the topographic pair needs to be unwrapped and scaled according to the baseline ratio between the two interferograms (Zebker et al., 1994b). Subtracting this unwrapped, scaled topographic pair from the deformation pair yields the differential pair, which should ideally reflect only deformation signal. The procedure is described mathematically in chapters 3 and 4.

2.5.12   Phase to height conversion

Phase to height conversion is the procedure that relates the unwrapped interferometric phase to topographic height, although the pixels remain in the range-azimuth radar coordinate system. The theoretical concept is already outlined in section 2.4.1, but a few practical aspects of the implementation are important to consider.

First, a theoretical situation such as the one sketched in figs. 2.10A and 2.10B will never occur, since it is not possible to derive the phase difference (and hence the difference in look angle $\partial\theta$) between a point on the ellipsoid and a point at the same range with a different height, because of the wrapped nature of the interferometric phase. Although this problem cannot be solved for a single pixel, the evaluation of the unwrapped phase gradients to height gradients is straightforward. Therefore the height gradients, or equivalently the relative heights, can be determined up to an arbitrary constant for the entire image.

Equation (2.4.12) needs to be evaluated for the phase to height conversion. Fixed parameters in this equation are λ, R_{1p}, $\sin\theta_p^0$, and $B_{\perp,p}^0$ for the imaginary point P' on the reference surface. However, the perpendicular baseline as well as the look angle will change as soon as the initial height H_p has been derived, see fig. 2.10B. Therefore, the procedure has to be repeated iteratively until H_p satisfies some accuracy criterion. Usually not more than three iterations are necessary to obtain mm-scale increments. Finally, the selection of one control point will fix the relative height differences to the coordinate system of the reference surface.

The evaluation of the perpendicular baseline per pixel is vital, not only because of the dependency on height. In fig. 2.17A, the variation of the baseline per pixel is shown for non-parallel orbits. Orbit 1 observes all pixels in one range line from one position, whereas the corresponding positions from orbit 2 are changing. For the second range line this behavior repeats for a slightly shifted position, leading to a sawtooth pattern. As a function of time (fast time and slow time combined), the behavior of the perpendicular baseline is as indicated in fig. 2.17B, assuming a reference surface without topography.

2.5.13   Geocoding

Geocoding refers to a coordinate transformation of the radar coordinates (range-/azimuth/height) to coordinates in a convenient geodetic reference system such as WGS84 (Φ, Λ, H), where we use Φ and Λ for the geographic latitude and longitude, respectively, and H for the height above the reference body (ellipsoid). For computational convenience, an intermediate earth-fixed Cartesian reference system is

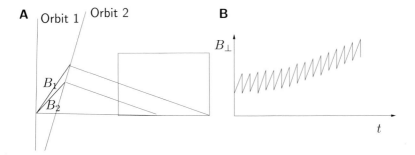

Fig. 2.17. Change in the baseline length for diverging orbits. **(A)** Top view of the baseline for two points in the same azimuth line, near-range and far range. **(B)** Sketch of the variation in the perpendicular baseline of diverging orbits as a function of time. The time-axis combines fast time (in range direction) and slow time (in azimuth direction).

used, in which the satellite orbits as well as the ellipsoid are given. The procedure is similar to the evaluation of the reference phase, see section 2.5.7, and is based on the following conditions.

1. For every pixel in the image, the range bin number is used to find the magnitude of the range vector, $R_1 = |\vec{R}_1|$, to the reference orbit.

2. The azimuth position along this orbit is retrieved from the line number, the azimuth starting time, and the PRF. This defines the state vector $\vec{X}_{s/c}$ for the satellite position.

3. The Doppler centroid, defining a possible squint angle, and the direction of the local velocity vector $\dot{\vec{X}}$ of the state vector, are used to define a plane. The plane is spanned by the origin of the coordinate system, the position of the satellite, and the squint angle with respect to the velocity direction. The range vector $\vec{R}_1$ must be situated in this plane.

4. Finally, the local earth radius ρ, defined by the ellipsoid, plus the topographic height H_p derived from the interferometric phase, define the length of the positioning vector $|\vec{X}_p| = |\rho + H_p|$.

These conditions uniquely define the position vector $\vec{X}_p$ in the geocentric Cartesian reference system. Latitude and longitude can easily be derived by a three-dimensional inverse Helmert transformation.

Applying this procedure for every pixel in the image yields a long list of (Φ, Λ, H) triplets, where H will be assigned a *not-a-number*-value if the height could not be derived, e.g., in decorrelated areas such as water, or in shadow or layover regions. A non-zero height will result in a horizontal displacement of approximately $H/\tan(\theta_{\mathrm{inc}})$ m, where θ_{inc} is the incidence angle. As a rule of thumb, for ERS conditions this horizontal displacement is approximately $2.3 \times H$. Therefore the positions

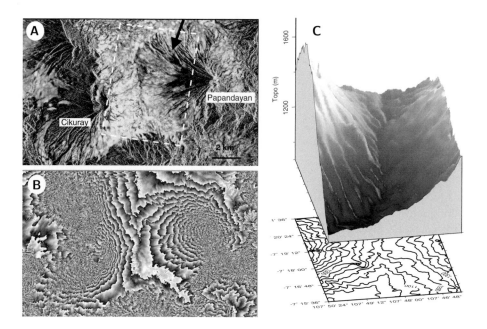

Fig. 2.18. Interferometric data from Mt. Cikuray and Mt. Papandayan, Java, Indonesia. **(A)** Interferogram magnitude. **(B)** Tandem interferogram with a perpendicular baseline of approximately 108 m. One cycle corresponds with 2π rad. **(C)** Geocoded (WGS84) DEM and 3D perspective of a 8×11 km subset of this interferogram. The location and point of view of the subset are indicated in (A).

of the (Φ, Λ, H) points are irregularly distributed, and interpolation is necessary to resample the data to a rectangular grid. Delaunay triangulation (Delaunay, 1934) followed by linear interpolation is an often applied method in this respect.

An example of geocoding a SAR derived DEM is shown in fig. 2.18. The interferogram (fig. 2.18A and B) covers an area of 16×10 km over central Java, Indonesia, showing two volcanoes, Mt. Papandayan in the west and Mt. Cikuray in the east. Clearly visible in the amplitude image are the foreshortening, layover, and shadow areas. The interferogram has a perpendicular baseline of ~108 m, corresponding with a height ambiguity of ~87 m. After a posteriori filtering using the method of Goldstein and Werner (1998), phase unwrapping, and phase-to-height transformation, the data were geocoded to WGS84. Local tiepoints were used to remove the bias in the height, to remove orbital errors, and to fine-tune the geocoding. Figure 2.18C shows a 3D perspective of the derived DEM.

Chapter 3

Functional model for radar interferometry

A generic Gauss-Markoff model is introduced and applied to describe the functional and stochastic relations for two-pass spaceborne radar interferometry. This chapter covers the functional part of this model, which describes the physical and geometric relations between the observations and the parameters. The model is refined for specific applications such as topography estimation, deformation mapping, and atmospheric mapping. An extended model is presented to treat three-pass and four-pass differential interferometry, as well as interferogram stacking.

key words: *Functional model, Topography, Deformation, Atmosphere*

The interferometric processing relates two registered signals y_1 and y_2 to the Hermitian product $y_1 y_2^*$, which has a phase equal to the phase difference of the two signals. The phase difference values for each pixel in the interferogram form the observation space, which is related to the parameter space by a functional relation. The following sections define a standard model to express this relation.

3.1 Gauss-Markoff model definition

The observations φ_k form a real stochastic vector of observations $\varphi \in \mathbb{R}^m$, characterized by the first moment $E\{\varphi\}$ and the second moment $D\{\varphi\}$. After linearization, the relation between the observations and the unknown parameters can be written as a Gauss-Markoff model (Gauss, 1809; Markoff, 1912), see Koch (1999):

$$
\underset{m \times 1}{E\{\varphi\}} = \underset{m \times n}{\mathbf{A}} \ \underset{n \times 1}{\mathbf{x}}; \quad \underset{m \times m}{D\{\varphi\}} = \underset{m \times m}{\mathbf{C}_\varphi} = \underset{1 \times 1}{\sigma^2} \ \underset{m \times m}{\mathbf{Q}_\varphi}, \tag{3.1.1}
$$

where $\mathbf{A}$ is the design matrix, $\mathbf{C}_\varphi$ is the variance-covariance matrix, $\mathbf{Q}_\varphi \in \mathbb{R}^{m \times m}$ the real positive-definite $m \times m$ cofactor matrix, and $\sigma^2 \in \mathbb{R}^+$ the a priori variance factor. The vector of parameters $\mathbf{x} \in \mathbb{R}^n$ is assumed to be real and non-stochastic.

The first part of eq. (3.1.1) is commonly referred to as the functional model or model of observation equations, whereas the second part is known as the stochastic model.

The latter is discussed in chapter 4. The functional model can also be written as

$$\varphi = \mathbf{A}\mathbf{x} + \varepsilon, \tag{3.1.2}$$

where $\varepsilon \in \mathbb{R}^m$ reflects the errors of φ with $E\{\varepsilon\} = 0$. Design matrix $\mathbf{A}$ should be of full rank to make the model uniquely solvable, which implies that $m \geq n$. Moreover, $\mathbf{Q}_\varphi$ should be positive definite, since in that case the *weight matrix* of the observations, $\mathbf{Q}_\varphi^{-1}$, will exist and will be positive definite as well (Koch, 1999). The best (minumum variance) linear unbiased estimator of the unknown parameters $\hat{\mathbf{x}}$ and the covariance matrix $\mathbf{C}_{\hat{\mathbf{x}}}$ is given by

$$\hat{\mathbf{x}} = (\mathbf{A}^T \mathbf{Q}_\varphi^{-1} \mathbf{A})^{-1} \mathbf{A}^T \mathbf{Q}_\varphi^{-1} \varphi \tag{3.1.3}$$

$$\mathbf{C}_{\hat{\mathbf{x}}} = \sigma^2 (\mathbf{A}^T \mathbf{Q}_\varphi^{-1} \mathbf{A})^{-1} \tag{3.1.4}$$

The advantage of describing the problem of parameter estimation in a systematic model such as the one outlined above is that different physical or geometrical aspects of the problem are reduced to relatively simple mathematical equations. Apart from providing a structural and perhaps elucidating approach, the formulation in a Gauss-Markoff model opens up a wealth of standard techniques for adjustment, quality description, and hypothesis testing. As a result, such techniques give a quantitative measure of the accuracy and reliability of the estimated parameters.

3.1.1 Observations

The primary observations are the raw radar echoes as received by the SAR antenna. Unfortunately, using this definition it is very difficult to obtain the fully linearized design matrix since SAR data processing is a highly non-linear operation (Bamler and Hartl, 1998). To simplify the model, we therefore define the focused interferogram as observations, and use system theoretical considerations to describe the stochasticity of the observations, see chapter 4. Some modifications of this definition are necessary to arrive at eq. (3.1.1), since (i) the interferogram consists of complex values, (ii) the interferometric phase is in first instance only known modulo 2π, and (iii) the interferometric phase might be altered due to the extraction of a reference and/or a topographic phase.[1] In fact, several definitions for the observation vector might be suitable, as long as the model in eq. (3.1.1) can be adopted. Here we define four types of observation vectors, based on (i) whether or not the influence of the reference body is accounted for, and (ii) whether or not the phase observations are wrapped. The four types are listed in table 3.1, where $W\{\cdot\}$ is the wrapping operator, with $W\{\phi\} = \mod(\phi + \pi, 2\pi) - \pi$. The reference phase ϑ is in first approximation regarded to be a deterministic parameter defined by

$$\vartheta_k = \frac{4\pi}{\lambda} B_k \sin(\theta_k^0 - \alpha_k), \tag{3.1.5}$$

---

[1]Another option is to define the gradient of the interferometric phase as observations, see Sandwell and Price (1998). This can be advantageous for identifying very localized gradients, e.g., due to surface deformations, and in the analysis of stacks of interferograms. Here we choose not to use this type of observation, as it relies on the assumption that the gradients are restricted to a $[-\pi, \pi)$ interval.

Table 3.1. The notation of four types of interferometric phase observations, including or without the reference phase and wrapped or unwrapped.

	absolute	wrapped
Including reference phase	ϕ	$\phi^{\mathrm{w}} = W\{\phi\}$
Reference phase subtracted	$\varphi = \phi - \vartheta$	$\varphi^{\mathrm{w}} = W\{\phi^{\mathrm{w}} - \vartheta\}$

where B_k and α_k are the baseline length and orientation (see fig. 3.2) for resolution cell k. These values are considered to be known, either by using precise orbit information or after an adjustment procedure involving tie-points, see also section 4.6.3. The look angle θ_k^0 follows from the definition of the reference body in the same datum as the orbits. The option to include the reference phase in the vector of observations, instead of including it in the design matrix, follows from its independence of the parameters to be estimated, resulting in a rank deficiency. Note that the four types of phase observation might be multilooked.

An interferogram of M_1 rows and M_2 columns consists of $m = M_1 M_2$ observations. A random (stochastic) wrapped interferometric phase observation for pixel k is written as ϕ_k^{w}, with $k = i + (j-1)M_1$, where $i = 1, \ldots, M_1$ and $j = 1, \ldots, M_2$ represent the row and column of the interferogram, respectively, see fig. 3.1. The vector of observations is written as ϕ^{w}. For the wrapped phase $\phi_k^{\mathrm{w}} \in [-\pi, \pi)$ and for the unwrapped (absolute) phase $\phi_k \in \mathbb{R}$.

3.1.2 Parameters and a priori information

Repeat-pass radar interferometry can be used for the recovery of geophysical parameters. Although many parameters affect the interferometric phase, at least five parameters are dominant. For every observation k these are

1. topographic height H_k [m],

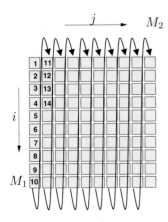

Fig. 3.1. Vectorization of the interferogram observations. The squares indicate the interferogram pixels at row i and column j, and are ordered columnwise to obtain observation vector ϕ_k^{w}, using $k = i + (j-1)M_1$, where $i = 1, \ldots, M_1$ and $j = 1, \ldots, M_2$.

2. deformation in slant direction D_k [m]
3. slant-atmospheric delay during acquisition 1: $S_k^{t_1}$ [m],
4. slant-atmospheric delay during acquisition 2: $S_k^{t_2}$ [m], and
5. the integer ambiguity number $w_k \in \mathbb{Z}$.

This results in a vector of parameters $\mathbf{x} \in \mathbb{R}^{5m}$, which is assumed to be real and non-stochastic. This formulation implies that every observation is related to at least five unknown parameters, which results in a rank defect of $4\,m$. Therefore, the problem is underdetermined. To solve this problem three strategies can be applied:

(i) more observations can be added to the model,
(ii) deterministic (a priori) information can be introduced, or
(iii) the model can be reformulated.

The first strategy is useful when a series of SAR acquisitions is available over the area of interest. More than one interferometric pair can be processed, and assuming that some parameters do not change for different interferograms the rank defect decreases. The second option can be used whenever one or more parameters are known, e.g., from different sources. A reference elevation model, a high likelihood that no deformation occurred, or confidence in the phase unwrapping might be used for that respect. The third strategy involves a conceptual change, for example by considering a category of parameters as stochastic variation of the observations instead of deterministic parameters. This approach will be used to model atmospheric signal variability in chapter 4. No strategy alone will be sufficient to resolve the rank defect. A combination of approaches is expected to be most effective.

The exact formulation of the parameter vector depends on the interferometric configuration, the amount of a priori information, and the goals of the study. In the remainder of this section we discuss several examples.

If the recovery of topographic information is the goal of the study, interferometric pairs with a short temporal baseline can exclude the presence of surface deformation signal, which reduces the rank defect by m. Another possibility is to use a second "topographic" interferogram: this will increase the number of observations to $2\,m$, whereas the number of parameters will be increased to $9\,m$, as the topographic height H_k will not change. Similarly, using the same SAR acquisitions in several interferometric pairs results in a reduction of the atmospheric parameters. Examples of this approach are presented in section 5.4.9. If the phase unwrapping can be assumed to be error free, e.g., for high coherence data with smooth topographic variation and/or small perpendicular baselines, the integer ambiguity number might be known, which reduces the rank defect by m.

For deformation studies, topographic information might be available from reference elevation models or other interferograms. In that case, an expression for the accuracy of these models needs to be included in the stochastic model, but the topographic parameter can be eliminated. A considerable improvement is the formulation of the deformation pattern in terms of a mathematical function. In that case, the number of unknown deformation parameters can often be reduced considerably.

A major source of uncertainty is formed by the atmospheric signal. Note that it is referred to as *signal* instead of disturbance, since the effects are spatially correlated and can be significant, see chapters 4 and 6. Moreover, using different interferometric combinations the atmospheric contribution at the time of some of the SAR acquisitions can be eliminated. A new application of repeat-pass interferometry, Interferometric Radar Meteorology, considers the atmospheric signal as the main parameter of interest. In that case, topography and deformation might be removed using a reference elevation and deformation model, respectively.

Since the practical definition of the model depends on the goals of the study, we define a generic model that involves all parameters discussed above.

3.1.3   Generic functional model

A generic linearized functional model for repeat-pass interferometry can be written as:

$$
E\left\{ \begin{bmatrix} \varphi_1^{\mathrm{w}} \\ \varphi_2^{\mathrm{w}} \\ \vdots \\ \varphi_m^{\mathrm{w}} \end{bmatrix} \right\} = \begin{bmatrix} \mathbf{A}_1 & & & \\ & \mathbf{A}_2 & & \\ & & \ddots & \\ & & & \mathbf{A}_m \end{bmatrix} \begin{bmatrix} \mathbf{x}_1 \\ \mathbf{x}_2 \\ \vdots \\ \mathbf{x}_m \end{bmatrix} \tag{3.1.6}
$$

where $\mathbf{A}_k$ is the part of the design matrix corresponding to observation k, defined as:

$$
\mathbf{A}_k = \left[-\frac{4\pi}{\lambda} \frac{B_{\perp,k}}{R_k \sin\theta_k}, \quad \frac{4\pi}{\lambda}, \quad \frac{4\pi}{\lambda}, \quad -\frac{4\pi}{\lambda}, \quad -2\pi \right], \tag{3.1.7}
$$

with $B_{\perp,k} = B\cos(\theta_k - \alpha)$. The observation vector $\varphi_k^{\mathrm{w}} = W\{\phi_k^{\mathrm{w}} - \vartheta_k\}$ is the phase contribution after subtracting the reference body, e.g., spheroid or ellipsoid, with:

$$
\vartheta_k = \frac{4\pi}{\lambda} B_k \sin(\theta_k^0 - \alpha_k), \tag{3.1.8}
$$

with the value for θ_k^0 obtained at the reference body. The parameter vector $\mathbf{x}_k$ corresponding to observation k is:

$$
\mathbf{x}_k = \left[H_k, \ D_k, \ S_k^{t_1}, \ S_k^{t_2}, \ w_k \right]^T. \tag{3.1.9}
$$

The dispersion of the vector of observations can be estimated using empirical coherence measurements or system theoretical considerations, and the diagonal variance-covariance matrix $\mathbf{C}_\varphi$ can be written as:

$$
\mathbf{C}_\varphi = \begin{bmatrix} \sigma_1^2 & & & \\ & \sigma_2^2 & & \\ & & \ddots & \\ & & & \sigma_n^2 \end{bmatrix} = \sigma^2 \mathbf{Q}_\varphi, \quad (\sigma^2 = 1), \tag{3.1.10}
$$

where σ_k^2 can be approximated using the estimated coherence values from the coregistered SAR data. As a first order approximation, we assume that there is no correlation between the phase observations in the interferogram, although a resolution cell strictly has about 50% overlap with the neighboring cells, see section 2.3.8.

In chapter 4 we propose to reformulate the model by lumping the atmospheric slant delay signal into one parameter $S_k = S_k^{t_1} - S_k^{t_2}$, transferring this parameter to the stochastic model, and replace $\mathbf{C}_\varphi$ by

$$\mathbf{C} \doteq \mathbf{C}_\varphi + \mathbf{C_s}, \tag{3.1.11}$$

where $\mathbf{C_s}$ is the variance-covariance matrix of the atmospheric delay parameter S_k. Using this approach, assuming ideal phase unwrapping and a priori knowledge of either topography or deformation, the design matrix $\mathbf{A}$ has full rank.

3.2 Topography estimation

The functional relation between the terrain height h_k above a reference body and the absolute interferometric phase φ_k has been defined in eq. (2.4.12), page 37, to be

$$\varphi_k = -\frac{4\pi}{\lambda} \frac{B_{\perp,k}^0}{R_k \sin \theta^0} H_k, \tag{3.2.1}$$

where the "naught" versions $B_{\perp,k}^0$ and θ^0 indicate the values for the reference body. These values change slightly for a point at some elevation. To refine the generic functional model for the needs of topography estimation it is necessary to reduce the rank defect in eq. (3.1.6). This could be achieved by eliminating the deformation component and the two atmospheric components. Regarding deformation, this could be achieved by choosing a time interval as short as possible, ideally simultaneous or single-pass acquisitions. For single-pass applications $\mathbf{A}_k$, cf. eq. (3.1.7), can be simplified to

$$\mathbf{A}_k = \left[\frac{2\pi}{\lambda} \frac{B_{\perp,k}}{R_k \sin \theta_k}, \quad -2\pi \right], \tag{3.2.2}$$

and

$$\mathbf{x}_k = \left[H_k, \ w_k \right]^T. \tag{3.2.3}$$

For repeat-pass topography estimation the ideal configuration is dependent on the time interval (which influences temporal decorrelation, surface deformation, vegetation), atmospheric variability, wavelength, baseline, and the amount and range of topography within the scene. Rules of thumb regarding atmospheric influence are given in section 4.7 and in chapter 6.

From fig. 3.2A, the difference between the geometric ranges from point P to satellite 1, R_p, and to satellite 2, R_p', can be expressed as

$$\Delta R_p = R_p' - R_p. \tag{3.2.4}$$

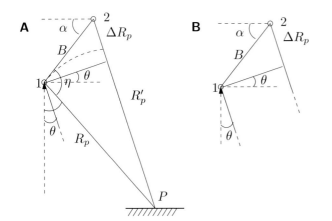

Fig. 3.2. Geometric configuration for pixel P and satellite positions 1 and 2. **(A)** Exact range difference. **(B)** Far Field Approximation.

This range difference can be expressed for each point P on the reference body from the baseline length B and angle α, the local look angle θ_p, and the geometric range R_p.

In fig. 3.2A, we find that angle η can be written as:

$$\eta = \frac{\pi}{2} - \theta + \alpha. \tag{3.2.5}$$

Using the cosine rule this enables an expression for R'_p as

$$R_p'^{\,2} = B^2 + R_p^2 + 2BR_p \sin(\alpha - \theta), \tag{3.2.6}$$

which gives an exact but non-linear expression for the range difference:

$$\Delta R_p = \sqrt{B^2 + R_p^2 + 2BR_p \sin(\alpha - \theta)} - R_p \tag{3.2.7}$$

Since for contemporary repeat-pass interferometric constellations the baseline length is typically less than 1000 m, the difference between the local look angle θ is less than 4 arc minutes. This enables an alternative expression for the range difference, the *far field approximation* (FFA) (Zebker and Goldstein, 1986). This approximation considers the local look angles for both sensors to be identical, as in fig. 3.2B. The parallel look directions provide the approximation for the range difference:

$$\Delta R_{p,\text{FFA}} = B \sin(\alpha - \theta). \tag{3.2.8}$$

Note that, compared to the exact expression in eq. (3.2.7), the absolute range R_p has been eliminated. Comparison of the exact and approximated range difference yields a measure for the approximation error.

$$
\begin{aligned}
\epsilon_{\text{FFA}} &= \Delta R_p - \Delta R_{p,\text{FFA}} \\
&= \sqrt{B^2 + R_p^2 + 2BR_p \sin(\alpha - \theta)} - R_p - B \sin(\alpha - \theta)
\end{aligned}
\tag{3.2.9}
$$

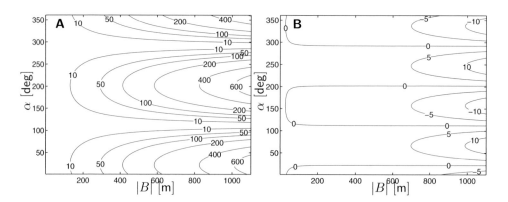

Fig. 3.3. **(A)** Approximation error of the range difference ΔR_p using the far field approximation in mm, for $\theta = 21°$ and $R_p = 850$ km. **(B)** Approximation error of the partial derivative of the range difference $\partial \Delta R_p / \partial \theta$ using the far field approximation in mm, for $\theta = 21°$ and $R_p = 850$ km.

Thus, for a given sensor, the approximation error is a function of two variables, the baseline length B and orientation α, see fig. 3.3A. Although relatively large approximation errors are apparent, up to 700 mm for large baselines, it needs to be stressed that the absolute range difference ΔR_p cannot be directly observed in the interferogram. This is due to orbit inaccuracies and the unknown phase ambiguity number. For a specific range R_p, rather the minute changes in the look angle θ are observed. These changes reflect the deviation between the look angle to a point on the reference body, θ_p^0, and to a point on the topography, cf. fig. 2.10A and 2.10B. Hence the partial derivatives to the look angle, for the true and the approximated range differences are

$$\frac{\partial}{\partial \theta} \Delta R_p = \frac{R_p}{\sqrt{B^2 + R_p^2 + 2BR_p \sin(\theta - \alpha)}} B \cos(\theta - \alpha) \qquad (3.2.10\text{a})$$

and

$$\frac{\partial}{\partial \theta} \Delta R_{p,\text{FFA}} = B \cos(\theta - \alpha). \qquad (3.2.10\text{b})$$

The approximation error derived from the difference between eq. (3.2.10a) and eq. (3.2.10b) is shown in fig. 3.3B. The values represent the range difference for a 1° change in look angle, which corresponds approximately with the height of Mt. Everest above the WGS84 ellipsoid. Although the approximation is already reasonably good, it can have a non-negligible effect on the sensitivity to height differences, expressed by the perpendicular baseline $B_\perp$. Therefore, it is always necessary to evaluate eq. (3.2.1) iteratively, updating the values for θ in every iteration. In practice, however, the iterations are initialized on the neighboring resolution cells, which makes one iteration sufficient to obtain a value to the cm-level.

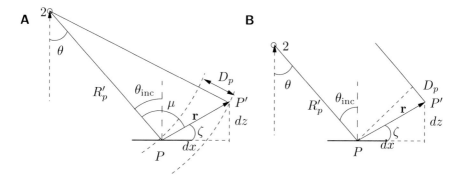

Fig. 3.4. Exact **(A)** and far field **(B)** representation of a deformation due to displacement vector **r**.

3.3   Deformation mapping

As a consequence of the time interval between the subsequent acquisitions, intermediate surface deformations will result in a relative range difference, here denoted by D_p. If we write the geometric range for the second acquisition as R'_p as a function of R_p, the constellation range difference ΔR, and the deformation D_p, we find

$$R'_p = R_p + \Delta R_p + D_p \tag{3.3.1}$$

Similar to the sensor constellation, an exact and an approximate expression can be derived. From fig. 3.4A, μ can be written as

$$\mu = \frac{\pi}{2} - \zeta + \theta_{\text{inc}}, \tag{3.3.2}$$

where θ_{inc} is the incidence angle at point P.

If resolution cell P has been subject to a deformation along displacement vector **r** in direction ζ, this would give a slant range deformation component D_p. Using the cosine rule the new range $R'_p + D_p$ can be expressed as

$$(R'_p + D_p)^2 = r^2 + R'^2_p - 2rR'_p \sin(\zeta - \theta_{\text{inc}}), \tag{3.3.3}$$

with $r = |\mathbf{r}|$, which yields the slant range deformation

$$D_p = \sqrt{r^2 + R'^2_p - 2rR'_p \sin(\zeta - \theta_{\text{inc}})} - R'_p. \tag{3.3.4}$$

In terms of horizontal and vertical deformation components we write

$$r^2 = dx^2 + dz^2, \tag{3.3.5}$$

$$\zeta = \arctan(\frac{dz}{dx}) \tag{3.3.6}$$

Using the far field approximation a parallel look direction can be assumed. Since the deformation in practice will be far less then a couple of meters, limited by a reduction in coherence for large deformations, this approximation can be safely applied, cf. fig. 3.4B.

A deformation $\mathbf{r}$ in direction ζ will result in a slant range deformation component

$$D_p = |\mathbf{r}| \cos(\theta_{\mathrm{inc}} - \zeta). \qquad (3.3.7)$$

Thus, only one component of the 3D deformation vector can be observed in the interferogram. In practice, the combination of ascending and descending orbit interferograms can be used to find a second component of the deformation vector. Due to the steep incidence angle of ERS, this satellite is especially sensitive to vertical deformation.

3.4 Atmospheric mapping

Radar interferometry can be applied to study vertically integrated atmospheric refractivity variations. The parameters $S_k^{t_1}$ and $S_k^{t_2}$ in eq. (3.1.9) reflect every type of slant-atmospheric phase delay or advancement with respect to the propagation velocity in vacuum c (Feynman et al., 1963). This can be due to the velocity variations along the geometric path between antenna and scatterer or due to the induced bending of the ray path (Bean and Dutton, 1968). For every point in three dimensional space, and for every moment in time, the electromagnetic properties concerning radio wave propagation are governed by the dimensionless refractive index $n(x, y, z, t)$ of the medium. The phase velocity is expressed by $v = c n^{-1}$.

Under normal atmospheric circumstances the refractive index varies only slightly from its value in a vacuum ($n = 1$), and therefore often the scaled-up refractivity N is used, with $N = (n - 1)10^6$. The slant-atmospheric delay for resolution cell k at acquisition time t_i can be written as (Hanssen, 1998)

$$S_k^{t_i} = 10^{-6} \underbrace{\int_0^H \frac{N}{\cos\theta_{\mathrm{inc}}} dh}_{S_{k,\mathrm{velocity}}^{t_i}} + \underbrace{\left(\int_0^H \frac{1}{\cos\theta_{\mathrm{inc}}(h)} dh - R_k \right)}_{S_{k,\mathrm{bending}}^{t_i}} \qquad (3.4.1)$$

where θ_{inc} is the incidence angle or zenith angle, H is the height of the satellite above the position of the scatterer, and R_k is the true slant range. In this equation we have separated the influence of propagation velocity and the bending of the ray path. For the propagation velocity part, measured along the true slant path the incidence angle θ_{inc} is a constant. For the bending part, the incidence angle $\theta_{\mathrm{inc}}(h)$ is a variable along the ray path. Bean and Dutton (1968) have shown that even for extreme refractivities, the ratio $S_k / S_{k,\mathrm{bending}}$ approaches zero for incidence angles less than $87°$. Therefore, for all SAR data available today the slant atmospheric delay can be considered a function of propagation velocity variations only, which reduces eq. (3.4.1) to

$$S_k^{t_i} = 10^{-6} \int_0^H \frac{N}{\cos\theta_{\mathrm{inc}}} dh. \qquad (3.4.2)$$

As a rule of thumb, we may use

$$S_k^{t_i} \ [\text{in} \quad \text{mm}] = \frac{N}{\cos \theta_{\text{inc}}} \ [\text{per} \quad \text{km}], \tag{3.4.3}$$

which is an approximation for the one-way path delay in mm, per km of signal path. Although eq. (3.4.2) combined with eqs. (3.1.6)–(3.1.9) describe the fundamental functional relation between the interferometric phase and the integrated refractivity, decomposition of N in a number of physical parameters can provide new insights in atmospheric behavior. This is mainly due to the fine resolution of the data, the high delay accuracy, and the all-weather capabilities. Using only one interferogram, an ambiguity will be introduced by the coherent superposition of two atmospheric states ($S_k^{t_1}$ and $S_k^{t_2}$). This ambiguity restricts the quantitative interpretation of the atmospheric signal to situations where events are dominant during one of the two SAR acquisitions. Nevertheless, several studies reported in chapter 6 clearly demonstrate the feasibility of meteorological interpretation. To circumvent the ambiguity problem, future applications of the technique can use "cascaded interferograms," in which every SAR image appears in two interferograms.

3.4.1 Decomposition of refractivity

For common radar frequencies, the refractivity N can be written as (Smith and Weintraub, 1953; Kursinski, 1997):

$$N = \underbrace{k_1 \frac{P}{T}}_{N_{\text{hyd}}} + \underbrace{\left(k_2' \frac{e}{T} + k_3 \frac{e}{T^2} \right)}_{N_{\text{wet}}} - \underbrace{4.03 \times 10^7 \frac{n_e}{f^2}}_{N_{\text{iono}}} + \underbrace{1.4W}_{N_{\text{liq}}}, \tag{3.4.4}$$

where P is the total atmospheric pressure in hPa, T is the atmospheric temperature in Kelvin, e is the partial pressure of water vapor in hPa, n_e is the electron number density per cubic meter, f is the radar frequency, and W is the liquid water content in g/m^3. The values $k_1 = 77.6$, $k_2' = 23.3$ and $k_3 = 3.75 \times 10^5$ are taken from Smith and Weintraub (1953), but also results from Thayer (1974) are commonly used. The four refractivity terms are referred to as hydrostatic term, wet term, ionospheric term, and liquid term, respectively. Resch (1984) indicated that the first two parts are accurate to 0.5 percent. Using this representation of the refractivity we decompose eq. (3.4.2) into

$$S_k^{t_i} = \frac{1}{10^6 \cos \theta_{\text{inc}}} \left\{ \int_0^H N_{\text{hyd}} \, dh + \int_0^H N_{\text{wet}} \, dh + \int_0^H N_{\text{iono}} \, dh + \int_0^H N_{\text{liq}} \, dh \right\}, \tag{3.4.5}$$

where the four integrals are labeled the hydrostatic delay, the wet delay, the ionospheric delay, and the liquid delay, respectively.

As the total signal delay can be up to several meters^2 it is evident that the interferometric data—which were originally wrapped to half the radar wavelength interval

---

^{2}The average delay due to a standard troposphere is approximately 2.5 m

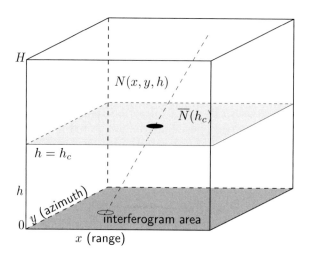

Fig. 3.5. The refractivity distribution $N(x, y, h)$ in a 3D Cartesian space above the interferogram can be described by eq. (3.4.6).

(2.8 cm for ERS)—can never capture the full integrated values, unless some sort of independent absolute calibration is possible. As a result, we can regard every integral in eq. (3.4.5) as the sum of two contributions. First, a fixed contribution (bias), which is constant for the whole scene. This value can be subtracted from the data. Second, a variable contribution dependent of the position in the image. In terms of the refractivity components, we can write

$$N(x, y, h) = \overline{N}(h_c) + \Delta N(x, y, h), \tag{3.4.6}$$

where $\overline{N}(h_c)$ is the mean refractivity for all values at height h_c within the interferogram area, see fig. 3.5. The variable $\Delta N(x, y, h)$, with $E\{\Delta N(x, y, h)\} = 0$, is the lateral variation of the refractivity. Consequently, with $S_k^{t_i} = \overline{S}^{t_1} + \Delta S_k^{t_1}$ we could write eq. (3.4.5) as

$$\Delta S_k^{t_i} = \frac{1}{10^6 \cos \theta_{\text{inc}}} \left(\int_0^H \Delta N_{\text{hyd}} \, dh + \int_0^H \Delta N_{\text{wet}} \, dh \right.$$
$$\left. + \int_0^H \Delta N_{\text{iono}} \, dh + \int_0^H \Delta N_{\text{liq}} \, dh \right). \tag{3.4.7}$$

As a result of this (relative) representation, it is obvious that the influence of the four refractivity components depends on their lateral variability within the scene. We will discuss these contributions in more detail in chapter 6. Vertical variation of refractivity is of importance if (i) there is topography in the image and if (ii) the mean vertical refractivity profiles differ between the two acquisitions, see section 4.8.

3.5 Generic model interferogram stacks

Recognition of the rank deficiency in the standard functional model, due to the superposition of several parameters per observation, has lead to the development of interferometric methods that combine more than two SAR images. These methods eliminate or suppress some parameters at the benefit of others. The three-pass and four-pass methods to eliminate topography for deformation studies are first examples of this approach, see Zebker et al. (1994b). Especially the interest in slow deformation processes and in suppressing atmospheric signal for topography as well as deformation studies have lead to the use of multiple interferograms, also referred to as stacking, multi-baseline interferometry, or time-series analysis, see e.g., Ferretti et al. (1997); Usai and Hanssen (1997); Hanssen and Usai (1997); Sandwell and Price (1998); Usai and Klees (1999b); Ferretti et al. (1999a).

For multiple interferograms, the generic functional model defined by eq. (3.1.6) for two-pass interferometry can be extended. When two interferograms are available, this results in:

$$
E\left\{
\begin{bmatrix}
\begin{bmatrix} \varphi_1^{\mathrm{w\,I}} \\ \varphi_2^{\mathrm{w\,I}} \\ \vdots \\ \varphi_m^{\mathrm{w\,I}} \end{bmatrix} \\
\begin{bmatrix} \varphi_1^{\mathrm{w\,II}} \\ \varphi_2^{\mathrm{w\,II}} \\ \vdots \\ \varphi_m^{\mathrm{w\,II}} \end{bmatrix}
\end{bmatrix}
\right\} =
\begin{bmatrix}
\mathbf{A}_1^{\mathrm{I}} & & & & & & \\
& \mathbf{A}_2^{\mathrm{I}} & & & & & \\
& & \ddots & & & & \\
& & & \mathbf{A}_m^{\mathrm{I}} & & & \\
& & & & \mathbf{A}_1^{\mathrm{II}} & & \\
& & & & & \mathbf{A}_2^{\mathrm{II}} & \\
& & & & & & \ddots \\
& & & & & & & \mathbf{A}_m^{\mathrm{II}}
\end{bmatrix}
\begin{bmatrix}
\mathbf{x}_1^{\mathrm{I}} \\ \mathbf{x}_2^{\mathrm{I}} \\ \vdots \\ \mathbf{x}_m^{\mathrm{I}} \\ \mathbf{x}_1^{\mathrm{II}} \\ \mathbf{x}_2^{\mathrm{II}} \\ \vdots \\ \mathbf{x}_m^{\mathrm{II}}
\end{bmatrix}, \qquad (3.5.1)
$$

where $\mathbf{A}_k^j$ is the part of the design matrix corresponding to observation k in interferogram j, defined as:

$$
\mathbf{A}_k^j = \left[-\frac{4\pi}{\lambda} \frac{B_{\perp,k}^j}{R_k^j \sin \theta_k^j}, \quad \frac{4\pi}{\lambda}, \quad \frac{4\pi}{\lambda}, \quad -\frac{4\pi}{\lambda}, \quad -2\pi \right], \qquad (3.5.2)
$$

and the parameter vector $\mathbf{x}_k^j$ corresponding to observation k and interferogram j is:

$$
\mathbf{x}_k^j = \left[H_k^j, \ D_k^j, \ S_k^{j_1}, \ S_k^{j_2}, \ w_k^j \right]^T. \qquad (3.5.3)
$$

Here we used the notation $S_k^{j_1}$ and $S_k^{j_2}$ to indicate the atmospheric delay for pixel k at the first SAR acquisitions j_1 and the second SAR acquisition j_2. In order to eliminate some of the parameters, both interferograms need to be resampled to the same (master) grid.

When both interferograms share a common SAR acquisition it is commonly referred to as *three-pass* interferometry. Say interferogram I consists of acquisition t_1 and t_2, whereas interferogram II consists of acquisition t_2 and t_3, and surface deformation doesn't affect topographic height to the accuracy level ($H_k^{\mathrm{I}} = H_k^{\mathrm{II}}$), we can reorganize eq. (3.5.1) to:

$$
E\left\{
\begin{bmatrix}
\begin{bmatrix} \varphi_1^{\mathrm{w\,I}} \\ \varphi_2^{\mathrm{w\,I}} \\ \vdots \\ \varphi_m^{\mathrm{w\,I}} \end{bmatrix} \\
\begin{bmatrix} \varphi_1^{\mathrm{w\,II}} \\ \varphi_2^{\mathrm{w\,II}} \\ \vdots \\ \varphi_m^{\mathrm{w\,II}} \end{bmatrix}
\end{bmatrix}
\right\} =
\begin{bmatrix}
\mathbf{A}_1^{\mathrm{I}} & & & & \\
& \mathbf{A}_2^{\mathrm{I}} & & & \\
& & \ddots & & \\
& & & \mathbf{A}_m^{\mathrm{I}} & \\
\mathbf{A}_1^{\mathrm{II}} & & & & \\
& \mathbf{A}_2^{\mathrm{II}} & & & \\
& & \ddots & & \\
& & & & \mathbf{A}_m^{\mathrm{II}}
\end{bmatrix}
\begin{bmatrix} \mathbf{x}_1 \\ \mathbf{x}_2 \\ \mathbf{x}_3 \\ \vdots \\ \mathbf{x}_m \end{bmatrix},
\tag{3.5.4}
$$

where $\mathbf{A}_k^{\mathrm{I}}$ and $\mathbf{A}_k^{\mathrm{II}}$ are defined as:

$$
\mathbf{A}_k^{\mathrm{I}} = \left[-\kappa \frac{B_{\perp,k}^{\mathrm{I}}}{R_k^{\mathrm{I}} \sin\theta_k^{\mathrm{I}}}, \quad \kappa, \quad 0, \quad \kappa, \quad -\kappa, \quad 0, \quad -2\pi, \quad 0 \right],
\tag{3.5.5a}
$$

$$
\mathbf{A}_k^{\mathrm{II}} = \left[-\kappa \frac{B_{\perp,k}^{\mathrm{II}}}{R_k^{\mathrm{II}} \sin\theta_k^{\mathrm{II}}}, \quad 0, \quad \kappa, \quad 0, \quad \kappa, \quad -\kappa, \quad 0, \quad -2\pi \right],
\tag{3.5.5b}
$$

with $\kappa = 4\pi/\lambda$. The parameter vector $\mathbf{x}_k$ corresponding to observation k in both interferograms is:

$$
\mathbf{x}_k = \left[H_k, \; D_k^{\mathrm{I}}, \; D_k^{\mathrm{II}}, \; S_k^{t_1}, \; S_k^{t_2}, \; S_k^{t_3}, \; w_k^{\mathrm{I}}, \; w_k^{\mathrm{II}} \right]^T.
\tag{3.5.6}
$$

Without a common SAR acquisition the two interferograms are referred to as *four-pass* interferometry. As long as the topography can be regarded to be constant, we can use eq. (3.5.4) with $\mathbf{A}_k^{\mathrm{I}}$ and $\mathbf{A}_k^{\mathrm{II}}$ redefined as:

$$
\mathbf{A}_k^{\mathrm{I}} = \left[-\kappa \frac{B_{\perp,k}^{\mathrm{I}}}{R_k^{\mathrm{I}} \sin\theta_k^{\mathrm{I}}}, \quad \kappa, \quad 0, \quad \kappa, \quad -\kappa, \quad 0, \quad 0, \quad -2\pi, \quad 0 \right],
\tag{3.5.7a}
$$

$$
\mathbf{A}_k^{\mathrm{II}} = \left[-\kappa \frac{B_{\perp,k}^{\mathrm{II}}}{R_k^{\mathrm{II}} \sin\theta_k^{\mathrm{II}}}, \quad 0, \quad \kappa, \quad 0, \quad 0, \quad \kappa, \quad -\kappa, \quad 0, \quad -2\pi \right],
\tag{3.5.7b}
$$

and the *four-pass* parameter vector $\mathbf{x}_k$ corresponding to observation k in both interferograms as:

$$
\mathbf{x}_k = \left[H_k, \; D_k^{\mathrm{I}}, \; D_k^{\mathrm{II}}, \; S_k^{t_1}, \; S_k^{t_2}, \; S_k^{t_3}, \; S_k^{t_4}, \; w_k^{\mathrm{I}}, \; w_k^{\mathrm{II}} \right]^T.
\tag{3.5.8}
$$

3.5.1 Three-pass differential interferometry

The concept of three-pass *differential* interferometry, as demonstrated by Zebker et al. (1994b), relies on the assumptions that (*i*) only one of the two interferograms is affected by surface deformation, and (*ii*) the deformation does not affect the topographic height to the accuracy level. Using the *flattened* phase $\varphi_k^j = \phi_k - \vartheta_k^j$ we extend eq. (3.5.4) to

$$
E\{\varphi_k^{\mathrm{I}}\} = -\kappa \frac{B_{\perp,k}^{\mathrm{I}}}{R_k^{\mathrm{I}} \sin\theta_k^{\mathrm{I}}} H_k + \kappa S_k^{t_1} - \kappa S_k^{t_2}
\tag{3.5.9}
$$

$$
E\{\varphi_k^{\mathrm{w\,II}}\} = -\kappa \frac{B_{\perp,k}^{\mathrm{II}}}{R_k^{\mathrm{II}} \sin\theta_k^{\mathrm{II}}} H_k + \kappa D_k^{\mathrm{II}} + \kappa S_k^{t_1} - \kappa S_k^{t_3} - w_k^{\mathrm{II}} 2\pi,
\tag{3.5.10}
$$

where interferogram I is considered to be the *topographic pair* (no deformation assumed, absolute phase) and interferogram II is the *deformation pair* (topography, deformation, atmosphere, wrapped phase). Resolving for H_k in (3.5.9) and inserting this in eq. (3.5.10) yields:

$$E\{\varphi_k^{\text{w II}}\} = \frac{B_{\perp,k}^{\text{II}}}{B_{\perp,k}^{\text{I}}} E\{\varphi_k^{\text{I}}\} + (1 - \frac{B_{\perp,k}^{\text{II}}}{B_{\perp,k}^{\text{I}}}) \kappa S_k^{t_1} + \frac{B_{\perp,k}^{\text{II}}}{B_{\perp,k}^{\text{I}}} \kappa S_k^{t_2} - \kappa S_k^{t_3} + \kappa D_k^{\text{II}} - w_k^{\text{II}} 2\pi,$$

$$(3.5.11)$$

where we safely assume that $R_k^j \sin \theta_k^j$ is equal for both interferograms. With

$$\Omega_k = B_{\perp,k}^{\text{II}} / B_{\perp,k}^{\text{I}} \tag{3.5.12}$$

(the baseline ratio) we reformulate the model of observation equations, cf. eq. (3.5.1), to a mixed model of form (Teunissen, 2000):

$$\begin{array}{cccc} \mathbf{B}^T & E\{\mathbf{y}\} & = & \mathbf{A} & \mathbf{x}, \\ m \times 2m & 2m \times 1 & & m \times 5m & 5m \times 1 \end{array} \tag{3.5.13}$$

assuming $\mathbf{A}$ and $\mathbf{B}$ have full rank.

$$\begin{bmatrix} -\Omega_1 & & & 1 & & \\ & \ddots & & & \ddots & \\ & & -\Omega_m & & & 1 \end{bmatrix} E\left\{ \begin{bmatrix} \begin{bmatrix} \varphi_1^{\text{I}} \\ \varphi_2^{\text{I}} \\ \vdots \\ \varphi_m^{\text{I}} \end{bmatrix} \\ \begin{bmatrix} \varphi_1^{\text{w II}} \\ \varphi_2^{\text{w II}} \\ \vdots \\ \varphi_m^{\text{w II}} \end{bmatrix} \end{bmatrix} \right\} = \begin{bmatrix} \mathbf{A}_1 & & \\ & \mathbf{A}_2 & \\ & & \ddots & \\ & & & \mathbf{A}_m \end{bmatrix} \begin{bmatrix} \mathbf{x}_1 \\ \mathbf{x}_2 \\ \vdots \\ \mathbf{x}_m \end{bmatrix},$$

$$(3.5.14)$$

where $\mathbf{A}_k$ is the differential design matrix corresponding to observation k, defined by:

$$\mathbf{A}_k = \begin{bmatrix} \kappa, & (1 - \Omega_k)\kappa, & \Omega_k \kappa, & -\kappa, -2\pi \end{bmatrix}, \tag{3.5.15}$$

and the parameter vector $\mathbf{x}_k$ corresponding to resolution cell k is:

$$\mathbf{x}_k = \begin{bmatrix} D_k^{\text{II}}, S_k^{t_1}, S_k^{t_2}, S_k^{t_3}, w_k^{\text{II}} \end{bmatrix}^T. \tag{3.5.16}$$

The three center elements of $\mathbf{A}_k$ are associated with the three atmospheric parameters. From eq. (3.5.15) it is clear that the baseline scaling operation Ω affects the atmospheric signal in the differential interferogram. In general, a large absolute baseline ratio $|\Omega|$ needs to be avoided as it amplifies the atmospheric signal in one of the two images contributing to the topographic pair. This is equivalent to choosing a topographic pair with a large baseline and a deformation pair with a relatively small baseline. In the limit, a zero baseline for the deformation pair would make the topographic pair unnecessary and has normal (non-scaled) atmospheric circumstances. The considerations for choosing the topographic pair are generally

Table 3.2. Amplification factor for atmospheric signal in the differential interferogram

	Ω	$(-\infty, -1]$	$[-1, 0]$	$[0, 1]$	$[1, 2]$	$[2, \infty)$
$S_k^{t_1}$ (common image):	$\lvert 1 - \Omega \rvert$	$[2, \infty)$	$[1, 2]$	$[0, 1]$	$[0, 1]$	$[1, \infty)$
$S_k^{t_2}$ (slave topo pair):	$\lvert \Omega \rvert$	$[1, \infty)$	$[0, 1]$	$[0, 1]$	$[1, 2]$	$[2, \infty)$

1. $B_\perp^{\text{topo}} \geq B_\perp^{\text{defo}}$ to obtain at least the same height accuracy as the deformation pair,

2. $B_\perp^{\text{topo}} < {\sim} 0.7 \, B_\perp^{\text{crit}}$ to obtain coherent data, and

3. terrain circumstances such as relief, vegetation, and temporal decorrelation.

Table 3.2 and fig. 3.6 show the consequence of the baseline scaling ratio Ω for the amplification or reduction of the atmospheric signals $S_k^{t_1}$ in the common image and $S_k^{t_2}$ in the slave of the topographic pair. A reduction of the atmospheric signal in both images can only be obtained for $0 < \Omega < 1$, all other baseline ratios result in the amplification of atmospheric signal in either one or both of the images of the topographic pair.

Regarding the atmospheric signal as stochastic—as will be proposed in chapter 4— implies that the baseline ratio must be used to weight the variance and covariance of the atmospheric signal . In spectral approximations, the spectra are multiplied by $(1 - \Omega)$ and Ω (Bracewell, 1986). The variance for a zero-mean signal will be amplified by $(1 - \Omega)^2$ and Ω^2.

3.5.2   Four-pass differential interferometry

In some cases no suitable SAR acquisition is available to form the topographic pair together with one acquisition from the deformation pair. However, there might be

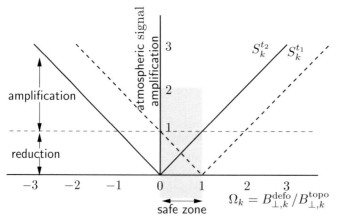

Fig. 3.6. Amplification/reduction of atmospheric signal in a three-pass or four-pass differential interferogram as a result of baseline scaling, see table 3.2. Only for $0 < \Omega_k < 1$ a reduction of the atmospheric signal in both images can be obtained.

another suitable interferometric combination that could serve as topographic pair. This situation occurs, e.g., when the perpendicular baselines for any combination with one of the acquisitions of the deformation pair are too large, or have a temporal baseline being too long. In those cases the two independent interferometric pairs can be used to apply four-pass differential interferometry.

A problem related to this technique is that the alignment procedures might fail. The correlation between the speckle pattern generally diminishes with increasing baseline. Nevertheless, there might be prominent reflectors that appear in both images. Since no coherent phase differences are necessary between both pairs, alignment criteria are less stringent. The most important factor for these applications is the roughness of the topography.

Assuming the topographic pair, labeled I, is flattened, unwrapped, and does not contain any deformation, the functional relations in eqs. (3.5.9) and (3.5.10) become

$$E\{\varphi_k^{\mathrm{I}}\} = -\kappa \frac{B_{\perp,k}^{\mathrm{I}}}{R_k^{\mathrm{I}} \sin \theta_k^{\mathrm{I}}} H_k + \kappa S_k^{t_1} - \kappa S_k^{t_2} \tag{3.5.17}$$

$$E\{\varphi_k^{\mathrm{w\,II}}\} = -\kappa \frac{B_{\perp,k}^{\mathrm{II}}}{R_k^{\mathrm{II}} \sin \theta_k^{\mathrm{II}}} H_k + \kappa D_k^{\mathrm{II}} + \kappa S_k^{t_3} - \kappa S_k^{t_4} - w_k^{\mathrm{II}} 2\pi. \tag{3.5.18}$$

Following the same procedure as for three-pass differential interferometry, we end up with the mixed model of eq. (3.5.14), with $\mathbf{A}_k$ defined as

$$\mathbf{A}_k = \begin{bmatrix} \kappa, & -\Omega_k\kappa, & \Omega_k\kappa, & \kappa, & -\kappa, & -2\pi \end{bmatrix}, \tag{3.5.19}$$

and the parameter vector $\mathbf{x}_k$ corresponding to resolution cell k as

$$\mathbf{x}_k = \begin{bmatrix} D_k^{\mathrm{II}}, & S_k^{t_1}, & S_k^{t_2}, & S_k^{t_3}, & S_k^{t_4}, & w_k^{\mathrm{II}} \end{bmatrix}^T. \tag{3.5.20}$$

Using table 3.2 the weighting of the atmospheric signal in the topographic pair can be determined. It is clear that a topographic pair with a long baseline is to be preferred.

3.5.3 Linear integer combinations

Linear combinations of interferograms as in eq. (3.5.14) can be used to decrease rank deficiency. However, an important limitation of scaling one of the interferograms with baseline ratio Ω_k is that this interferogram needs to be unwrapped before scaling, introducing errors that propagate into the final results. The necessity for the phase unwrapping is nevertheless limited to $\Omega_k \in \mathbb{R} \setminus \mathbb{Z}$. In these situations, scaling a *wrapped* interferogram will introduce fractional, hence erroneous, phase jumps in the data. Massonnet et al. (1996c) showed that this condition does not apply for $\Omega_k \in \mathbb{Z} \setminus \{0\}$, hence linear *integer* combinations. In these cases, the original fringes remain at their position but new fringes may be introduced between them.

Since an integer scaled interferogram corresponds with an integer scaled perpendicular baseline, $a B_{\perp}^{\mathrm{I}}$, with $a \in \mathbb{Z} \setminus \{0\}$, it is possible to form the difference with an interferogram with perpendicular baseline $B_\perp^{II}$. The resulting difference interferogram will have a perpendicular baseline of $B_\perp^{III} = B_\perp^{II} - a\,B_\perp^{I}$. By choosing factor a so that $B_\perp^{III}$ is very small, the difference interferogram will have much less fringes, and might be easier to unwrap. Although the accuracy of this unwrapped (first-order) difference interferogram is not optimal—the small baseline results in a large height ambiguity—it can be subtracted from the original interferograms. This yields an interferogram with residual fringes, which might be more successfully unwrapped than the original one. Adding the residual unwrapped fringes to the unwrapped fringes of the linear combination consequently yields the final results in which possible errors in the linear integer combination are eliminated. Figure 3.7 shows an example of this procedure for an area in Cornwall, United Kingdom. It is important to note that the integer scaling of one of the baselines amplifies noise in that interferogram as well. In practice, the scaling factor proves to be limited to $|a| \leq 3$.

It might be elucidating to express the generic functional model in terms of topographic height only, assuming no deformation or atmospheric signal. In that case we can write the functional relation between height H_k and the flattened, wrapped interferometric phase in interferogram I and II as

$$
\begin{aligned}
H_k &= \frac{\lambda}{4\pi}\frac{R_k^{I}\sin\theta_k^{I}}{B_{\perp,k}^{I}}\;\varphi_k^{w,I} \;+\; \frac{\lambda}{2}\frac{R_k^{I}\sin\theta_k^{I}}{B_{\perp,k}^{I}}\;w_k^{I} \\
&= \frac{\lambda}{4\pi}\frac{R_k^{II}\sin\theta_k^{II}}{B_{\perp,k}^{II}}\;\varphi_k^{w,II} \;+\; \frac{\lambda}{2}\frac{R_k^{II}\sin\theta_k^{II}}{B_{\perp,k}^{II}}\;w_k^{II},
\end{aligned}
\tag{3.5.21}
$$

which can be simplified when denoting the height ambiguity for the two interferograms by $h_{2\pi}^{I}$ and $h_{2\pi}^{II}$:

$$
\begin{aligned}
H_k &= \tfrac{1}{2\pi}h_{2\pi}^{I}\;\varphi_k^{w,I} \;+\; h_{2\pi}^{I}\;w_k^{I} \\
&= \tfrac{1}{2\pi}h_{2\pi}^{II}\;\varphi_k^{w,II} \;+\; h_{2\pi}^{II}\;w_k^{II},
\end{aligned}
\tag{3.5.22}
$$

with $\varphi_k^{w} \in [-\pi,\pi)$, $w_k \in \mathbb{Z}$, and $H_k \in \mathbb{R}$. The linear integer combination $\Delta\varphi_k^{w} = W\{a\,\varphi_k^{w,I} + b\,\varphi_k^{w,II}\}$, with $(a,b) \in \mathbb{Z}\setminus\{0\}$, can be written as

$$
H_k = \tfrac{1}{2\pi}h_{2\pi}^{III}\;\Delta\varphi_k^{w} \;+\; h_{2\pi}^{III}\;w_k^{III}
\tag{3.5.23}
$$

with

$$
\frac{1}{h_{2\pi}^{III}} = a\,\frac{1}{h_{2\pi}^{I}} + b\,\frac{1}{h_{2\pi}^{II}}.
\tag{3.5.24}
$$

Therefore, by choosing factors a and b such that $h_{2\pi}^{III}$ is large, phase gradients correspond with large height gradients. As a result, at the expense of increased phase noise, the height gradients can be such that the likelihood for estimating the correct integer value for w_k^{III} is increasing considerably. Therefore, this methodology can be used to reduce the solution space in a way comparable to the two carrier frequencies (L1,L2) approach in GPS ambiguity fixing.

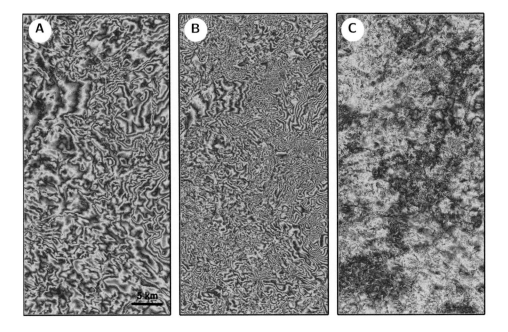

Fig. 3.7. Two interferograms and their integer linear combination for an area in Cornwall, United Kingdom. (Size: 21×42 km, posting: 60×60 m). **(A)** Interferogram I with $B_\perp^I = 170$ m. **(B)** Interferogram II with $B_\perp^{II} = -366$ m. **(C)** Linear combination III=2I + II with $B_\perp^{III} = -26$ m, showing a dramatic decrease in the number of fringes. Noise and atmospheric signal in interferogram I has doubled in this linear combination.

Chapter 4

Stochastic model for radar interferometry

This chapter presents the stochastic part of the Gauss-Markoff model for radar interferometry. It revisits the necessary theoretical concepts before focusing on the single-point statistics and multiple-point statistics. The quality of an interferogram is described in terms of the first and second moments. Error sources introduced by scattering, geometry, instrument, orbit, atmosphere, and signal processing are discussed. Models are proposed to describe atmospheric signal due to turbulent mixing and vertical stratification. Error propagation is used to derive a generic stochastic model.

key words: *Stochastic model, Quality analysis, Error sources, Error propagation, Atmosphere*

In the previous chapter, the functional part of the mathematical model has been defined. The functional model relates the interferometric observations to the unknown parameters of interest, such as topography, deformation, or atmospheric signal. In this chapter we will focus on the stochastic part of the mathematical model, which describes the dispersion of the observations in terms of variances and covariances. A key aspect in the definition of this model is the analysis of the error sources and the way these errors propagate into the stochastic model. As the sub-title of this book *"Data Interpretation and Error Analysis"* suggests, the analysis, propagation, and mathematical formulation of errors is indissoluble from a correct interpretation of the data.

In this chapter, we will derive and discuss the most important error sources in repeat-pass radar interferometry, starting with a concise theoretical background of the used concepts in section 4.1. Section 4.2 covers the single point observation statistics in the absence of biases or correlations between resolution cells. The concept of coherence and its relation with the signal-to-noise ratio is discussed in section 4.3, followed by a systematic treatment of the error sources affecting coherence and single point statistics. This evaluation includes instrument noise, processing induced errors, and temporal decorrelation. Sections 4.5–4.8 describe errors affecting the correlation between different resolution cells in the interferogram, starting with integer phase ambiguities, followed by orbit induced errors in 4.6 and atmospheric induced errors in section 4.7 and 4.8. In section 4.9 all error sources are combined in a joint stochastic model, and the propagation of phase errors into heights, deformations or atmospheric parameters is derived.

4.1   Theory

In this introduction a brief review is presented of the main statistical parameters used to describe the behavior of radar interferometric data.

4.1.1   First-order moment or expectation value

The random variable $\varphi(\mathbf{x})$ at location $\mathbf{x} = (x, y)$ in an image has an expectation value

$$E\{\varphi(\mathbf{x})\} = \mu(\mathbf{x}) = \int_{-\infty}^{\infty} \varphi \, pdf(\varphi) dx, \qquad (4.1.1)$$

where $pdf(\varphi)$ is the probability density function of $\varphi(\mathbf{x})$. The expectation value $E\{\varphi(\mathbf{x})\}$ can be a function of the location $\mathbf{x}$.

4.1.2   Second-order moments

The dispersion of the observations about their expectation value is described by the second-order moments. Three second-order moments are the variance, the covariance and the structure function or variogram. Note that structure function is a term frequently used in turbulence literature, while variogram originates from geostatistics. Covariance and structure function are often referred to as the two structural functions.

The variance is defined as (Box et al., 1994)

$$\sigma_{\varphi(\mathbf{x})}^2 = E\{[\varphi(\mathbf{x}) - \mu(\mathbf{x})]^2\} = \int_{-\infty}^{\infty} (\varphi - \mu(\mathbf{x}))^2 \, pdf(\varphi) dx, , \qquad (4.1.2)$$

and the covariance as

$$C_\varphi(\mathbf{x}_1, \mathbf{x}_2) = E\{[\varphi(\mathbf{x}_1) - \mu(\mathbf{x}_1)] \, [\varphi(\mathbf{x}_2) - \mu(\mathbf{x}_2)]\}. \qquad (4.1.3)$$

which is a function of the locations $\mathbf{x}_1$ and $\mathbf{x}_2$.

The third second-order moment is the structure function, which is defined as the variance of the increment $\Delta\varphi(\mathbf{x}_1, \mathbf{x}_2) = [\varphi(\mathbf{x}_1) - \varphi(\mathbf{x}_2)]$:

$$D_\varphi(\mathbf{x}_1, \mathbf{x}_2) = 2\Gamma_\varphi(\mathbf{x}_1, \mathbf{x}_2) = E\{[\Delta\varphi(\mathbf{x}_1, \mathbf{x}_2) - E\{\Delta\varphi(\mathbf{x}_1, \mathbf{x}_2)\}]^2\}, \qquad (4.1.4)$$

where $D_\varphi(\mathbf{x}_1, \mathbf{x}_2)$ is the structure function or variogram and $\Gamma_\varphi(\mathbf{x}_1, \mathbf{x}_2)$ is the semi-variogram (Journel and Huijbregts, 1978).

Autocorrelation

The autocorrelation function ρ is a standardized version of the covariance function, bounded to the interval $[-1, 1]$, with

$$\rho_\varphi(\mathbf{x}_1, \mathbf{x}_2) = \frac{E\{[\varphi(\mathbf{x}_1) - \mu(\mathbf{x}_1)][\varphi(\mathbf{x}_2) - \mu(\mathbf{x}_2)]\}}{\sqrt{E\{(\varphi(\mathbf{x}_1) - \mu(\mathbf{x}_1))^2\}E\{(\varphi(\mathbf{x}_2) - \mu(\mathbf{x}_2))^2\}}}, \qquad (4.1.5)$$

or

$$\rho_\varphi(\mathbf{x}_1, \mathbf{x}_2) = \frac{E\{[\varphi(\mathbf{x}_1) - \mu][\varphi(\mathbf{x}_2) - \mu]\}}{\sigma^2_{\varphi(\mathbf{x})}}, \qquad (4.1.6)$$

for a (second-order) stationary process. In contrast to the covariance function, the autocorrelation function is independent of the magnitude of the data values.

4.1.3 Positive definiteness

A covariance function, empirical or analytical, is positive (semi)-definite whenever it fulfills the following three conditions (Journel and Huijbregts, 1978):

- $C_\varphi(0) = \sigma^2_{\varphi(\mathbf{x})} \geq 0$,

- $C_\varphi(\mathbf{r}) = C_\varphi(-\mathbf{r})$, this is always true for $\mathbf{r} \geq 0$, and

- $|C_\varphi(\mathbf{r})| \leq C_\varphi(0)$, Schwarz's inequality.

Often the covariance function will reach zero values for a distance larger than a specific value, indicating that there is no correlation for two values which are further apart than that distance. This distance is referred to as the *range* of the covariance function. This implies with eq. (4.1.12) that the value of the structure function after that range will be $2C_\varphi(0)$ or $2\sigma^2_{\varphi(\mathbf{x})}$, which is called the *sill* value.

If the covariance function is positive semi-definite, the variance of any linear combination of variables $\varphi(\mathbf{x})$ must be positive (Wackernagel, 1995). Therefore, the covariance matrix C is positive semi-definite if and only if the covariance function fulfills these three requirements.

4.1.4 Stationarity

To apply specific statistical measures it is necessary to determine whether the underlying physical process can be considered statistically stationary. Unfortunately, unless infinitely many sample functions are available of this process, which reach to infinity, it is never possible to answer this question (Newland, 1993). Therefore, it is often tried to apply the mathematics under certain assumptions of the physical behavior of the process, or to relax the stationarity conditions. In terms of nested sets, we define the set of *strictly stationary* random functions as a subset of the set of *second-order stationary* random functions, which is a subset of *intrinsic* random functions, see fig. 4.1. All these sets are contained in the general set of random functions (Journel and Huijbregts, 1978) (Chilès and Delfiner, 1999).

Proceeding from the large sets to the smaller subsets we find:

- Intrinsic random functions. A random function $\varphi(\mathbf{x})$ is intrinsic when (i) the expectation value $E\{\varphi(\mathbf{x})\}$ is constant and exists for every location $\mathbf{x}$:

$$E\{\varphi(\mathbf{x})\} = m, \quad \forall \mathbf{x}, \qquad (4.1.7)$$

and (ii) for all vectors $\mathbf{r}$ the increment $\Delta\varphi(\mathbf{x}+\mathbf{r}, \mathbf{x}) = [\varphi(\mathbf{x}+\mathbf{r}) - \varphi(\mathbf{x})]$ has a constant finite variance:

$$\begin{aligned} D_\varphi(\mathbf{r}) &= E\{[\Delta\varphi(\mathbf{x}+\mathbf{r}, \mathbf{x}) - E\{\Delta\varphi(\mathbf{x}+\mathbf{r}, \mathbf{x})\}]^2\}, \quad \forall\mathbf{x}, \\ &= E\{[\Delta\varphi(\mathbf{x}+\mathbf{r}, \mathbf{x})]^2\}, \quad \forall\mathbf{x}. \end{aligned} \tag{4.1.8}$$

Therefore, this set of functions is also referred to as *random functions with stationary increments*. This does not imply that the variance and covariance of $\varphi(\mathbf{x})$ can be uniquely determined. Written in matrix notation the expectation of the increment and its variance are

$$E\{\Delta\varphi(\mathbf{x}+\mathbf{r}, \mathbf{x})\} = \begin{bmatrix} 1 & -1 \end{bmatrix} E\left\{ \begin{bmatrix} \varphi(\mathbf{x}+\mathbf{r}) \\ \varphi(\mathbf{x}) \end{bmatrix} \right\} = 0; \tag{4.1.9}$$

$$\sigma^2_{\Delta\varphi(\mathbf{x}+\mathbf{r}, \mathbf{x})} = D_\varphi(\mathbf{r}) = \begin{bmatrix} 1 & -1 \end{bmatrix} \begin{bmatrix} \sigma^2_{\varphi(\mathbf{x}+\mathbf{r})} & \sigma_{\varphi(\mathbf{x}+\mathbf{r}), \varphi(\mathbf{x})} \\ \sigma_{\varphi(\mathbf{x}), \varphi(\mathbf{x}+\mathbf{r})} & \sigma^2_{\varphi(\mathbf{x})} \end{bmatrix} \begin{bmatrix} 1 \\ -1 \end{bmatrix}. \tag{4.1.10}$$

From (4.1.10) it follows that there are more combinations of variances and covariances of $\varphi(\mathbf{x})$ which result in the same value for $\sigma^2_{\Delta\varphi(\mathbf{x}+\mathbf{r}, \mathbf{x})}$. Note that the definition of the structure function $D_\varphi(\mathbf{r})$ does not require the existence of a constant mean and finite variance of $\varphi(\mathbf{x})$ (Goovaerts, 1997).

- Second-order stationarity.

A random function $\varphi(\mathbf{x})$ is stationary of order 2 when the first two moments are stationary. In that case, the two characteristics of the intrinsic hypothesis hold, and (iii) for each pair $[\varphi(\mathbf{x}), \varphi(\mathbf{x}+\mathbf{r})]$ the covariance exists and is only dependent on the distance $\mathbf{r}$:

$$C_\varphi(\mathbf{r}) = E\{\varphi(\mathbf{x}+\mathbf{r})\varphi(\mathbf{x})\} - E\{\varphi(\mathbf{x}+\mathbf{r})\} E\{\varphi(\mathbf{x})\}, \quad \forall\mathbf{x}. \tag{4.1.11}$$

The existence of the covariance implies that the right-hand side of eq. (4.1.10) is known. Therefore, it is always possible for a second-order stationary function to derive the structure function from the covariance function

$$\begin{aligned} \sigma^2_{\Delta\varphi(\mathbf{x}+\mathbf{r}, \mathbf{x})} = D_\varphi(\mathbf{r}) &= 2(\sigma^2_{\varphi(\mathbf{x})} - \sigma_{\varphi(\mathbf{x}+\mathbf{r}), \varphi(\mathbf{x})}) \\ &= 2(C_\varphi(0) - C_\varphi(\mathbf{r})), \quad \forall\mathbf{x}. \end{aligned} \tag{4.1.12}$$

Therefore, under the hypothesis of second-order stationarity, the covariance and the structure function are in fact equivalent (Journel and Huijbregts, 1978). Note that there are physical phenomena, such as Brownian motion, which have an infinite dispersion. For such phenomena $C_\varphi(0)$ does not exist, and although they might fulfill the intrinsic hypothesis they cannot be considered second-order stationary. On the other hand, since practical measurements span a finite range, second-order stationarity can be assumed within this range. This corresponds to an hypothesis of quasi-stationarity.

- Strict stationarity. A random function $\varphi(\mathbf{x})$ is strictly stationary if the whole probability density function (i.e., all statistical moments) is invariant under translation. Note that strict stationarity is a severe requirement, and usually the simple term *stationarity* refers to second-order stationarity.

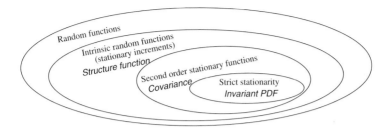

Fig. 4.1. Schematic visualization of the four sets: random functions, random functions with stationary increments (intrinsic random), second-order stationary functions, and stationary functions. The covariance function is defined only for second-order stationary functions, whereas the structure function is defined for the larger subset of intrinsic random functions.

Often the covariance or structure function are only used in a limited region. By limiting the hypothesis of second-order stationarity to this region, the function can be considered quasi-stationary.

It is important to note that stationarity is a property of the random function or process, not the data, due to the finite and discrete nature of the data. Therefore, in practice, stationarity is a decision made by the user, not a hypothesis that can be proved or rejected by data. Often it is even only a characteristic of the used model and not of the physical properties under study (Goovaerts, 1997).

Homogeneity and isotropy

Homogeneity is the equivalent of second-order stationarity and is used especially for 2D and 3D random fields (Balakrishnan, 1995). A second-order stationary random field is isotropic if it is homogeneous and if the covariance function only depends on the length of the increment and not on its orientation (Chilès and Delfiner, 1999), (Balakrishnan, 1995):

$$C_\varphi(\mathbf{r}) = C_\varphi(r), \quad r = |\mathbf{r}|. \tag{4.1.13}$$

Ergodicity

For an ergodic signal, the spatial (or temporal) and ensemble averages are equivalent. For example, interferometric coherence estimation using a spatial window relies on the assumption of ergodicity.

4.1.5 Scale-invariance and fractal dimension

For a scale-invariant (scaling) signal there is no characteristic scale in the process, so that power laws prevail over a large range of scales. For a scaling process $\varphi(x)$ the power spectrum $P_\varphi(k)$ exhibits a power-law behavior:

$$P_\varphi(k) \propto k^{-\beta}, \tag{4.1.14}$$

for all discrete wavenumbers k within the scaling range. Note that the expression power spectrum is also commonly referred to as wavenumber spectrum, energy spectrum, variance spectrum, periodogram, and spectral density. The spectral exponent β contains information on the degree of stationarity of the signal (Mandelbrot, 1983). If $\beta < 1$, the signal is stationary, i.e. $\varphi(x)$ is statistically invariant for translation. For $1 < \beta < 3$, the signal is non-stationary with stationary increments, so the structure function will be stationary (Ishimaru, 1978; Agnew, 1992; Davis et al., 1994). Many geophysical processes belong to the latter category over a bounded range of scales, which restrict the process to its physically accessible values.

Fractal dimension

Scale-invariant processes can be conveniently described using power-law statistics or fractal statistics. Since many processes in nature are scale-invariant within a certain range, the fractal dimension D is an elegant parameterization of the behavior of a process, equivalent to the spectral exponent β. A fractal that is statistically *self-similar* exhibits the same behavior in all dimensions, and is by definition isotropic. For example, the isoline curves (horizontal planar sections) of a topography map are self-similar. The fractal surfaces that are important in this study are *self-affine* fractals. For such fractals the scale in the vertical plane differs from the scale in the horizontal plane (Chilès and Delfiner, 1999).

For a one-dimensional signal or two-dimensional random field the relationship between the spectral exponent β, Haussdorff measure H, and fractal dimension D_1 or D_2 is (Turcotte, 1997):

$$\beta = 2H + 1 = \begin{cases} 5 - 2D_1, & \text{for a one-dimensional signal, and} \\ 7 - 2D_2, & \text{for a two-dimensional field.} \end{cases} \quad (4.1.15)$$

For a self-affine fractal ($0 < H < 1$, $1 < D_1 < 2$) we find $1 < \beta < 3$. This category of processes is referred to as fractional Brownian motion (fBm). *Standard* fBm or *random walk* is a subset within this category with $H = 0.5$, $D_1 = 1.5$, or $\beta = 2$ (Mandelbrot and Ness, 1968).

The fractal dimension or power-law exponent is a measure for the smoothness of a signal or surface. Figure 4.2 shows three characteristic fractal simulations: "white noise," "$1/f$-noise" (where f is the continuous wavenumber), and "standard fBm" in a one-dimensional realization (left column), two-dimensional realization (middle column), and the power spectral shape in a log-log plot (right column). Notice that a lower fractal dimension or steeper slope of the power spectrum (higher value for β) corresponds with a smoother surface.

The three examples in fig. 4.2 demonstrate the relationship between stationarity and the spectral exponent or fractal dimension. White noise ($\beta = 0$) is strictly stationary: $E\{\varphi(x)\} = 0$ and $E\{\varphi(x)^2\} = \text{constant}$ for any position x. For $1 < \beta < 3$ the signal is not strictly stationary: taking $\varphi(0) = 0$ for convenience, the variance $E\{\varphi(x)^2\}$ is proportional to x, even though $E\{\varphi(x)\} = 0$. Since the structure function $D_\varphi(\mathbf{r}) = E\{[\varphi(\mathbf{x} + \mathbf{r}) - \varphi(\mathbf{x})]^2\}$ is independent of position $\mathbf{x}$, the signal has stationary increments and is intrinsic. Nevertheless, for a limited range such as depicted here, the signal can be considered second-order stationary.

The concepts of scale-invariance will be used in section 4.7 to describe and model the general behavior of atmospheric signal in the SAR interferograms.

4.2   Single-point observation statistics

Radar interferometry can be regarded as a parameter estimation problem, where the observations consist of complex-valued numbers for all resolution cells and the parameters depend on the goal of the study, as discussed in chapter 3. In that chapter, the standard Gauss-Markoff model, defined in eq. (3.1.1), was proposed to obtain optimal estimates of the parameters and their corresponding covariance matrix, see eqs. (3.1.3) and (3.1.4), respectively. The functional relation between parameters and observations has been defined. Nevertheless, correct assessment of the dispersion of the observations in the stochastic model is the key issue for evaluating adjustment procedures and assessing the final data quality. Let us look at this dispersion in more detail.

Many authors have evaluated the (complex) variance of single-look or multilook interferometric observations (Li and Goldstein, 1990; Tough et al., 1995; Joughin, 1995; Touzi et al., 1999; Lee et al., 1994; Just and Bamler, 1994; Bamler and Just, 1993; Davenport and Root, 1987; Rodriguez and Martin, 1992). In these approaches, however, the covariance between observations was considered to be restricted to neighboring resolution cells. This hypothesis was mainly supported by the fact that the bandwidth of the SAR in range and azimuth direction is smaller that the frequency range determined by the Nyquist frequency. This effect results in resolution cells that are slightly larger than the posting of the observations, and is accounted for, e.g., by taking an *effective* number of looks instead of the nominal amount of looks. Correlations between resolution cells that are not neighboring are generally not accounted for. Ignoring (in first approximation) the correlations between neighboring resolution cells results in a diagonal variance-covariance matrix, cf. eq. (3.1.10):

$$\mathbf{C}_\varphi = \begin{bmatrix} \sigma_1^2 & & & \\ & \sigma_2^2 & & \\ & & \ddots & \\ & & & \sigma_n^2 \end{bmatrix} = \sigma^2 \mathbf{Q}_\varphi, \quad (\sigma^2 = 1), \qquad (4.2.1)$$

where σ_k^2 can be approximated using the estimated coherence values from the aligned SAR data. This derivation will be discussed in section 4.2.2.

There are two reasons why this approach is generally insufficient for parameter estimation from interferometric radar data. First, the interferometric observation is inherently relative. This implies that, instead of observing single (multilooked) resolution cells, one is generally interested in the difference between two spatially separated resolution cells, as implicitly visualized in an interferogram. This introduces the need to analyze covariances between the data. Secondly, for all repeat-pass interferometric configurations the influence of spatially varying signal propagation effects cannot be ignored. As we will see in section 4.7, atmospheric signal generally

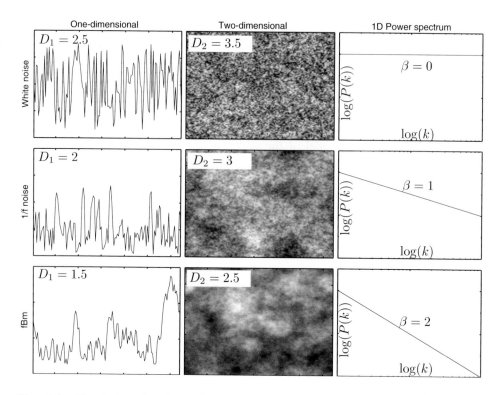

Fig. 4.2. Simulations for three characteristic fractal models: white noise, $1/f$ noise, and standard fractional Brownian motion. The three columns are a 1D realization, a 2D realization, and the corresponding shape of the 1D power spectrum.

obeys a power-law behavior. This implies that a larger separation between resolution cells results in a larger variance of the difference. Hence, the varying correlation between resolution cells, dependent on their spatial separation, cannot be ignored, and the variance-covariance matrix $\mathbf{C}_\varphi$ in eq. (4.2.1) will become a full matrix, as:

$$\mathbf{C} \doteq \mathbf{C}_\varphi + \mathbf{C}_\mathbf{s}, \qquad (4.2.2)$$

where $\mathbf{C}_\mathbf{s}$ is the variance-covariance matrix of the atmospheric delay parameter S_k, defined by lumping the atmospheric slant delay signals $S_k^{t_1}$ and $S_k^{t_2}$ during the two acquisitions into one parameter $S_k = S_k^{t_1} - S_k^{t_2}$. Of course, it depends on the parameter of interest (topography, deformation, or atmosphere) whether this approach should be followed. If the main objective of the analysis of the interferometric data is atmosphere studies, it is obvious that the atmospheric delay parameter should be included in the functional model, not in the stochastic model. As a consequence, for atmospheric studies the absolute variance of the data is much better than for topography or deformation studies.

In the following subsections we summarize the analysis of SAR and InSAR data statistics for single (multilooked) resolution cells.

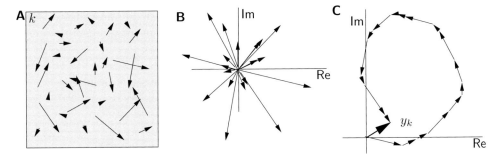

Fig. 4.3. Summation of scatterers within resolution cell k. **(A)** The distribution of N individual phasors within a resolution cell.**(B)** All phasors translated to the same origin (scaled). **(C)** The phasor sum of all scatterers. The fat arrow indicates the sum, which is the complex observable y_k (scaled).

4.2.1 Complex PDF SAR resolution cell

The response of a resolution cell on earth to the arriving radar pulse is strongly dependent on the scattering mechanisms involved. Although this is an extended field of study, only two mechanisms will be discussed here: distributed scattering (also referred to as Gaussian or Rayleigh scattering) and point scattering (Madsen, 1986; Bamler and Hartl, 1998).

For SAR systems where the size of a resolution cell is many times larger than the radar wavelength, many terrain elements or scatterers contribute to the response for one resolution cell (Krul, 1982; Curlander and McDonough, 1991). In practice it is therefore not possible to determine the response of individual scatterers within a resolution cell. This results in the well known, highly unpredictable speckle characteristics of SAR magnitude images (Goodman, 1975). The measured reflection is rather written as a sum of many (sub) scatterers, see fig. 4.3. Using a defined set of assumptions, it is possible to apply the central limit theorem, which defines the observations y_k as *complex (circular) Gaussian random variables*. These assumptions are (Madsen, 1986; Bamler and Hartl, 1998):

1. No single scatterer should dominate the others in a resolution cell. This assumption generally holds for, e.g., agricultural fields, forests, deserts, and many other natural scatterers.

2. The phase of every individual scatterer must be *uniformly distributed* between $-\pi$ and $+\pi$. When the slant-range resolution is much larger than the radar wavelength, a very large range of phase shifts is apparent. Folded into the interval $[-\pi, \pi)$, this leads to a uniform distribution.

3. The phases of the individual scatterers must be uncorrelated. This is reasonable as correlated scatterers are gathered into several (uncorrelated) scatter centers.

4. The amplitude and the phase of every scatterer must be uncorrelated. This assumption holds because the phase delay due to the propagation of the signal is independent of the scattering magnitude.

The probability density function of a complex circular Gaussian variable y (or the *joint* PDF of its real and imaginary component) is written as (Dainty, 1975; Davenport and Root, 1987):

$$pdf(y) = pdf(\text{Re}\{y\}, \text{Im}\{y\}) = \frac{1}{\pi 2\sigma^2} \exp\left(-\frac{(\text{Re}\{y\})^2 + (\text{Im}\{y\})^2}{2\sigma^2}\right), \qquad (4.2.3)$$

where

$$\sigma^2 = \sigma_y^2 = \sigma_{\text{Re}\{y\}}^2 = \sigma_{\text{Im}\{y\}}^2. \qquad (4.2.4)$$

In the context of radar observations, the factor $2\sigma^2$ in (4.2.3) is often substituted by (Bamler and Hartl, 1998)

$$E\{p\} = 2\sigma^2, \qquad (4.2.5)$$

where $p = |y|^2$ is the intensity (or power) of the resolution cell and $E\{p\}$ is its expectation value. [1] From equation (4.2.3) it is clear that the real and imaginary parts of a complex circular Gaussian variable are uncorrelated. From (4.2.3) and

$$\text{Re}\{y\} = a \cos \psi \qquad (4.2.7a)$$
$$\text{Im}\{y\} = a \sin \psi, \qquad (4.2.7b)$$

we find the Jacobian $|J| = a$ and derive the joint probability density function for the amplitude a and phase ψ random variables:

$$pdf(a, \psi) = \begin{cases} \frac{a}{2\pi\sigma^2} \exp\left(-\frac{a^2}{2\sigma^2}\right) & \text{for } a \geq 0 \text{ and } -\pi \leq \psi < \pi \\ 0 & \text{otherwise.} \end{cases} \qquad (4.2.8)$$

The marginal PDF of a is obtained integrating ψ out, between $-\pi$ and π. This yields

$$pdf(a) = \begin{cases} \frac{a}{\sigma^2} \exp\left(-\frac{a^2}{2\sigma^2}\right) & \text{for } a \geq 0 \\ 0 & \text{otherwise.} \end{cases} \qquad (4.2.9)$$

---

[1] A system theoretical description of $E\{p\}$ was given by Madsen (1986),

$$E\{p\} = N \, E\{\sqrt{\sigma_j^{0\,2}}\} E\{h_j^2\}, \qquad (4.2.6)$$

where N is the number of independent scatterers within the resolution cell, σ_j is the complex backscatter coefficient per scatterer, and h is the coherent system impulse response function. Similar expressions are discussed by Bamler and Schättler (1993); Bamler and Hartl (1998).

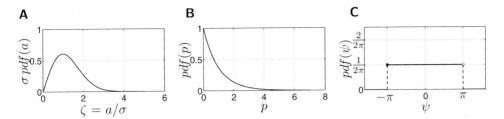

Fig. 4.4. Probability density functions for amplitude, power, and phase. **(A)** The Rayleigh distribution is used as a model for the amplitude of the radar data (Davenport and Root, 1987), whereas **(B)** the exponential distribution describes its intensity. Here the exponential distribution is shown for $E\{p\} = 1$. **(C)** The phase of the radar data is uniformly distributed.

Equation (4.2.9) is the Rayleigh distribution (Papoulis, 1968), see fig. 4.4A. The marginal PDF of ψ is found integrating (4.2.8) over a, between 0 and ∞:

$$pdf(\psi) = \begin{cases} \frac{1}{2\pi} & \text{for } -\pi \leq \psi < \pi \\ 0 & \text{otherwise.} \end{cases} \tag{4.2.10}$$

Equation (4.2.10) describes a *uniform distribution*. From (4.2.8), (4.2.9), and (4.2.10) it follows that a and ψ are uncorrelated, since

$$pdf(a, \psi) = pdf(a)pdf(\psi). \tag{4.2.11}$$

The pixel intensity variation for a distributed scene, expressed by the exponential PDF, is known as *speckle*. Assuming ergodicity, averaging of resolution cells is often applied to reduce the effect of speckle. As a result, the PDF of the intensity value of N averaged resolution cells can be described by a χ^2-distributed PDF with $2N$ degrees of freedom (Raney, 1998):

$$pdf(p_{2N}) = \frac{p^{N-1}}{E\{p\}^N \Gamma(N)} \exp\left(-\frac{p}{E\{p\}}\right). \tag{4.2.12}$$

Note that for $N = 1$ this equals the exponential PDF, while for $N \to \infty$ it equals a Gaussian PDF.

In table 4.1 the probability density functions of the intensity p, magnitude a, and phase ψ of a resolution cell are summarized (Madsen, 1986; Bamler and Hartl, 1998), (Raney, 1998):

4.2.2 Complex PDF interferometric resolution cell

In section 4.2.1 the behavior of a single SAR resolution cell, under the assumption of a distributed scattering mechanism, was described by circular Gaussian statistics, see eq. (4.2.3). Although the PDF of the phase has a uniform distribution, see table 4.1, the phase of the complex product of two circular Gaussian signals is not necessarily uniform, as long as the two signals have some degree of correlation.

Table 4.1. Overview of the PDF, expectation, and variance for the phase, amplitude, and intensity (power) of a single SAR image.

parameter	PDF	expectation	variance
Intensity	$pdf(p) = \frac{1}{E\{p\}} \exp\left(-\frac{p}{E\{p\}}\right)$	$E\{p\}$	$\sigma_p^2 = E\{p\}^2$
Amplitude	$pdf(a) = \frac{2a}{E\{p\}} \exp\left(-\frac{p}{E\{p\}}\right)$	$E\{a\} = \frac{1}{4}\sqrt{2\pi}E\{p\}$	$\sigma_a^2 = (1 - \frac{\pi}{4})E\{p\}$
Phase	$pdf(\psi) = \frac{1}{2\pi} \ (-\pi \leq \psi < \pi)$	$E\{\psi\} = 0$	$\sigma_\psi^2 = \pi^2/3$

The joint PDF of two circular Gaussian signals y_1 and y_2, with $E\{y_1\} = E\{y_2\} = 0$ and $E\{y_1^2\} = E\{y_2^2\}$ can be written as (Goodman, 1963; Hannan and Thomson, 1971; Lee et al., 1994; Just and Bamler, 1994; Tough et al., 1995; Bamler and Hartl, 1998)

$$pdf\{y_1, y_2\} = \frac{1}{\pi^2 |C_y|} \exp\{-\begin{bmatrix} y_1^* & y_2^* \end{bmatrix} C_y^{-1} \begin{bmatrix} y_1 \\ y_2 \end{bmatrix}\}, \qquad (4.2.13)$$

where C_y is the complex covariance matrix, defined by

$$C_y = E\{\begin{bmatrix} y_1 \\ y_2 \end{bmatrix} \begin{bmatrix} y_1^* & y_2^* \end{bmatrix}\} = \begin{bmatrix} E\{|y_1|^2\} & \gamma\sqrt{E\{|y_1|^2\}E\{|y_2|^2\}} \\ \gamma^*\sqrt{E\{|y_1|^2\}E\{|y_2|^2\}} & E\{|y_2|^2\} \end{bmatrix}, \qquad (4.2.14)$$

and $|C_y|$ is its determinant (Tough et al., 1995):

$$|C_y| = E\{|y_1|^2\}\, E\{|y_2|^2\}\, (1 - |\gamma|^2). \qquad (4.2.15)$$

The complex coherence (or complex correlation coefficient) γ is defined as

$$\gamma = \frac{E\{y_1 y_2^*\}}{\sqrt{E\{|y_1|^2\}\, E\{|y_2|^2\}}} = |\gamma| \exp(j\phi_0). \qquad (4.2.16)$$

As the interferometric combination of y_1 and y_2 can be written as the Hermitian product:

$$v = y_1 y_2^* = |y_1||y_2| \exp(j(\psi_1 - \psi_2)) = |y_1||y_2| \exp(j\phi_0), \qquad (4.2.17)$$

we find that $E\{v\}$ is equal to the numerator of the complex coherence. Since the denominator is real-valued, the phase of the complex correlation is the expected phase ϕ_0 of the interferogram. The magnitude $|\gamma|$, with $0 \leq \gamma \leq 1$, is a measure of the phase noise. Applying a multilook procedure with N looks, $N \geq 1$, during this multiplication, eq. (4.2.17) can be written as

$$v = \sum_{n=1}^{N} y_1^{(n)} y_2^{(n)*} = a \exp(j\phi_0). \qquad (4.2.18)$$

Rodriguez and Martin (1992) showed that the maximum likelihood estimator of the interferometric phase, for distributed scattering mechanisms, is

$$\hat{\phi} = \phi_0 = \arctan\left(\frac{\text{Im}\sum_{n=1}^{N} y_1^{(n)} y_2^{*(n)}}{\text{Re}\sum_{n=1}^{N} y_1^{(n)} y_2^{*(n)}}\right), \tag{4.2.19}$$

which is unbiased modulo 2π.

The joint probability density function of the interferometric amplitude a and phase ϕ for L independent looks ($L < N$ for SAR systems with a limited bandwidth) can be shown to be (Goodman, 1963; Barber, 1993b; Lee et al., 1994; Tough et al., 1995):

$$pdf(a, \phi) = \frac{2L(L\,a)^L}{\pi \zeta^{L+1}(1 - |\gamma|^2)\Gamma(L)} \exp\left(\frac{2|\gamma|L\,a\cos(\phi - \phi_0)}{\zeta(1 - |\gamma|^2)}\right) K_{L-1}\left(\frac{2L\,a}{\zeta(1 - |\gamma|^2)}\right), \tag{4.2.20}$$

where $\zeta = \sqrt{E\{y_1\}E\{y_2\}}$ is the average power of the two image segments. $K_{L-1}(\cdot)$ is the modified Bessel function of the third kind (also referred to as Hankel's function (Gradshteyn et al., 1994)), and γ is the complex correlation coefficient. The gamma function is defined as:

$$\Gamma(L) = \int_0^\infty t^{L-1} e^{-t} dt, \quad \text{for} \quad L \in \mathbb{R}, \tag{4.2.21}$$

or as

$$\Gamma(L) = (L-1)! \quad \text{for} \quad (L-1) \in \mathbb{N}. \tag{4.2.22}$$

The marginal probability density function for the interferometric phase ϕ is obtained by integrating over all amplitudes a (Tough et al., 1995):

$$\begin{aligned}
pdf(\phi; \gamma, L, \phi_0) = \frac{(1 - |\gamma|^2)^L}{2\pi} &\left\{ \frac{\Gamma(2L - 1)}{[\Gamma(L)]^2 \, 2^{2(L-1)}} \right. \\
&\times \left[\frac{(2L - 1)\beta}{(1 - \beta^2)^{L+\frac{1}{2}}} \left(\frac{\pi}{2} + \arcsin\beta\right) + \frac{1}{(1 - \beta^2)^L} \right] \\
&+ \left. \frac{1}{2(L-1)} \sum_{r=0}^{L-2} \frac{\Gamma(L - \frac{1}{2})}{\Gamma(L - \frac{1}{2} - r)} \frac{\Gamma(L - 1 - r)}{\Gamma(L - 1)} \frac{1 + (2r + 1)\beta^2}{(1 - \beta^2)^{r+2}} \right\}
\end{aligned} \tag{4.2.23}$$

with $\beta = |\gamma|\cos(\phi - \phi_0)$. Note that the finite summation in (4.2.23) disappears for the single look case ($L = 1$).

Barber (1993a), Lee et al. (1994), and Joughin and Winebrenner (1994) found an equivalent expression, using a hypergeometric function:

$$\begin{aligned}
pdf(\phi; \gamma, L, \phi_0) = &\frac{\Gamma(L + 1/2)\,(1 - \gamma^2)^L\,|\gamma|\cos(\phi - \phi_0)}{2\sqrt{\pi}\,\Gamma(L)\,(1 - \gamma^2\cos^2(\phi - \phi_0))^{L+1/2}} \\
&+ \frac{(1 - \gamma^2)^L}{2\pi} \, _2F_1(L, 1; 1/2; \gamma^2\cos^2(\phi - \phi_0)),
\end{aligned} \tag{4.2.24}$$

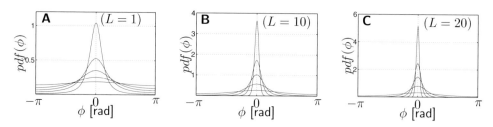

Fig. 4.5. Probability density functions for coherence levels $|\gamma| = 0.1$, 0.3, 0.5, 0.7, and 0.9 (smallest curves correspond with highest coherence). **(A)** Single look, **(B)** Multilook level 10, **(C)** Multilook level 20.

where the hypergeometric function is defined as (Oberhettinger, 1970; Gradshteyn et al., 1994):

$$
\begin{aligned}
{}_2F_1\left(L, 1; \frac{1}{2}; \gamma^2 \cos^2 \phi\right) &= F\left(1, L; \frac{1}{2}; \gamma^2 \cos^2 \phi\right) \\
&= \sum_{i=0}^{\infty} \frac{(L)_i (L)_i}{\left(\frac{1}{2}\right)_i} \frac{(\gamma^2 \cos^2 \phi)^i}{i} \\
&= \frac{\Gamma(\frac{1}{2})}{\Gamma(L)\Gamma(1)} \sum_{i=0}^{\infty} \frac{\Gamma(L+i)\,\Gamma(1+i)}{\Gamma(\frac{1}{2}+i)} \frac{(\gamma^2 \cos^2 \phi)^i}{i!}.
\end{aligned}
\tag{4.2.25}
$$

For single-look data, $L = 1$, eqs. (4.2.23) or (4.2.24) reduce to (Just and Bamler, 1994; Tough et al., 1995):

$$
\begin{aligned}
pdf(\phi; \gamma, \phi_0) =\; &\frac{(1 - |\gamma|^2)}{2\pi} \frac{1}{1 - |\gamma|^2 \cos^2(\phi - \phi_0)} \\
&\left(\frac{|\gamma| \cos(\phi - \phi_0) \arccos(-|\gamma| \cos(\phi - \phi_0))}{\sqrt{(1 - |\gamma|^2 \cos^2(\phi - \phi_0))}} + 1\right)
\end{aligned}
\tag{4.2.26}
$$

Figure 4.5 shows the shape of the probability density functions for different coherence levels. Figures 4.5A, B, and C correspond with multilook levels of 1, 10, and 20, respectively.

With the PDF's defined in eqs. (4.2.23) and (4.2.24), the phase variance resulting from $\gamma < 1$ can be easily derived using eq. (4.1.2):

$$
\sigma_\phi^2 = \int_{-\pi}^{+\pi} [\phi - E\{\phi\}]^2 \, pdf(\phi)d\phi.
\tag{4.2.27}
$$

For single-look data the phase variance can be expressed in a closed form using (Bamler and Hartl, 1998)

$$
\sigma_{\phi, L=1}^2 = E\{[\phi - E\{\phi\}]^2\} = \frac{\pi^2}{3} - \pi \arcsin(|\gamma|) + \arcsin^2(|\gamma|) - \frac{\mathrm{Li}_2(|\gamma|^2)}{2},
\tag{4.2.28}
$$

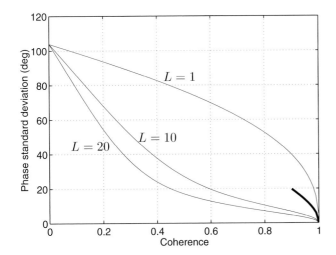

Fig. 4.6. Standard deviation of the interferometric phase as a function of coherence, for three multilook levels. The short bold line between $0.9 < |\gamma| \leq 1$ is the standard deviation of the interferometric phase for a *point scatterer*, using eq. (4.2.31), see Just and Bamler (1994). This equation only applies for $|\gamma|$ close to 1.

where Li_2 is Euler's dilogarithm, defined as (Abramowitz and Stegun, 1970)

$$\text{Li}_2(|\gamma|^2) = \sum_{k=1}^{\infty} \frac{|\gamma|^{2k}}{k^2}, \tag{4.2.29}$$

and

$$E\{\phi\} = \phi_0. \tag{4.2.30}$$

For numerical evaluations, eq. (4.2.23) appears to be faster and more accurate than eq. (4.2.24), since it does not involve an infinite summation. In fig. 4.6, the standard deviation of the interferometric phase is evaluated as a function of the coherence level, for three different multilook levels.

Statistics for point scatterers For point scatterers, which can be regarded as a constant, deterministic signal plus noise, the variance σ_ϕ^2 according to eq. (4.2.27) is overestimated. For those cases, where $|\gamma|$ is close to 1, the variance of the phase is given by (Bendat and Piersol, 1986), (Just and Bamler, 1994):

$$\sigma_\varphi^2 = \frac{1 - \gamma^2}{2\gamma^2} \quad [\text{rad}^2], \tag{4.2.31}$$

which follows from the Cramér-Rao bound for the phase variance, as derived by Rodriguez and Martin (1992):

$$\sigma_\varphi^2 = \frac{1 - \gamma^2}{2\gamma^2 L} \quad [\text{rad}^2]. \tag{4.2.32}$$

In fig. 4.6 the behavior of eq. (4.2.31) is indicated by the short bold line between $0.9 < |\gamma| \leq 1$. If a stack of coregistered interferograms is available, a pixel with a relatively large amplitude in every interferogram is an indication of possibly stable phase behavior. This can be a first criterion for the selection of *permanent scatterers*, see Ferretti et al. (2000). Simply said, for point scatterers the amplitude behavior is an indication for the phase variance.

4.3 Coherence and SNR

The complex coherence γ between two zero-mean circular Gaussian variables y_1 and y_2 is defined as (Born et al., 1959; Foster and Guinzy, 1967; Papoulis, 1991)

$$\gamma = \frac{E\{y_1 y_2^*\}}{\sqrt{E\{|y_1|^2\}E\{|y_2|^2\}}}, \quad 0 \leq \gamma \leq 1. \tag{4.3.1}$$

In this definition a coherence value can be assigned to every pixel in the interferogram. The coherence can be used as a measure for the accuracy of the interferometric phase, as discussed in section 4.2.2, or as a tool for image classification. Ideally, the expectation values in eq. (4.3.1) are obtained using a suite of observations for every single pixel, i.e., a large number of interferograms acquired simultaneously and under exactly the same circumstances. In this case, an ensemble average could be used to obtain the expectation values in eq. (4.3.1). Unfortunately, this procedure is not feasible, as every full-resolution pixel is observed only once during each SAR acquisition.

In practical situations the accuracy of phase observations of a uniform region is assumed to be stationary. Under the assumption of ergodicity it is possible to exchange the ensemble averages with spatial averages, obtained over a limited area surrounding the pixel of interest. This assumption is used to obtain the maximum likelihood estimator of the coherence magnitude $|\hat{\gamma}|$ over an estimation window of N pixels (Seymour and Cumming, 1994),

$$|\hat{\gamma}| = \frac{|\sum_{n=1}^{N} y_1^{(n)} y_2^{*(n)}|}{\sqrt{\sum_{n=1}^{N} |y_1^{(n)}|^2 \sum_{n=1}^{N} |y_2^{(n)}|^2}}. \tag{4.3.2}$$

Since $|\hat{\gamma}|$ (for simplicity often denoted as γ) is used in many of the equations defined in the previous section, it is important to consider its statistics.

Touzi and Lopes (1996) have demonstrated that the probability density function of the coherence magnitude estimator can be expressed as a function of the absolute value of the true coherence $|\gamma|$ and the number of independent samples, $L > 2$ as

$$pdf(|\hat{\gamma}|; |\gamma|, L) = 2(L-1)(1-|\gamma|^2)^L |\hat{\gamma}|(1-|\hat{\gamma}|^2)^{L-2} \,_2F_1(L, L, 1, |\gamma|^2|\hat{\gamma}|^2). \tag{4.3.3}$$

This equation holds for distributed scattering mechanisms. The expectation of $|\hat{\gamma}|$ is derived as

$$E\{|\hat{\gamma}|\} = \frac{\Gamma(L)\Gamma(3/2)}{\Gamma(L+1/2)} \,_3F_2(3/2, L, L; L+1/2, 1; |\hat{\gamma}|^2)(1-|\hat{\gamma}|^2)^L \tag{4.3.4}$$

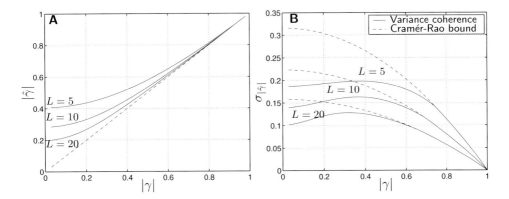

Fig. 4.7. (A) The expectation value $|\hat{\gamma}|$ as a function of $|\gamma|$ for a various number of looks L. **(B)** The standard deviation of coherence estimator $|\hat{\gamma}|$ as a function of the "true" coherence $|\gamma|$, for various L.

which is plotted in fig. 4.7A.

It appears that the estimate $E\{|\hat{\gamma}|\}$ is biased towards higher values, in particular for low coherence and/or small estimation windows (Joughin and Winebrenner, 1994; Tough et al., 1995; Touzi et al., 1996, 1999). The variance of the coherence magnitude estimator is derived from eq. (4.3.3) as well:

$$D\{|\hat{\gamma}|\} = \sigma_{|\hat{\gamma}|}^2 = \left[\frac{\Gamma(L)\Gamma(2)}{\Gamma(L+1)} \, _3F_2(2, L, L; L+1, 1; |\hat{\gamma}|^2)(1 - |\hat{\gamma}|^2)^L \right] - E\{|\hat{\gamma}|\}^2,$$
(4.3.5)

which is plotted in fig. 4.7B. The Cramér-Rao bound, which is a lower bound for the variance (assuming unbiased estimation) is defined as

$$\sigma_{|\hat{\gamma}|,\mathrm{CR}}^2 = \frac{(1 - |\gamma|^2)^2}{2L} \leq \sigma_{|\hat{\gamma}|}^2,$$
(4.3.6)

and indicated as a reference by the dashed line in fig. 4.7B. From figs. 4.7A and 4.7B it can be observed that the Cramér-Rao bound is a good approximation as long as the estimates are unbiased (Priestley, 1981). Removal of the bias in $E\{|\hat{\gamma}|\}$ by inverting eq. (4.3.4) is only possible when $\sigma_{|\hat{\gamma}|}^2$ is low or when L is sufficiently large (Touzi et al., 1999). For biased estimates, eq. (4.3.5) is to be preferred over eq. (4.3.6).

Phase corrected coherence The coherence as determined using eq. (4.3.2) is affected by interferometric phase variation due to (i) noise in the data and/or to (ii) systematic phase variation within the coherence estimation window. Systematic phase variation is a result of effective path length variations over the scene, which can be attributed to topographic, atmospheric, or deformation gradients. If the coherence estimation is solely used for deriving the phase statistics, eq. (4.3.2) is correct. However, if the coherence is used for, e.g., classification of terrain types, derived from the noise level of the data, the systematic phase variation needs to be excluded. This necessitates an adapted definition of the coherence, the "phase-corrected" coherence $\hat{\gamma}_{\mathrm{PC}}$ defined as (Hagberg et al., 1995; Monti Guarnieri and Prati, 1997; Dammert, 1997)

$$|\hat{\gamma}_{\mathrm{PC}}| = \frac{|\sum_{n=1}^{N} y_1^{(n)} y_2^{*(n)} \exp(-j\phi^{(n)})|}{\sqrt{\sum_{n=1}^{N} |y_1^{(n)}|^2 \sum_{n=1}^{N} |y_2^{(n)}|^2}}, \tag{4.3.7}$$

where $\phi^{(n)}$ is the systematic phase component for each pixel.

Relation with SNR The absolute value of the coherence $|\gamma|$, which ranges between 0 and 1, is equivalently described as a function of the signal to noise ratio (SNR) (Foster and Guinzy, 1967; Prati and Rocca, 1992; Zebker and Villasenor, 1992; Bamler and Just, 1993):

$$|\gamma| = \frac{SNR}{SNR+1}. \tag{4.3.8}$$

4.4 Sources of decorrelation

The definition of decorrelation can be rather arbitrary. Here we define decorrelation as the noise caused by error sources that have a correlation length smaller than a regular coherence estimation window. This implies that orbit errors and errors due to atmospheric heterogeneities will be discussed separately.

In the previous section we have shown that phase noise in interferograms can be expressed as a function of the absolute value of the complex coherence. The complex coherence can either be estimated from the interferometric data, based on ergodicity assumptions, or derived theoretically based on the known sensor characteristics, acquisition circumstances, and signal processing algorithms. Several sources of decorrelation, expressed by their correlation terms, can be distinguished, such as

- baseline or geometric (surface) decorrelation (γ_{geom}), caused be the difference in the incidence angles between the two acquisitions;
- Doppler centroid (surface) decorrelation (γ_{DC}), caused by the differences in the Doppler centroids between the two acquisitions;
- volume decorrelation (γ_{vol}), caused by penetration of the radar wave in the scattering medium;
- thermal or system noise ($\gamma_{\mathrm{thermal}}$), caused by the characteristics of the system, including gain factors and antenna characteristics;
- temporal terrain decorrelation ($\gamma_{\mathrm{temporal}}$), caused by physical changes in the terrain, affecting the scattering characteristics of the surface, and
- processing induced decorrelation ($\gamma_{\mathrm{processing}}$), which results from the chosen algorithms, e.g., for coregistration and interpolation.

It can be shown that the listed correlation terms are multiplicative (Zebker and Villasenor, 1992), which results in a total correlation or coherence:

$$\gamma_{tot} = \gamma_{geom} \times \gamma_{DC} \times \gamma_{vol} \times \gamma_{thermal} \times \gamma_{temporal} \times \gamma_{processing}. \qquad (4.4.1)$$

In the following sections these correlation factors will be discussed in more detail.

4.4.1 Thermal decorrelation

The influence of thermal noise on the interferometric phase can be derived theoretically by determining the signal-to-noise ratio of a specific system (Zebker et al., 1994c; Zebker, 1996). In table 4.2 the SNR for ERS-1 and ERS-2 is derived.

The average received signal power $\overline{P}_r$ can be determined using the radar equation, which can be written in the following form (Ulaby et al., 1982; Zebker, 1996):

$$\overline{P}_r = \frac{P_t}{4\pi R^2} G A_{scat} \sigma^0 \frac{A}{4\pi R^2}, \qquad (4.4.2)$$

where

P_t,	the transmit power [Watts],
R,	the distance between the antenna and the resolution cell [m] $(R = h/\cos\theta_{inc})$,
h,	the altitude of the satellite [m],
θ_{inc},	the incidence angle,
G,	the directivity gain $(G = 4\pi A/\lambda^2)$ [dimensionless],
A,	the antenna area $(L_a \times D_a)$ [m^2],
λ,	the radar wavelength [m],
A_{scat},	the scattering area [m^2], and
σ^0,	the normalized radar cross section [dimensionless].[2]

The scattering area A_{scat} is defined as the size of the area which is illuminated by a single pulse. In azimuth direction, this area is determined by the width of the antenna beam β multiplied with the average range between the antenna and the ground R:

$$\beta R = \frac{\lambda R}{L_a}, \qquad (4.4.3)$$

where L_a is the length of the antenna. In range direction, the range of a single pulse is determined by pulse length τ and the speed of light c, followed by a projection from slant-range to ground-range:

$$\frac{c\tau}{2\sin\theta_{inc}}. \qquad (4.4.4)$$

---

[2]The normalized radar cross section is the fraction of the power intercepted by the scatterer and the power returned to the radar. The parameters that affect σ^0 are primarily surface geometry and moisture content. The surface geometry is determined by the roughness, slope, and vertical or horizontal heterogeneity. The electrical properties of soil and vegetation, expressed by the dielectric constant ε, are strongly influenced by the moisture content.

With eqs. (4.4.3) and (4.4.4) we find

$$A_{\text{scat}} = \frac{\lambda R}{L_a} \frac{c\tau}{2\sin\theta_{\text{inc}}}. \tag{4.4.5}$$

The power of the thermal noise P_n in the receiver system is (Curlander and Mc-Donough, 1991; Zebker et al., 1992; Zebker, 1996)

$$P_n = kT_{\text{sys}}B_R, \tag{4.4.6}$$

with

k,   Boltzmann's constant [J/K],

T_{sys},   the receiver noise temperature [K],[3]

B_R,   the system bandwidth [Hz].

The signal-to-noise ratio is found dividing the signal power by the power of the noise,

$$SNR = \overline{P}_r/P_n, \quad \text{or} \quad \log(SNR) = \log(\overline{P}_r) - \log(P_n), \tag{4.4.7}$$

and is calculated for ERS-1/2 in table 4.2. It shows that the SNR value is influenced by two classes of parameters, those which are determined by the design of the radar system and those determined by the scene. The latter is only expressed by σ^0. Therefore, we can determine the signal-to-noise ratio as a function of σ^0, see fig. 4.8. The correlation coefficient between two complex signals $y_1 = c + n_1$ and $y_2 = c + n_2$, consisting of a common part c and two thermal noise parts n_1 and n_2 can be written as (Foster and Guinzy, 1967; Zebker and Villasenor, 1992):

$$
\begin{aligned}
\gamma_{\text{thermal}} &= \frac{E\{y_1 y_2^*\}}{\sqrt{E\{y_1 y_1^*\}E\{y_2 y_2^*\}}} \\
&= \frac{E\{cc^* + 2cn_2^* + 2c^*n_1 + n_1 n_2^*\}}{\sqrt{E\{cc^* + 2cn_1^* + n_1 n_1^*\}E\{cc^* + 2cn_2^* + n_2 n_2^*\}}}.
\end{aligned} \tag{4.4.8}
$$

The two noise components are assumed to be uncorrelated, and the signal is uncorrelated with the noise, we can write

$$\gamma_{\text{thermal}} = \frac{E\{cc^*\}}{\sqrt{E\{(cc^* + nn^*)^2\}}} = \frac{|c|^2}{|c|^2 + |n|^2}, \tag{4.4.9}$$

assuming $E\{n\} = E\{n_1\} = E\{n_2\}$. Using the definition of the signal-to-noise ratio $|c|^2/|n|^2$, we can obtain (Zebker and Villasenor, 1992)

$$\gamma_{\text{thermal}} = \frac{1}{1 + SNR^{-1}}, \quad \text{or}$$

$$= \frac{1}{\sqrt{(1 + SNR_1^{-1})(1 + SNR_2^{-1})}}, \tag{4.4.10}$$

---

[3]The temperature T is determined by matching the product kT with the correct noise power spectral density (Curlander and McDonough, 1991).

Table 4.2. Determination of the SNR for ERS-1/2 system parameters.

Signal parameter	Unit	Value	dB	dB
Peak transmit power[†] P_t	[W]	4800	36.8	
Altitude h	[km]	785	–	
Mid incidence angle θ_{inc}		23°	–	
Distance $R = h/\cos\theta_{inc}$	[km]	852	–	
Wavelength	[m]	0.0566	–	
Antenna width D_a	[m]	1.0	–	
Antenna length L_a	[m]	10.0	–	
Antenna area	[m²]	10.0	–	
Antenna gain[*]		3.923×10^4	45.9	
Antenna efficiency		50%	−3.0	
$1/4\pi$		0.08	−11.0	
$1/R^2$	[m⁻²]	1.375×10^{-12}	−118.6	
Pulse length	[μs]	37.12 ± 0.06	–	
Illuminated area	[km²]	68.7	78.4	
σ^0		4%	−14.0	
$1/R^2$	[m⁻²]	1.375×10^{-12}	−118.6	
$1/4\pi$		0.08	−11.0	
Antenna area	[m²]	10	10.0	
Antenna efficiency		50%	−3.0	
Cable/system losses		50%	−3.0	
Oversampling gain (18.96/15.55)		1.23^2	1.8	
Signal power			$-109.3 \rightarrow$	-109.3
Boltzmann's constant	[JK⁻¹]	1.38×10^{-23}	−228.6	
Receiver noise temperature	[K]	3700 K	35.7	
Range bandwidth	[MHz]	15.55 ± 0.1	71.9	
Noise power			$-121.0 \rightarrow$	-121.0
SNR				11.7

Parameters are listed in the order in which they appear in eqs. (4.4.2) and (4.4.6).   [†] Transmit power at power amplifier out.   [*] The antenna gain for ERS-2 is different from the ERS-1 gain, to compensate for a minor saturation observed with ERS-1.   [dB = 10 × log₁₀(Value)].   Data are obtained from Zebker et al. (1994c); European Space Agency (1999); Alaska SAR Facility (1997). The value for σ^0 is varying and dependent of the terrain characteristics.

if the SNR values are different for the two sensors, as with ERS-1 and ERS-2. Thus, combining eqs. (4.4.7) and (4.4.10) we found a direct relationship between the radar system parameters and the thermal correlation coefficient. Note that this relationship cannot be regarded separately from the scene's radar cross section σ^0.

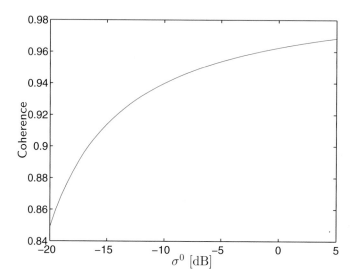

Fig. 4.8. Coherence based on thermal noise, using ERS settings as in table 4.2, as a function of the normalized radar cross section σ^0.

4.4.2 Geometric decorrelation

Geometric (baseline) decorrelation is a result of a difference in the angle of incidence between the two sensors at the earth's surface (Gatelli et al., 1994). The main concepts of geometric decorrelation have been described in section 2.5.5, see page 47. There we found that the amount of decorrelation increases linearly with the amount of spectral shift between the two SAR acquisitions, assuming a rectangular spectral window, see fig. 2.14. Although the amount of spectral shift, i.e., the fringe frequency, can be due to deformation or atmosphere as well as topography, we focus here on the latter.

The critical baseline $B_{\perp,\mathrm{crit}}$ is the baseline causing a spectral shift equal to the bandwidth B_R. It is a function of the wavelength λ, the incidence angle θ_{inc}, and the topographic slope ζ, defined positive away from the satellite:

$$B_{\perp,\mathrm{crit}} = \lambda(B_R/c)R_1 \tan(\theta_{\mathrm{inc}} - \zeta). \qquad (4.4.11)$$

The geometric decorrelation can now be simply defined as

$$|\gamma_{\mathrm{geom}}| = \begin{cases} \frac{B_{\perp,\mathrm{crit}}-B_\perp}{B_{\perp,\mathrm{crit}}}, & |B_\perp| \leq B_{\perp,\mathrm{crit}} \\ 0, & |B_\perp| > B_{\perp,\mathrm{crit}}, \end{cases} \qquad (4.4.12)$$

and is plotted for three topographic slopes in fig. 4.4.2. For a flat horizontal terrain, the ERS critical baseline is approximately 1.1 km. A rectangular spectrum was assumed in this derivation.

A priori filtering can eliminate the geometric decorrelation to a considerable extent, at the expense of spatial resolution. Slope-adaptive spectral shift filtering needs an estimate of the topographic slopes, either from an a priori DEM or from the local fringe frequency. It is obvious that fringe frequency estimation needs a window of a considerable size to operate, which will limit the filtering for very rugged terrain.

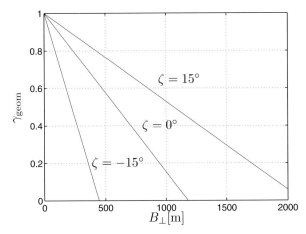

Fig. 4.9. Geometric decorrelation for ERS-1/2 as a function of the perpendicular baseline $B_\perp$, and the slope of the terrain ζ. The critical baseline $B_{\perp,\mathrm{crit}}$ for ERS for a flat terrain is approximately 1.1 km, dependent of the incidence angle.

4.4.3 Doppler centroid decorrelation

The azimuth equivalent of geometric decorrelation is caused by the difference in Doppler centroid frequencies, Δf_{DC}, between both acquisitions, as discussed in section 2.5.5. Assuming uniform spectral weighting, the coherence factor γ_{DC} is decreasing linearly with increasing Δf_{DC} as in

$$|\gamma_{\mathrm{DC}}| = \begin{cases} 1 - \Delta f_{\mathrm{DC}}/B_A & |\Delta f_{\mathrm{DC}}| \leq B_A \\ 0 & |\Delta f_{\mathrm{DC}}| > B_A \end{cases}, \qquad (4.4.13)$$

where B_A is the bandwidth in azimuth direction. For the yaw-steered SAR onboard ERS, this effect is minimal. However, for interferograms between ERS-1 and ERS-2 acquisitions, as well as ERS-2 interferograms with images acquired after 7 February 2000,[4] the effect can be significant (Swart, 2000). If the final interferometric combination is known, it is common to use the average Doppler of the two images during the SAR processing. In that case, azimuth filtering is not strictly necessary. However, if pre-processed SLC images are used, or if interferograms need to be stacked and Doppler centroids cover a wide range of values, azimuth filtering can improve the interferogram quality significantly.

4.4.4 Temporal and volume decorrelation

Temporal decorrelation occurs when the distribution of wavelength-scale scatterers within a resolution cell, or their electrical characteristics, differs as a function of time between the first and the second data acquisition, uncorrelated with other resolution cells. The measured interferometric phase ϕ is the difference between the phase values of corresponding aligned resolution cells during the acquisitions, hence

$$\phi = \psi_{t_1} - \psi_{t_2}. \qquad (4.4.14)$$

---

[4]At 7 February 2000, an error in two out of three gyroscopes of ERS-2 reduced its Doppler steering capabilities.

Although considerably simplified, it is helpful to consider both initial phase observations consisting of two deterministic components and a stochastic or noise component

$$\psi_{t_1} = \psi_{\text{geom},t_1} + \psi_{\text{scat},t_1} + n_{t_1} \qquad (4.4.15)$$

$$\psi_{t_2} = \psi_{\text{geom},t_2} + \psi_{\text{scat},t_2} + n_{t_2}, \qquad (4.4.16)$$

where ψ_{geom,t_i} is the deterministic phase component due to the geometry of the sensor and the targets on earth at acquisition time t_i. In this expression, signal delay contributions are also considered in the geometric component of the phase. The deterministic phase contribution of all scatterers within the resolution cell is denoted by ψ_{scat,t_i}, and n_{t_i} is additional noise. The scattering component of the phase is a function of the distribution of scatterers and the viewing direction of the radar. It is a deterministic value, since a repeated observation under exactly the same circumstances will yield the same phase value. Temporal decorrelation effectively changes the contribution of ψ_{scat,t_i} for different t_i. Useful coherent interferometric phase observations are only possible whenever the scattering components ψ_{scat,t_i} are more or less the same during each of the two acquisitions, expressed by the temporal coherence γ_{temporal}, cf. eq. (4.4.1). High temporal coherence values are generally obtained for areas without lush vegetation (i.e., arid areas or polar areas). Especially for vegetated areas longer wavelength radar data, for instance L-band, are preferred over C-band or X-band data, since these instruments are less sensitive to small changes in the scattering characteristics.

Presenting analytical models for temporal decorrelation has proven to be unsuccessful so far, since the range of possible temporal changes is too wide. For example, anthropogenic changes caused by farming activities or construction works cannot be modeled quantitatively due to its unpredictable and often discrete nature. Specific weather conditions can have identical effects. On the other hand, for temporal decorrelation caused by natural processes such as vegetation growth or seasonal effects rules of thumb might be applied. For example, complete decorrelation due to snow cover might change to good correlation as soon as the snow disappears. Experiments with 1-day and 3-day period interferograms show that rates of decorrelation differ considerably. Assuming that the temporal decorrelation is caused by the movement of scatterers only, Zebker and Villasenor (1992) obtain an approximation based on the rms of the displacements.

Quantitative analysis of temporal decorrelation, for example for classification purposes, is not straightforward. Additional effects such as geometric decorrelation and volume decorrelation need to be quantified separately in order to obtain meaningful measures (Hoen and Zebker, 2000). Volume decorrelation, related to the penetration of the radar waves, depends significantly on the radar wavelengths and the scattering medium, see Hoen and Zebker (2000) for a review.

4.4.5 Interpolation errors

One of the first steps in SAR interferogram processing is the resampling of one complex SAR image y_1 to map it onto a second image y_2 to within an accuracy of about a tenth of a resolution element.

Although the implementation may be different, resampling can be viewed as consisting of two steps:

1. reconstruction of the continuous signal from its sampled version by convolution with an interpolation kernel $i(x, y)$;
2. sampling of the reconstructed signal at the new sampling grid.

This scheme holds even in many cases, where the convolution [step 1] is not obvious. For example, nearest neighbor and Lagrange-type interpolation of equidistantly sampled data can be considered as a convolution with particular kernels. The choice of the interpolation kernel (especially its length) requires a tradeoff between interpolation accuracy and computational efficiency. This section[5] shows that straightforward system theoretical considerations give objective criteria for choosing or designing interpolation kernels for interferometric processing.

Theory of interpolation errors   The following analysis starts from the classical Fourier domain description of interpolation errors in stationary signals. Often these errors are quantified in terms of an L^2-norm. We will instead employ coherence theory for SAR interferograms (Just and Bamler, 1994; Cattabeni et al., 1994) to predict the effect of interpolation on interferogram phase quality. Figure 4.10A shows (for the one-dimensional case) how the Fourier transform $I(f)$ of a kernel $i(x)$ acts as a transfer function on the periodically repeated signal power spectral density $|H(f)|^2$. The two classes of errors to be considered are the distortion of the useful spectral band $|f| \leq B/2$, and the insufficient suppression of its replicas $|f - nf_s|\|_{n \neq 0} \leq B/2$, where f_s is the sampling frequency. Hence, the interpolated signal will not be strictly low-pass limited and the subsequent new sampling creates aliasing terms. If in the resampling process all interpixel positions are equally probable, the aliasing terms are superposed incoherently and can be treated as noise with a signal-to-noise ratio of:

$$SNR = \frac{S}{N} = \frac{\int_{-B/2}^{+B/2} |H(f)|^2 |I(f)|^2 df}{\sum_{n \neq 0} \int_{nf_s - B/2}^{nf_s + B/2} |H(f - nf_s)|^2 |I(f)|^2 df} \qquad (4.4.17)$$

In the following we will quantify interpolation errors in terms of interferogram decorrelation and associated phase noise. We assume that the original data y_1 have been sampled at least at the Nyquist rate and that the sampling distance after resampling is similar to the original one.

The system model of fig. 4.10B is sufficient for our analysis: consider a perfect and noise-free interferometric data pair of coherence $\gamma = 1$, before interpolation. Both y_1 and y_2 have passed the SAR imaging and processing system described by a transfer function $H(f)$. Signal y_1 additionally suffers from the interpolation transfer function $I(f)$ and alias noise n. It can be derived from Just and Bamler (1994)

---

[5]This section has been published by Hanssen and Bamler (1999), and is here extended with the analytic Fourier transform of the 6-point cubic convolution kernel.

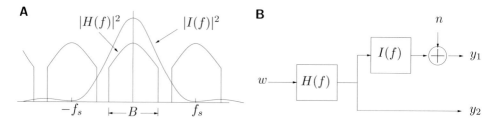

Fig. 4.10. (A) Fourier transform $I(f)$ of interpolator $i(x)$ acting on the replicated signal spectrum $H(f)$. **(B)** System model for evaluating interpolation errors (w is a white circular Gaussian process.

and Cattabeni et al. (1994) that, for circular Gaussian signals (i.e., for distributed targets), the coherence of such a system is given by

$$\gamma = \frac{1}{\sqrt{1 + \frac{N}{S}}} \frac{\int |H(f)|^2 I(f) df}{\sqrt{\int |H(f)|^2 df \ \int |H(f)|^2 |I(f)|^2 df}}. \tag{4.4.18}$$

These equations are readily extended to two dimensions, azimuth and range. If both $H(f_a, f_r)$ and $I(f_a, f_r)$ are separable, we find:

$$\gamma = \gamma_a \times \gamma_r. \tag{4.4.19}$$

The phase noise resulting from $\gamma < 1$ is known to be (in the N-look case):

$$\sigma_\phi^2 = \int_{-\pi}^{+\pi} (\phi - E\{\phi\})^2 \mathrm{pdf}(\phi) d\phi, \tag{4.4.20}$$

where

$$\mathrm{pdf}(\phi; \gamma, N) = \frac{\Gamma(N + 1/2)\,(1 - \gamma^2)^N \, \gamma \cos\phi}{2\sqrt{\pi}\,\Gamma(N)\,(1 - \gamma^2 \cos^2\phi)^{N+1/2}}$$
$$+ \frac{(1 - \gamma^2)^N}{2\pi}\, {}_2F_1(N, 1; 1/2; \gamma^2 \cos^2\phi). \tag{4.4.21}$$

Examples for interpolators   The interpolators and their spectra evaluated here are (assuming unity sample grid distance) as follows (Keys, 1981; Park and Schowengerdt, 1983).

- Nearest neighbor:[6]

$$i(x) = \mathrm{rect}(x) = \begin{cases} 0, & |x| > \frac{1}{2} \\ \frac{1}{2}, & |x| = \frac{1}{2} \\ 1 & |x| < \frac{1}{2} \end{cases} \tag{4.4.22a}$$

$$I(f) = \mathrm{sinc}(f). \tag{4.4.22b}$$

---

[6]Note that the rect function needs to be 0.5 at the discontinuities to meet the Dirichlet derivative conditions (Bracewell, 1986). In practice, however, it is often defined asymmetrically.

- Piecewise linear interpolation:

$$i(x) = \mathrm{tri}(x) = \begin{cases} 0, & |x| > 1 \\ 1 - |x|, & |x| < 1 \end{cases} \tag{4.4.23a}$$

$$I(f) = \mathrm{sinc}^2(f). \tag{4.4.23b}$$

- Four-point cubic convolution $(\alpha = -1)$:

$$i(x) = \begin{cases} (\alpha + 2)|x|^3 - (\alpha + 3)|x|^2 + 1, & 0 \leq |x| < 1 \\ \alpha|x|^3 - 5\alpha|x|^2 + 8\alpha|x| - 4\alpha, & 1 \leq |x| < 2 \\ 0, & 2 \leq |x| \end{cases} \tag{4.4.24a}$$

$$\begin{aligned} I(f) &= \frac{3}{(\pi f)^2} \left[\mathrm{sinc}^2(f) - \mathrm{sinc}(2f) \right] \\ &+ \frac{2\alpha}{(\pi f)^2} \left[3\mathrm{sinc}^2(2f) - 2\mathrm{sinc}(2f) - \mathrm{sinc}(4f) \right]. \end{aligned} \tag{4.4.24b}$$

- Six-point cubic convolution $(\alpha = -\frac{1}{2}, \beta = \frac{1}{2})$:

$$i(x) = \begin{cases} (\alpha - \beta + 2)|x|^3 - (\alpha - \beta + 3)|x|^2 + 1, & 0 \leq |x| < 1 \\ \alpha|x|^3 - (5\alpha - \beta)|x|^2 + (8\alpha - 3\beta)|x| - (4\alpha - 2\beta), & 1 \leq |x| < 2 \\ \beta|x|^3 - 8\beta|x|^2 + 21\beta|x| - 18\beta, & 2 \leq |x| < 3 \\ 0, & 3 \leq |x| \end{cases} \tag{4.4.25a}$$

$$\begin{aligned} I(f) &= -\frac{3}{4} \frac{\beta \cos(6\pi f)}{\pi^4 f^4} - \frac{1}{2} \frac{\beta \sin(6\pi f)}{\pi^3 f^3} + \frac{3}{4} \frac{(\beta - a)\cos(4\pi f)}{\pi^4 f^4} \\ &- \frac{1}{2} \frac{(\alpha + 3\beta)\sin(4\pi f)}{\pi^3 f^3} + \frac{3}{4} \frac{(\beta - 2)\cos(2\pi f)}{\pi^4 f^4} \\ &- \frac{1}{2} \frac{(4\alpha + 3 - 3\beta)\sin(2\pi f)}{\pi^3 f^3} + \frac{3}{4} \frac{\alpha - \beta + 2}{\pi^4 f^4}. \end{aligned} \tag{4.4.25b}$$

- Truncated sinc

$$i(x) = \mathrm{sinc}(x)\mathrm{rect}(x/L), \qquad \text{with} \quad L = 6, 8, 16 \tag{4.4.26a}$$

$$I(f) = \frac{1}{\pi}(\mathrm{Si}(\pi(f + 0.5)L) - \mathrm{Si}(\pi(f - 0.5)L)), \tag{4.4.26b}$$

where $\mathrm{Si} = \int (\sin(x)/x)dx$.

Table 4.3 lists the theoretically derived coherence and one-look phase noise introduced by the first three and the last of these interpolators for one and two dimensions. ERS range signal parameters have been used for both dimensions with uniformly weighted spectrum and oversampling ratio of $f_s/B = 18.96\,\mathrm{MHz}/15.5\,\mathrm{MHz} = 1.223$

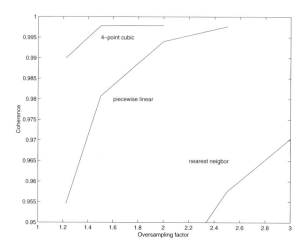

Fig. 4.11. Improvement of 2-D interpolation for oversampled input data (same oversampling factor in range and azimuth).

(in real systems, azimuth oversampling is slightly higher than in range), i.e., $H(f) = \text{rect}(f/B)$. Often, SAR data are oversampled by a higher factor before an interferogram is computed, be it either to avoid undersampling of the interferogram or as a consequence of baseline dependent spectral shift filtering. In these cases the requirement on the interpolator is relaxed. Figure 4.11 shows how decorrelation reduces with oversampling.

Experimental results   Using uniformly-distributed random generators R_i, a 1-D white circular Gaussian signal w is computed with amplitude $|w| = \sqrt{-\ln R_1[\cdot]}$ and phase $\arg(w) = 2\pi R_2[\cdot]$. Low pass filtering yields a correlated random signal. An oversampling ratio of 12.23 is used to create the reference signal u, whereas the test signal y_s is a subsampled version thereof. Using a subsampling ratio of $1/10$ reduces the oversampling ratio of the test signal to 1.223 to resemble ERS conditions. The test signal y_s is then interpolated using the kernels under investigation, yielding an estimate $\hat{y}$ of the reference signal. The interpolation kernels *nearest neighbor*, *piecewise linear*, *four-point* and *six-point cubic convolution* and *six-point, eight-point*, and *16-point truncated sinc* are created using eqs. (4.4.22a)–(4.4.26a). For every kernel the interferometric phase error $\phi = \arg[\hat{y}\,u^*]$, the phase error histogram, the total coherence γ, and the standard deviation of the interferometric phase error σ_ϕ are evaluated. Single experiment results of the interferometric phase error are shown for the first four evaluated kernels in fig. 4.12. It can be seen that the variation of the interpolated signal decreases considerably as the kernel contains more sample points. Nevertheless, spurious spikes up to $\pm\pi$ still cause residues in the interferogram. The histogram is depicted in fig. 4.13A. The total coherence γ and the standard deviation of the interferometric phase σ_ϕ is studied using averaged values from 500 simulation loops. The results are given in Table 4.3 to allow comparison with the theoretical findings. Coherence has been estimated as the sample correlation coefficient of the reference signal u and the interpolated signal $\hat{y}$. Figure 4.13B shows the mean

Table 4.3. Influence of interpolators on interferogram coherence and phase noise.

Interpolator	One-Dim.				Two-Dim.	
	γ		σ_ϕ*		γ	σ_ϕ*
	theory	simu	theory	simu	theory	theory
nearest neighbor	0.9132	0.9042	37.4	36.4	0.8345	48.7
piecewise linear	0.9773	0.9757	21.4	20.1	0.9551	28.5
4-point cubic convolution	0.9949	0.9946	11.3	11.3	0.9898	15.2
6-point cubic convolution	0.9988	0.9988	—	5.6	0.9976	—
6-point truncated sinc	0.9975	0.9973	8.3	8.2	0.9950	11.2
8-point truncated sinc	0.9980	0.9979	7.4	7.0	0.9961	10.1
16-point truncated sinc	0.9995	0.9995	4.1	3.8	0.9990	5.6

*Phase standard deviation (deg) without multilooking.   The influence of the different interpolators is derived from theory and simulations.

standard deviation of the phase as a function of the coherence for the four shortest kernels.

Discussion of results The spurious spikes in the interferogram, as shown in fig. 4.12, appear at those positions in the signal where the amplitude is extremely low. The signal-to-noise ratio at these interpolation points is therefore dominated by the interpolation noise. This makes a sudden phase jump at low-amplitude areas likely. Due to the small amplitude, multilooking suppresses these spikes and considerably diminishes the phase noise.

The cubic spline interpolation kernels used here can be referred to as parametric cubic convolution (PCC) (Park and Schowengerdt, 1983). The parameters α and β for the four- and six-point kernels chosen here have proved to be close to optimal for this particular configuration. Optimization for specific purposes can be performed by evaluating eq. (4.4.17).

A comparison of cubic splines and truncated sincs underlines the necessity of careful interpolator design: note that the cubic splines used here are special cases of *weighted* truncated sincs. The six-point cubic convolution kernel showed better quality than an eight-point unweighted truncated sinc.

Interpolation errors are due to the aliasing of repeated signal spectra and the cutoff of the signal spectra's corners. Hence, the choice of an optimal interpolator will always depend on the correlation properties of the signal. However, a subjective recommendation can be given for ERS conditions, where temporal decorrelation dominates the interferogram quality anyway. In these cases, a four-point cubic convolution with $\alpha = -1$ proved to be sufficient. For high resolution applications of high coherence single-pass interferometers, in which multilooking is not desirable, longer interpolation kernels, like the optimized six-point cubic convolution presented here, are recommended.

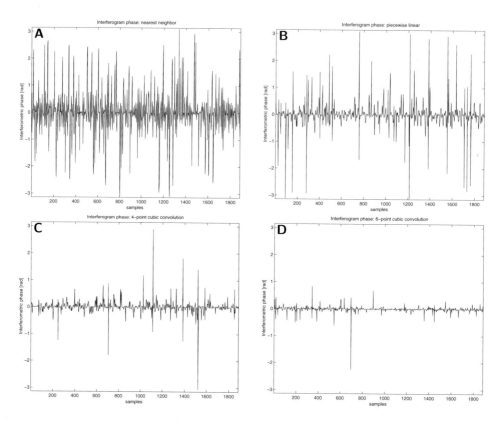

Fig. 4.12. Simulated interferograms using four kernels: **(A)** nearest neighbor, **(B)** piecewise linear, **(C)** four-point cubic convolution, and **(D)** six-point cubic convolution.

4.4.6   Coregistration errors

Coherent interferometric SAR products rely on an optimal coregistration of the two complex images. Errors in the alignment introduce loss of coherence, and therefore phase noise. It is obvious that a shift of a full resolution cell results in complete decorrelation for distributed scattering mechanisms, since there is no physical correspondence between the scatterers in both images left. Therefore, subpixel coregistration accuracy is necessary to obtain coherent interferometric products. A relative shift μ_r as a fraction of the range resolution Δ_r results in a coherence of, see Just and Bamler (1994):

$$|\gamma_{\mathrm{coreg},r}| = \begin{cases} \mathrm{sinc}(\mu_r) = \frac{\sin(\pi\mu_r)}{\pi\mu_r} & 0 \le \mu_r \le 1 \\ 0 & \mu_r > 1. \end{cases} \qquad (4.4.27)$$

This equation holds for a range as well as an azimuth misregistration. Application of this equation results in fig. 4.14, where the effect of a misregistration is visualized for coherence and phase standard deviation. Note that in this diagram it

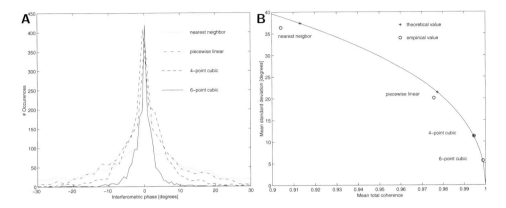

Fig. 4.13. (A) Histogram of the phase errors for four kernels: nearest neighbor, piecewise linear, 4-point cubic convolution, and 6-point cubic convolution. **(B)** Phase standard deviation phase and coherence for four interpolation kernels.

is assumed that both signals are identical (SNR=∞), and only misregistration influences the coherence. The top axis shows the pixel cell size (posting), which is approximately a factor 1.22 times the resolution cell size, both in range and azimuth direction. The figure shows that the coherence does not improve significantly when the coregistration accuracy is better than 0.1 resolution cell (1/8th of a pixel). For this coregistration accuracy we find a coherence $|\gamma_{\mathrm{coreg},r}| = 0.98$ and a phase standard deviation of $\sim 19°$. Since an identical result can be obtained in the azimuth direction, we find the total coherence $|\gamma_{\mathrm{coreg}}|$ as a result of the coregistration to be the product of both directions:

$$|\gamma_{\mathrm{coreg}}| = |\gamma_{\mathrm{coreg},r}|\,|\gamma_{\mathrm{coreg},a}| = 0.98^2 = 0.96. \qquad (4.4.28)$$

4.5 Integer phase ambiguities

An analysis of the quality, efficiency, and statistics of phase unwrapping algorithms is outside the scope of this study. Such analyses or reviews can be found, e.g., in Fornaro et al. (1996b); Ghiglia and Pritt (1998); Zebker and Lu (1998), or Bamler and Hartl (1998). Nevertheless, some general remarks are in order to stress the importance of phase unwrapping and its consequence for the quality of the final results.

The phase statistics discussed in the previous sections can be regarded as *single-point* statistics. They provide estimates for the phase variance—either through the coherence or through the SNR—but do not influence the expectation value of the phase data. In contrast, if only a single SAR interferogram is available, phase unwrapping is only possible by considering the phase differences between neighboring pixels, making it essentially a *multiple-point* statistic. Errors induced by the integration of the phase gradients in one part of the interferogram can propagate into other parts easily. Therefore, the quality of the solution at a certain location cannot be es-

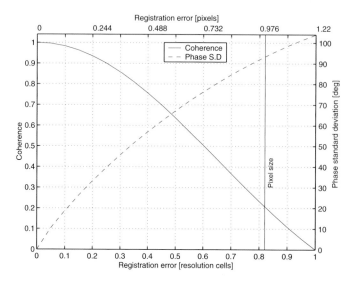

Fig. 4.14. The effect of a registration error μ as a fraction of the resolution cell size, expressed in the coherence $|\gamma_{\text{coreg}}|$ as well as the phase standard deviation σ_φ using eq. (4.2.26). The upper x-axis shows the fractional pixel cell size.

timated only by analyzing its direct surroundings. Figure 4.15 shows a section of the probability density function of the absolute phase, which is known up to a 2π ambiguity. Due to this ambiguity, the absolute position of the PDF is unknown, showing the integer nature of the unwrapped phase solutions. If the expectation value of the wrapped phase is denoted as $E\{\phi^{\text{w}}\}$, the expectation value of the unwrapped phase should be $E\{\phi\} = E\{\phi^{\text{w}}\} + k2\pi$, with $k \in \mathbb{Z}$. Figure 4.16 shows an example of phase unwrapping errors, caused by a least-squares algorithm which doesn't account for the integer nature of the phase data. The result of phase unwrapping (fig. 4.16A) is shown in fig. 4.16B. Although the unwrapping seems successful—the unwrapped phase looks continuous—"spikes" in the result (e.g., at the white arrows) indicate residues that have not been unloaded. If the algorithm would obey the $k \in \mathbb{Z}$-condition, such spikes would not show up. Moreover, noisy areas such as the lake in the center of the image have been simply interpolated, hereby loosing all possible phase information. Another indication of failing the $k \in \mathbb{Z}$-condition is shown in fig. 4.16C, where the unwrapped phases are "re-wrapped" to the $[-\pi, \pi)$ interval, and subtracted from the original data (fig. 4.16A). The $k \in \mathbb{Z}$-condition would imply that this difference needs to be 0 rad (gray) for the full image. It is clearly visible that some areas have been biased after unwrapping. Areas with decreased coherence, such as the hills in the south-western part of the image, have been interpolated.

Fortunately, most of the currently used algorithms obey the integerness-condition. This implies that the errors introduced by phase unwrapping have a discrete ($k2\pi$) distribution. It is obvious that this characteristic of the data constrains a proper quality description of the final interferometric products (topography/deformation), in the sense that the PDF of these products is ambiguous as well.

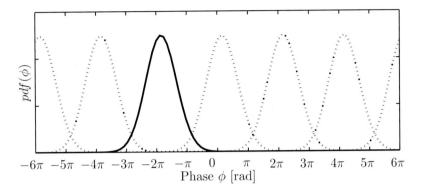

Fig. 4.15. A section of the probability density function of the unwrapped interferometric phase, for different $k2\pi$ ambiguity numbers. Although the shape of the PDF is known, its absolute position is not.

4.6 Influence and modeling of orbital errors

Spaceborne synthetic aperture (SAR) interferometry requires orbital satellite ephemerides or state vectors in order to determine the interferometric baseline vector and to refer the interferometric products to a reference datum. As such, orbit errors propagate directly into errors in topographic height or deformation maps. For global applications such as potential field modeling or oceanography, orbit errors are conveniently parameterized as "once per revolution" errors, and much effort is expended to improve gravity field models, dynamic models, and the quality of geodetic tracking measurements. The way orbit errors propagate into SAR interferograms differs from this conventional approach, in the sense that the absolute orbit accuracy is less important than the relative accuracy between the state vectors during the few seconds of image acquisition. The required absolute accuracy for orbit determination would need to be on the order of 1 mm to fully correct the residual interferometric fringes, which is far below the current value of around 5–10 cm.

In the following it is described how orbit errors affect the interferometric baseline and how a baseline error influences the so-called "reference phase," that is, the expected interferometric phase for a reference body such as an ellipsoid. After a brief review of precise orbit determination in section 4.6.1, the interferometric geometry is discussed in section 4.6.2. In the range direction the reference phase can be expressed as a function of the baseline geometry and the range pixel number, which is described in section 4.6.3. Section 4.6.4 covers how the predicted errors in the state vectors propagate into baseline errors and residual reference phase signal. Different methods to model and reduce the residual reference phase are discussed in section 4.6.5. This section also evaluates the remaining errors after applying different correction methods.

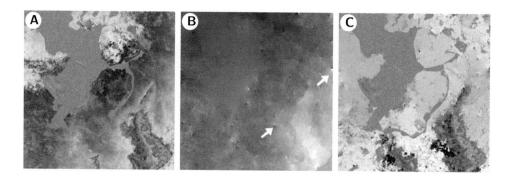

Fig. 4.16. The effect of phase unwrapping errors using a least-squares algorithm. Interferogram of IJssel lake, The Netherlands. **(A)** Relative (wrapped) interferometric phase. **(B)** Absolute (unwrapped) phase. The white arrows show some "spike"-effects due to uncompensated residues. **(C)** Difference between re-wrapped absolute phase and original wrapped phase. Errors have propagated over large areas of the interferogram.

4.6.1 Precise orbit determination

For ERS-1 and ERS-2, precise orbit determination (POD) is based on tracking data, a gravity model, and several dynamical models. Tracking data include satellite laser ranging (SLR) for ERS-1 and a combination of SLR and data from the Precise Range and Range-Rate Equipment (PRARE) for ERS-2. As the tracking data from SLR can only be acquired in the vicinity of an SLR station, which are unevenly distributed over earth, gravity field information (a dynamic model) is needed to determine or interpolate the position of the satellite at other positions. Gravity models, such as GEM-T2, PGM035, JGM 3, and DGM-E04 are used for this purpose. Over the oceans, altimeter crossover height differences can be used as additional tracking data. Dynamical models are used to include nonconservative forces such as atmospheric drag and solar radiation. Precise orbit determination incorporates a data adjustment using observations and models, and produces a set of state vectors consisting of time, velocity vectors, and position vectors (Scharroo and Visser, 1998).

Geographically correlated orbit errors, such as errors in the gravity field model, are repetitive along the same ground track (Scharroo and Visser, 1998). For repeat-pass SAR interferometry, where collinear tracks are a requisite, such errors are identical and cancel. The remaining error sources include nonconservative forces such as time variable atmospheric drag, time variable solar radiation pressure, and other unmodeled effects.

4.6.2 Interferometric baseline geometry

The reference phase variation is due to the geometric configuration of the two satellites and the variation of this configuration in the azimuth direction of the SAR image. The geometric configuration is sufficiently modeled by the interferometric

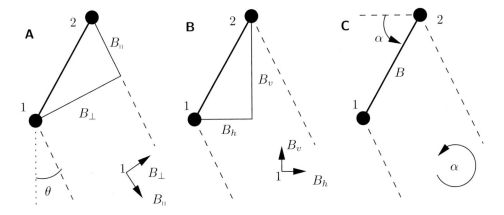

Fig. 4.17. Three representations for the interferometer baseline. **(A)** parallel/perpendicular, **(B)** horizontal/vertical, and **(C)** length/orientation. Position 1 is the reference position. $B_\| > 0$ when $R_1 > R_2$, where R_1 and R_2 are the corresponding slant-ranges to position 1 and 2. Angle α is defined positive counterclockwise from the reference satellite (1), starting from the horizontal at the side of the look direction. In (C), $\alpha > 0$.

baseline vector, which can be described conveniently using three different representations, see fig. 4.17A the parallel/perpendicular representation, fig. 4.17B the horizontal/vertical representation, and fig. 4.17C the representation in baseline length and angle, written in vector notation as

$$\vec{B} = (B_\|, B_\perp) = (B_h, B_v) = (B, \alpha), \qquad (4.6.1)$$

see fig. 4.17. The parallel/perpendicular representation can be readily used to describe the sensitivity of the configuration for topographic heights and the amount of geometric (baseline) decorrelation. The perpendicular baseline, $B_\perp$, can be regarded as a scaling factor for the height ambiguity. A large $B_\perp$ usually implies geometric (baseline) decorrelation, which decreases the SNR of the interferogram. The parallel baseline, $B_\|$, is in fact the basic path length difference

$$B_\| = R_1 - R_2 \qquad (4.6.2)$$

that is inferred directly from the absolute interferometric phase. R_1 and R_2 correspond with the geometric range between a resolution cell on earth and antenna positions 1 and 2, respectively. An important aspect of the parallel/perpendicular representation is its dependence on the look angle θ (fig. 4.17A). This implies that both baseline parameters depend on the relative position of a pixel in the interferogram, which is related to (i) orbit convergence/divergence, (ii) the geometric difference between near-range and far-range, and (iii) the contribution of topography.

In contrast to the parallel/perpendicular baseline, the horizontal/vertical representation is equal for all pixels in the same range line, as it is independent of the look angle (fig. 4.17B). The components are written as B_h and B_v, respectively. The advantage of this representation is that the horizontal baseline can be directly coupled to across-track orbit errors, while the vertical baseline is directly related to the radial component of the state vector. This representation will be used in the error analysis in section 4.6.4.

The third baseline representation—in terms of baseline length B and orientation α— can be useful because it can be easily related to the look angle θ, which is necessary to explain certain peculiar orbital phase effects, see the examples in section 4.6.3.

Sign of the baseline components

Regarding the sign of the baseline parameters we need to account for the interferogram formation equations. Let y_1 and y_2 be the complex values of image 1 and 2. In terms of amplitude and phase components we can write:

$$
\begin{aligned}
y_1 &= a_1 \exp(j\psi_1) = a_1 \exp(j[-2kR_1]) \\
y_2 &= a_2 \exp(j\psi_2) = a_2 \exp(j[-2kR_2]),
\end{aligned}
\tag{4.6.3}
$$

with $k = 2\pi/\lambda$. Here we only regard the geometric part of the phase observation, hence no scattering components. The interferogram is formed by complex multiplication:

$$
\begin{aligned}
y_1 y_2^* &= a_1 a_2 \exp(i(\psi_1 - \psi_2)) \\
&= a_1 a_2 \exp(i[-2k(R_1 - R_2)]),
\end{aligned}
\tag{4.6.4}
$$

and therefore the relation between the interferometric phase $\phi = \psi_1 - \psi_2$ and the range difference to both satellites is

$$
\phi = -\frac{4\pi}{\lambda}(R_1 - R_2).
\tag{4.6.5}
$$

The range difference $(R_1 - R_2)$ is equal to the parallel baseline $B_\parallel$, see eq. (4.6.2), so for $R_1 < R_2$, as sketched in (fig. 4.17A), $B_\parallel$ has a negative value.

The sign of $B_\perp$ is defined indirectly by the definition of orientation angle α. Here we define α positive counterclockwise (consistent with the look angle θ) from the reference satellite (number 1), starting from the horizontal at the side of the look direction (or from the horizontal opposite to the look direction for satellite 2, as in fig. 4.17C). In this definition, $B_\perp$ is positive whenever satellite 2 is located to the right of the slant-range line of satellite 1. For convenience: when $B_\perp > 0$, then $B_\parallel = R_1 - R_2$ will increase from near-range to far-range, or from the foot to the top of a mountain, which results in a decreasing phase. In table 4.4, the conversion formulas between the three representations are listed.

Knowledge of the baseline parameters is necessary to predict the interferometric phase variation for an arbitrary reference surface. We can subtract this reference phase from the observed data, and interpret the phase residuals as, e.g., topographic height, deformation, or atmospheric signal. Frequently used reference surfaces are locally best fitting ellipsoids, so that ellipsoidal heights can be obtained, but in fact any suitable reference body is possible, even a geoid or a sphere.

Table 4.4. Conversion formulas for the three baseline representations.

	$[B_h, B_v]$	$[B, \alpha]$	$[B_\perp, B_{||}]$
$[B_h, B_v]$	—	$B_h = B \cos \alpha$	$B_h = B_\perp \cos \theta + B_{||} \sin \theta$
	—	$B_v = B \sin \alpha$	$B_v = B_\perp \sin \theta - B_{||} \cos \theta$
$[B, \alpha]$	$\alpha = \arctan(B_v/B_h)$	—	$\alpha = \theta - \arctan(B_{||}/B_\perp)$
	$B = \sqrt{B_h^2 + B_v^2}$	—	$B = \sqrt{B_\perp{}^2 + B_{||}^2}$
$[B_\perp, B_{||}]$	$B_{||} = B_h \sin \theta - B_v \cos \theta$	$B_{||} = B \sin(\theta - \alpha)$	—
	$B_\perp = B_h \cos \theta + B_v \sin \theta$	$B_\perp = B \cos(\theta - \alpha)$	—

Note that the arctangent is the four-quadrant arctangent (`atan2`).

Ideally, in order to predict the reference phase, we would need to determine $\vec{B}(i,j)$ for every azimuth and range resolution cell i and j. However, often it appears that the variation of $\vec{B}$ with time, denoted by $d\vec{B}/dt$, can be approximated by a linear drift for the acquisition period of a SAR image, typically 15.4 seconds for ERS-1/2. In range direction, the phase of the reference surface can be predicted by $\vec{B}$ directly. Thus, we need only $\vec{B}$ and $d\vec{B}/dt$ to compute the reference phase in two dimensions.

4.6.3   Reference phase in range

To derive a convenient analytical expression for the reference interferometric phase, a spherical approximation is used, see fig. 4.18, where the earth radius R_e is adapted to the latitude of the spacecraft. The slant range R_1 will be used as input parameter, depending on the range pixel number. Using straightforward trigonometry, an expression for location angle μ is derived, which is related to the interferometric reference phase φ_{ref}.

The horizontal and vertical baseline components are readily derived from the state vectors $\vec{\rho_1}$ and $\vec{\rho_2}$, using

$$\vec{B} = \vec{\rho_2} - \vec{\rho_1}, \tag{4.6.6a}$$

$$B_v = (\vec{\rho_2} - \vec{\rho_1})\frac{\vec{\rho_1}}{\rho_1}, \quad \text{or} \quad \vec{B_v} = B_v\frac{\vec{\rho_1}}{\rho_1}, \quad \text{and} \tag{4.6.6b}$$

$$B_h = \pm\sqrt{B^2 - B_v^2}, \quad \text{or} \quad \vec{B_h} = \vec{B} - \vec{B_v}. \tag{4.6.6c}$$

The slant-range value R_1 is derived from the roundtrip travel time to the first range pixel, τ_1, the sampling frequency ($f_s = 18.96$ MHz), and the velocity of electromagnetic waves c. The range increments dR can be found using

$$dR = \frac{c}{2f_s}, \tag{4.6.7}$$

and the range to the first range pixel is

$$R_1(1) = \frac{1}{2}\tau_1 c. \tag{4.6.8}$$

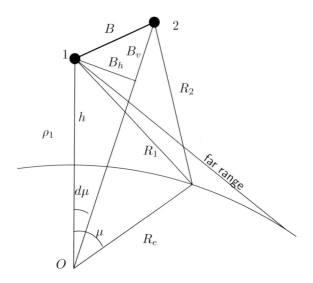

Fig. 4.18. Geometric configuration in two dimensions. R_1 and R_2 are the two range vectors to the same resolution element. The state vector to the reference satellite is denoted ρ_1, the sum of earth radius R_e and satellite height h. The location angle of a coregistered resolution cell in the interferogram is denoted by μ.

Using (4.6.7) and (4.6.8) the range as function of pixel i can be expressed by

$$R_1(i) = R_1(1) + (i-1)dR. \tag{4.6.9}$$

The relation between the measured range R_1 and the unknown location angle μ is

$$R_1^2 = \rho_1^2 + R_e^2 - 2\rho_1 R_e \cos\mu, \tag{4.6.10}$$

and, using the same configuration for the second satellite position, we find

$$R_2^2 = (\rho_1 + B_v)^2 + R_e^2 - 2(\rho_1 + B_v)R_e \cos(\mu - B_h/\rho_1) \tag{4.6.11}$$

using the far-field approximation—assuming parallel state vectors $\vec{\rho_1}$ and $\vec{\rho_2}$. In range direction, B_h, B_v, ρ_1, and R_e are constant. Therefore, as R_1 changes from near-range to far-range, only μ will change. For a coregistered image, we want to express the values of R_2 as a function of R_1. From the interferometric measurements we find that, cf. eq. (4.6.5),

$$R_2 = R_1 + \frac{\lambda}{4\pi}\varphi. \tag{4.6.12}$$

Combining eqs. (4.6.11) and (4.6.12), we find for the reference phase

$$\varphi_{\text{ref}} = -\frac{4\pi}{\lambda}\left(R_1 - \sqrt{(\rho_1 + B_v)^2 + R_e^2 - 2(\rho_1 + B_v)R_e \cos(\mu - B_h/\rho_1)}\right), \tag{4.6.13}$$

and from (4.6.10) follows

$$\mu = \arccos\left(\frac{\rho_1^2 + R_e^2 - R_1^2}{2\rho_1 R_e}\right). \tag{4.6.14}$$

Note that in fig. 4.18, μ is largely exaggerated: in reality μ is only a few degrees. Using

$$\cos(\alpha + \beta) = \cos\alpha\cos\beta - \sin\alpha\sin\beta, \tag{4.6.15}$$

and

$$\sin(\arccos(a/b)) = \sqrt{b^2 - a^2}/b, \tag{4.6.16}$$

we find an expression of φ_{ref} as a function of R_1 only:

$$\varphi_{\text{ref}}(R_1) = -\frac{4\pi}{\lambda}\left(R_1 - \sqrt{(\rho_1 + B_v)^2 + R_e^2 - \frac{\rho_1 + B_v}{\rho_1}(D\cos\frac{B_h}{\rho_1} + \sqrt{E}\sin\frac{B_h}{\rho_1})}\right), \tag{4.6.17}$$

where

$$D = (\rho_1^2 + R_e^2 - R_1^2)$$
$$E = 4\rho_1^2 R_e^2 - D^2.$$

Using (4.6.9), where $R_1 = R_1(i)$, equation (4.6.17) describes the behavior of the reference phase φ_{ref} for range pixel i. The reference phase depends on the baseline vector $\vec{B}$, here expressed using the components B_h and B_v. In the next section, it will be shown that this parameterization enables a direct relation to the radial and across-track orbit errors.

Figure 4.19 shows three baseline configurations with their corresponding reference phase in range direction. The figure shows how the phase variation is directly related to the parallel baseline. Moving from near-range ($\theta_{\text{nr}} \approx 18°$) to far-range ($\theta_{\text{fr}} \approx 24°$) $B_\parallel$ increases in fig. 4.19A and 4.19B. Since $\varphi \propto -B_\parallel$, the values are positive-decreasing in fig. 4.19A and negative-decreasing in situation fig. 4.19B. Obviously, reversing the two satellite positions will yield increasing values in both situations.[7] Although the reference phase variation seems nearly linear in fig. 4.19A and 4.19B, a special situation occurs when the baseline orientation α is in the range

$$\theta_{\text{nr}} + \frac{\pi}{2} < \alpha < \theta_{\text{fr}} + \frac{\pi}{2}, \quad \text{or}$$
$$\theta_{\text{nr}} - \frac{\pi}{2} < \alpha < \theta_{\text{fr}} - \frac{\pi}{2}, \tag{4.6.18}$$

as shown in fig. 4.19C. In this interval the reference phase will be both increasing and decreasing. The right column of fig. 4.19 shows the absolute phase in radians as a function of the range direction R_1, corresponding to the sketches in the left column. The insets show the corresponding wrapped phase equivalents.

---

[7]Note that the behavior of the reference phase (increasing or decreasing) is caused by the behavior of the parallel baseline $B_\parallel$. The latter, however, can also be directly derived from the sign of the perpendicular baseline $B_\perp$.

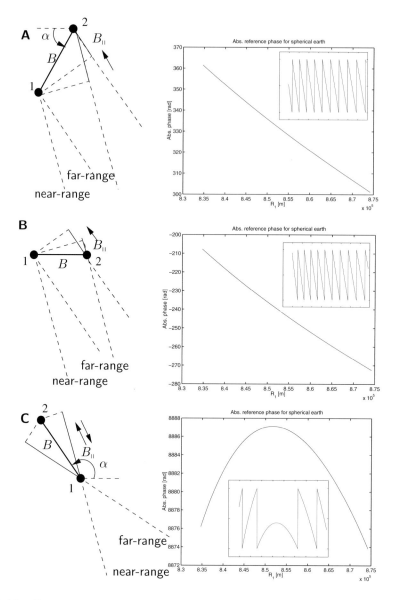

Fig. 4.19. The variation of the parallel baseline $B_{||}$ as a function of range, shown for three baseline configurations. **(A)** $B = 3$ m, and $\alpha = 51°$. The parallel baseline is negative, increasing in range, resulting in a positive phase, decreasing over range. **(B)** $B = 3$ m, $\alpha = 0°$. A positive baseline, increasing in range, resulting in a negative phase, decreasing in range. **(C)** $B = 40$ m, $\alpha = 111°$. The phase increases until the look angle reaches $\theta = \alpha - \frac{\pi}{2}$. Nearly halfway the image, ϕ_{ref} starts to decrease, resulting in a hyperbolic-shaped reference phase in the interferogram.

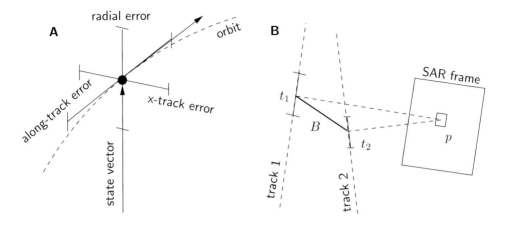

Fig. 4.20. (A) Three-dimensional representation of an orbital arc with state vector and velocity vector. Included are the across-track, along-track, and radial error bars. **(B)** Top view of two converging tracks. Resolution element p is observed at $t = t_1$ from the reference track, and at $t = t_2$ from the secondary track. The along-track error is indicated by the error bars along the track.

4.6.4 Description of orbit errors

Errors in the orbital state vectors can be represented as errors in the along-track, the across-track, and the radial directions, as sketched in fig. 4.20A. For SAR interferometry, along-track errors are usually sufficiently corrected for during the coregistration of the two images, see fig. 4.20B. In fact, it is sufficient to perform the alignment of the images with an accuracy of 1/10th of the resolution size, see section 4.4.6, which results in a derived along-track accuracy of approximately 40 cm for ERS. Note that along-track positioning errors can also be regarded as timing errors.

As only radial and across-track errors will propagate as systematic phase errors in the interferogram, the problem is effectively two-dimensional. From the SAR image coordinates, range and azimuth pixel positions, these effects can be separated in a nearly instantaneous component in range direction, and a time-dependent component in azimuth direction. The effects will be discussed in both directions.

Errors in the two state vectors need to be propagated to the baseline vector $\vec{B}$ rather than analyzed separately, since an interferogram is an inherently relative measure. A quasi-3D plot is shown in fig. 4.21A, indicating the change in the baseline vector $\vec{B}$ and the influence of radial and across-track orbit errors. Using the DEOS precise orbits, radial and across-track rms errors are on the order of 5 and 8 cm respectively (Visser et al., 1997; Scharroo and Visser, 1998). Although geographically correlated errors cannot be resolved using this approach—which makes the estimate often too optimistic in an absolute sense—this is of no consequence for the interferometric baseline.

The relationship between baseline vector errors and radial/across-track errors is

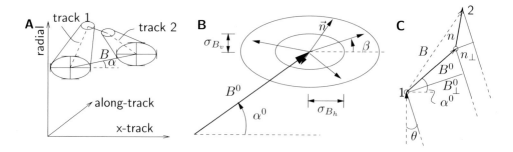

Fig. 4.21. (A) Quasi-3D view of the two orbits, their error ellipses based on radial and across-track errors, and their relation with baseline length B and orientation α. Note that we apply the far-field approximation here, assuming the radial directions are parallel for both satellite positions. **(B)** Radial and across-track rms orbit error approximations result in an error ellipse centered at the end of $\vec{B}$, using equation (4.6.19a). With a 95% confidence interval, the error vector $\vec{n}$ is situated somewhere within the outer ellipse. **(C)** The true baseline vector $\vec{B}$ is the vector sum of baseline estimate $\vec{B}^0$ and error baseline $\vec{n}$.

given by:

$$\sigma_{B_v} = \sqrt{\sigma_{\text{radial},1}^2 + \sigma_{\text{radial},2}^2} \qquad (4.6.19a)$$

$$\sigma_{B_h} = \sqrt{\sigma_{\text{xtrack},1}^2 + \sigma_{\text{xtrack},2}^2} \qquad (4.6.19b)$$

assuming uncorrelated errors between orbit 1 and 2. This implies that the orbit errors can be regarded as a single "noise vector" $\vec{n}$ superposed on the baseline estimates B_0, as sketched in fig. 4.21B.

Representing orbit errors as a noise vector $\vec{n}$ on top of the baseline vector leads to a convenient way of interpreting residual orbit fringes. The noise vector can be regarded as a second baseline vector. As we subtract the expected (a priori) reference phase, a residual reference interferogram will remain, with the unknown a posteriori "baseline" vector $\vec{n}$. Unfortunately, the only available information we have on $\vec{n}$ is that its length is less than two times the largest value of σ_{B_v} and σ_{B_h}, with a probability of 95%. However, as no information is available on the orientation of $\vec{n}$, denoted by β in fig. 4.21B, the residual reference interferogram can have any kind of behavior. Substituting B, α, and $B_\parallel$ by n, β and $n_\parallel$ respectively in fig. 4.19 illustrates this relation. Obviously, since $n \ll B$, the amount of fringes decreases considerably. Figure 4.22 illustrates the behavior of the reference phase in range direction for a radial rms of 5 cm and an across-track rms of 10 cm, for a range of angles β. Experiments with similar radial and across-track rms values show that the largest effects occur when $\beta \approx \pm\theta$ whereas when $\beta \approx \theta \pm \frac{\pi}{2}$, the error can be almost neglected. However, since there is no way of knowing β, it seems wise to account for the largest errors. In fig. 4.23 the maximum amount of residual fringes in an interferogram is shown, as a function of the a priori radial and across-track rms values of a single state vector. With a 95% likelihood, the graph shows the maximum amount of residual fringes in a 100×100 km interferogram.

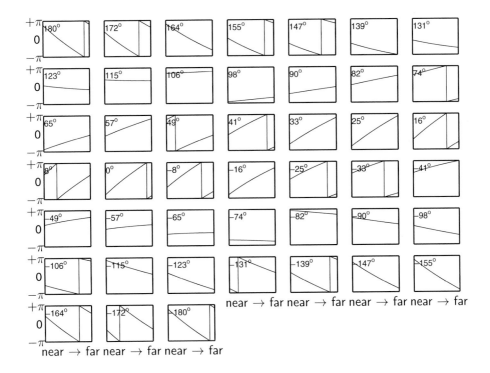

Fig. 4.22. The wrapped maximum residual reference phase as a function of range R_1 for a radial orbit rms of 5 cm and an across-track orbit rms of 10 cm, using ERS characteristics. The corresponding axes of the error ellipse are 15 and 30 cm respectively. The wrapped phase is shown for 45 values of β, equally distributed along the boundary of the error ellipse. On the horizontal axis, the slant range is shown in meters. The vertical axis covers the wrapped phase interval.

From this relation, the approximate orbit accuracy can be derived from the number of residual fringes in the interferogram. One fringe or less corresponds to rms values of 5 and 10 cm in radial and across-track direction, and keeping the ratio between the two error sources constant, the errors scale nearly linear with the number of fringes. For example, when observing 16 residual fringes in the interferogram, which cannot be contributed to, e.g., topography, expected maximal rms values are approximately 80 and 160 cm in radial and across-track direction.

4.6.5 Correcting the residual reference phase

Knowing how residual fringes originate from orbit errors, it needs to be studied how they can be eliminated from the interferogram. Strictly, the only way to do this is to try to estimate the total orbit error vector $\vec{n}$. The parallel component $n_{\parallel}$ is needed to correct for the residual reference phase, whereas the perpendicular component

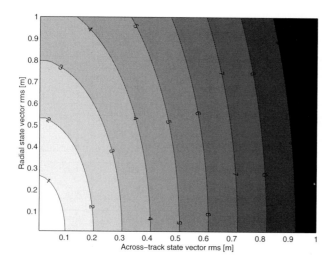

Fig. 4.23. Maximum amount of residual fringes in an interferogram, as a function of the reported radial and across-track state vector accuracies. With a 95% likelihood, the maximum amount of fringes is less then shown here.

$n_\perp$ needs to be added to the perpendicular baseline $B_\perp$ to obtain the correct height conversion factor.

If the interferogram is used for topography height estimation, it seems suitable to use tie-points to constrain the reference phase at selected locations in the interferogram. A similar approach is used by Massonnet and Feigl (1998), see fig. 4.24A. In this approach, one satellite position is fixed, while the second one, indicated in the figure, is changed in two steps. First, since the range distance $R_{2,\mathrm{nr}}$ to the first pixel in range can be kept constant, the position of the satellite needs to be somewhere on curve 1. By counting the number of fringes between near-range and far-range, range vector $R_{2,\mathrm{fr}}$ needs to be changed in length, denoted as step 1. Keeping this value for $R_{2,\mathrm{fr}}$, the position of the satellite needs to be somewhere on curve 2. Therefore, in step 2, the position is moved to the cross-section of curve 1 and 2. Using the same approach in azimuth direction, residual azimuth fringes can be eliminated.

Although it seems that this approach solves for the error vector $\vec{n}$ this is in fact not totally true. In fact, the method corrects only for the *change* in the parallel component of the error vector, $n_\parallel$, see equation (4.6.20), which causes the residual fringes in the interferogram. The method does not correct for the perpendicular component of the error vector, $n_\perp$, and still leaves an error in the height conversion factor. In fig. 4.24B it is shown how this method fails finding the correct position of the satellite. The a priori position of the satellite is indicated as position 1. The true position of the satellite is indicated as position 2. The error vector $\vec{n}$ points from position 1 to 2. As B is often much larger than n, the major part of the reference phase is decreasing with range, indicated by the boldly drawn parallel component of $\vec{B}$. The reference phase from this baseline can be eliminated totally. The influence of the error vector $\vec{n}$ is not known, though. From the figure it is clear how $n_\parallel$ is also decreasing with range. The length change of $n_\parallel$ from near-range to far-range is directly related to the number of residual fringes in the interferogram. This length

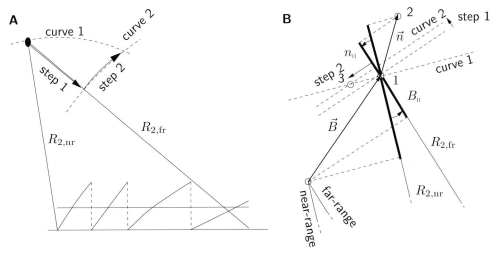

Fig. 4.24. (A) Correction of orbit error for the first range line in two steps, see Massonnet and Feigl (1998). **(B)** Difference between the true error vector $\vec{n}$ (position 2), and the position obtained with the method in (A) (position 3). The parallel components of $\vec{B}$ and $\vec{n}$ are denoted by the thick lines. Position 1 indicates the a priori position found by using (precise) orbital state vectors.

change will be denoted as

$$\delta n_{\|}(R_2) = n_{\|}(R_2) - n_{\|}(R_{2,\mathrm{nr}}). \tag{4.6.20}$$

Since $n_{\|}(R_{2,\mathrm{nr}})$ is unknown, we can only observe the residual fringes in the interferogram, which are equivalent to $\delta n_{\|}(R_2)$. In fact, we observe from the interferogram that $R_{2,\mathrm{fr}}$ is too short by $\delta n_{\|}(R_2)$. Correcting for this error is indicated as step 1 in fig. 4.24B. Keeping $R_{2,\mathrm{nr}}$ constant, the position that Massonnet and Feigl (1998) would find will be the intersection of curve 1 and curve 2, indicated as position 3. It is clear from the figure that this method might eliminate the residual reference phase, but it decreases the perpendicular baseline $B_\perp$, whereas the correct solution increases $B_\perp$. With an angle of approximately 6° between the near-range and far-range look angle, a correction for two residual fringes (approximately 6 cm) would lead to a decrease in $B_\perp$ of 57 cm. Such an error would result in an error of 53 cm per fringe for a baseline of 100 meters, and 9 to 56 meters per fringe for a baseline of 25 to 10 meters, respectively.

A different problem connected with using tie-points is caused by atmospheric phase variation. To translate the height difference between two tiepoints to a phase difference, it has to be assumed that there are no additional phase gradients in the interferogram, for example caused by atmospheric heterogeneities. Previous studies have shown that the phase difference between a set of tiepoints within a SAR image can easily exceed several fringes. Thus, without any prior information on atmospheric behavior during the image acquisitions it seems cumbersome to correct for residual reference phase. It could be tried to diminish this problem by using a large number of equally spread tiepoints over the whole image, assuming that atmospheric phase signal will average out in the adjustment procedure. However, for interferograms which have deformation as well as topographic signal, the tiepoint approach cannot be used.

In practical applications, a number of phase interpretation problems accumulate. Phase gradients can be due to atmospheric signal, errors in the topographic elevation model, or long wavelength deformation signal. In some cases, the specific behavior of the phase gradients enables identification of the driving mechanisms. Unfortunately, for largely decorrelated interferograms as used for this study, phase gradients cannot be followed over an extended spatial range. Therefore, they can be due to any of the mentioned sources. Moreover, phase unwrapping between the coherent patches, here referred to as "patch unwrapping," can cause severe misinterpretation of phase gradients, see section 5.4.

For this reason it is valuable to have a first order approximation of the type, behavior, and magnitude of residual reference phase errors, only based on the a priori radial and across-track orbit accuracies. In range direction, the residual reference phase is approximated by a linear and quadratic polynomial. The linear model corresponds with fitting a linear plane through the interferogram. Using the configuration shown in fig. 4.21B, and the a priori radial and across-track orbit rms errors, a set of noise vectors $\vec{n}$ is simulated, with lengths corresponding to the 95% level under 192 equally spaced orientation angles β. For each of these combinations the values of the polynomial coefficients for the 1st and 2nd degree approximations are determined, as well as the error between the true signal and the approximation. The error is defined as the maximum residual phase value minus the minimum residual phase value. The results are shown in fig. 4.25. It can be observed that using a linear correction for the residual reference phase, a maximal phase error of approximately 0.35 rad (6% of a cycle) remains in the interferogram after correction. These maximum errors occur when the noise vector $\vec{n}$ is nearly horizontal. Scaling the radial and across-track rms errors with a factor f will linearly scale the maximum error. Since the linear trend is removed, the shape of the remaining error in the interferogram will be approximately hyperbolic. From this distribution, without knowledge of β, the probability of a residual error larger than a certain threshold can be obtained.

Using the quadratic polynomial approximation, the ratio between the coefficient of the linear term and the coefficient of the quadratic term is approximately 10^{-6}, indicating that the signal is only weakly curved. This value is determined by the sensor height and viewing direction. The error approximated with the quadratic polynomial approach is shown in the lower right panel of fig. 4.25. This error is a scaled-down version of the linear approximation error with a factor 0.1. Maximal errors reach 0.035 rad (0.6% of a cycle). The shape of this small remaining error is a sine-like curve over the whole range line.

The severity of remaining errors depends on the application. For example, for topographic mapping using a small perpendicular baseline $B_{\perp}$, the residual error after correction with a linear trend can be significantly (6% of the height ambiguity). In those cases, it can be recommended to remove a quadratic polynomial surface from the interferogram. It has to be noted that the ratio between the magnitude of the first and the second degree coefficient in this case has to be restricted to approx-

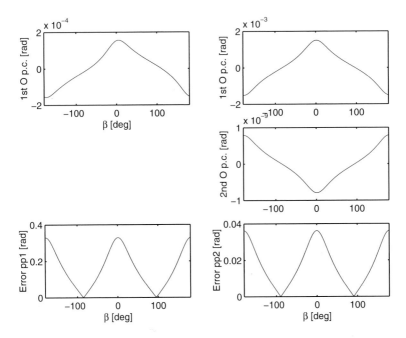

Fig. 4.25. Approximations of the residual reference phase using a linear (left column) and a quadratic (right column) polynomial model. In the left column the polynomial coefficient for the linear term as a function of orientation angle β, and the maximum error between the true residual phase and the approximation are shown. The right column shows the 1st degree and 2nd degree polynomial coefficients used for the approximation. In the lower right panel, the error between the true residual phase and the quadratic approximation is shown. A radial rms of 5 cm and a across-track rms of 10 cm are used.

imately 10^{-6}, when approximating these coefficients from the interferogram. The shape of the error signal in fig. 4.25 is dependent on the ratio between the radial and across-track error. For equally sized errors, the curve can be described by $|\cos\beta|$. A relatively larger across-track error will make the curves more pointed.

4.6.6 Note on the reference system

A spherical approximation has been used for convenience in obtaining analytical equations. The equations governing the behavior of the reference phase can, however, be easily extended to other reference systems, such as ellipsoidal systems or the geoid. In order to do this, the term expressing the earth radius R_e needs to be replaced by, e.g., $R'_e = R_e + H$, where H expresses the height of the chosen reference surface over the spherical earth. This height might be expressed in a spherical harmonics series, as a function of latitude and longitude. For example, using the geoid as reference surface, the height N between ellipsoid and geoid needs to be known, as well as the height ζ between that ellipsoidal position and the sphere. Both height levels are not parallel, as the mass distribution of the gravity anomalies cause deviations, see fig. 2.9 at page 33. Nonetheless, since an interferogram is a relative measure, these effects will only be of influence when the error due to this approximation is significantly different between, e.g., the two opposite sides of the interferogram. For example, the total bias between, e.g., ellipsoidal height ζ with respect to the sphere is not important: only gradients can be measured.

4.6.7 Conclusions

Errors in the orbital state vectors result in a residual reference phase in the radar interferogram, which can be erroneously interpreted as topography, deformation, or atmospheric signal. These errors are not related to the baseline length B and orientation angle α. In fact, it is not possible to predict whether the residual reference phase is increasing, decreasing, or both in range direction, as this is dictated by the unknown orientation β of the error vector $\vec{n}$. The total residual reference phase is restricted to the range $\frac{4\pi}{\lambda}\{0,\max(\delta n_{\shortparallel}(R_2))\}$, where $\max(\delta n_{\shortparallel}(R_2))$ is determined by the magnitude of the radial and across-track orbit errors.

Correction of the residual reference phase can be based on tiepoints, but this is strongly dependent on the amount and extent of atmospheric or deformation signal in the interferogram. Especially for interferograms with extended decorrelated regions, this technique cannot be applied. A method proposed by Massonnet and Feigl (1998) can be used for eliminating the residual reference phase, but it may introduce new errors in the perpendicular baseline $B_\perp$.

Approximations of the residual reference phase, based on a priori estimations of the radial and across-track orbit errors, are valuable to suppress the residual reference phase, and to assess the magnitude and behavior of the maximum error after correction. For rms values of 5 and 10 cm in radial and across-track direction respectively, it is found that a linear approximation will suppress the residual reference phase to a maximum error of 0.35 rad, in a hyperbolic shaped plane. A quadratic polynomial model further decreases the maximum error to 0.035 rad, in a sine-like plane. In order to apply a quadratic polynomial model, as a rule of thumb, the ratio between the magnitude of the second degree term and the first degree term needs to be restricted to 10^{-6}.

4.6.8 Propagation of orbit errors to height errors

Assume that the orbit errors, expressed as an rms in radial and across-track direction, result in an error baseline $\vec{n}$, with length $n = |\vec{n}|$ and orientation β, see fig. 4.21B. Since the topographic height is related to the interferometric phase with, see eq. (3.2.1),

$$H = -\frac{\lambda}{4\pi}\frac{R\sin\theta}{B_\perp}\varphi, \qquad (4.6.21)$$

where

$$B_\perp = B_\perp^0 + n_\perp = |\vec{B}|\cos(\theta - \alpha) + |\vec{n}|\cos(\theta - \beta), \qquad (4.6.22)$$

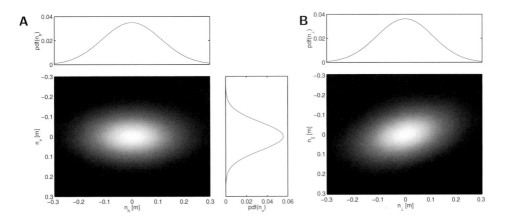

Fig. 4.26. (A) Joint probability density function for the horizontal/vertical error baseline (n_h, n_v). **(B)** Rotated probability density function for the parallel/perpendicular error baseline $(n_\perp, n_\parallel)$, and the conditional PDF for the perpendicular error baseline.

see fig. 4.21C. Here, $B_\perp^0$ is the perpendicular baseline derived from the observed state vectors and $n_\perp$ is the residual (error) perpendicular baseline caused by the error in the state vectors. Hence the topographic error is

$$dH = \frac{\lambda}{4\pi} \frac{R\sin\theta\, n_\perp}{(B_\perp^0)^2} \varphi = H^0 \frac{-n_\perp}{B_\perp^0}, \qquad (4.6.23)$$

where H^0 is the initial height derived using the available orbit information. Hence, the baseline error results in an erroneous scaling of the topographic height with factor $-n_\perp/B_\perp^0$. For perpendicular baselines larger than 20 m, the scaling factor is typically less than 1%, using ERS precise orbits. To derive the absolute error in an interferogram with an arbitrary state vector accuracy and perpendicular baseline we need to evaluate the propagation of the radial and across-track orbit PDF to the PDF of the perpendicular error baseline $n_\perp$.

Assuming a Gaussian PDF for the radial as well as the across-track orbit error, see eq. (4.6.19a), we find the joint probability density function for the baseline error to be

$$pdf(n_h, n_v) = \frac{1}{\sqrt{2\pi}\sigma_{n_h}} \exp\left(\frac{-n_h^2}{2\sigma_{n_h}^2}\right) \frac{1}{\sqrt{2\pi}\sigma_{n_v}} \exp\left(\frac{-n_v^2}{2\sigma_{n_v}^2}\right), \qquad (4.6.24)$$

where $\sigma_{n_h} = \sigma_{B_h}$ and $\sigma_{n_v} = \sigma_{B_v}$, cf. eq. (4.6.19a). In fig. 4.26A the marginal PDF's for the horizontal and vertical baseline errors are shown in combination with the joint PDF, based on a 5 cm radial rms and an 8 cm across-track rms. A simple coordinate transformation from the horizontal/vertical baseline error to the perpendicular/parallel baseline error can be written as:

$$\begin{bmatrix} n_h \\ n_v \end{bmatrix} = \begin{bmatrix} \cos\theta & -\sin\theta \\ \sin\theta & \cos\theta \end{bmatrix} \begin{bmatrix} n_\perp \\ n_\parallel \end{bmatrix}, \qquad (4.6.25)$$

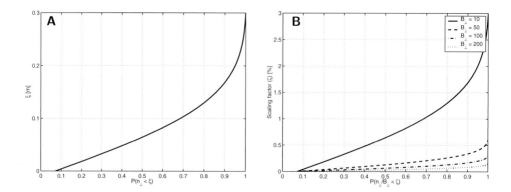

Fig. 4.27. **(A)** Probability of an error perpendicular baseline $n_\perp$ to be less than a threshold of ξ, using accuracies of 5 cm for the radial and 8 cm for the across-track state vector error. **(B)** Probability of a scaling factor $n_\perp/B_\perp$ to be less than a threshold ζ, for four baseline situations and identical state vector accuracies as in fig. 4.27A.

and the marginal PDF for $n_\perp$ is consequently the integral over all parallel baselines $n_\parallel$

$$
\begin{aligned}
pdf(n_\perp) &= \int_{-\infty}^{\infty} pdf(n_h, n_v) dn_\parallel \\
&= \frac{1}{2\pi \sigma_{n_h} \sigma_{n_v}} \int_{-\infty}^{\infty} \exp\left(\frac{-n_h^2}{2\sigma_{n_h}^2} + \frac{-n_v^2}{2\sigma_{n_v}^2} \right) dn_\parallel.
\end{aligned}
\tag{4.6.26}
$$

As a result, the probability of an error perpendicular baseline $n_\perp$ to be less or larger than a certain threshold value can now be numerically derived, based on the rms values of the state vectors. In fig. 4.27A the probability of an error baseline to be less than a threshold ξ, in meters, is given. Figure 4.27B shows for four perpendicular baseline lengths how this influences the scaling factor $-n_\perp/B_\perp^0$. For example, for $B_\perp = 10$ m, there's a probability of 70% that the erroneous scaling is less than 1%. Planning suitable SAR acquisitions for topographic mapping is possible using eq. (4.6.26) or from fig. 4.27A. Suppose we want to map a mountain with an elevation of 1 km with an accuracy better than 1 m. With a 90% likelihood this yields a maximum perpendicular baseline error of 0.17 m, hence $n_\perp < 0.17$. With eq. (4.6.23) we find for $dH = 1$ m and $H^0 = 1000$ m that we need a perpendicular baseline $B_\perp > 170$ m.

4.7　Atmospheric signal: turbulence

First indications that atmospheric delay variations might cause problems for the interpretation of interferograms have been reported by Goldstein (1995); Massonnet and Feigl (1995a); Tarayre and Massonnet (1996). There is currently no alternative instrumentation to measure this delay with sufficient accuracy, spatial resolution, and temporal sampling to eliminate its influence entirely from the interferograms.

Neither is it possible to measure the refractivity distribution causing the delay, cf. eq. (3.4.2), in three dimensions simultaneously with the SAR acquisitions. In appendix A we show that attempts to eliminate atmospheric influence from interferograms using GPS (time-series or networks) can give a reasonable approximation of the long wavelength features, but cannot eliminate signal with wavelengths shorter than, say, 10 km. Moreover, it introduces interpolation errors in the modified interferogram.

Since it is not possible to correct for atmospheric errors in a deterministic way, it is important to develop a mathematical model that describes the behavior of atmospheric delay in interferograms stochastically. A stochastic model is necessary for quality control, for example to prevent erroneous interpretation, and is imperative for data adjustment using series of interferograms.

We distinguish two types of atmospheric signal, based on their physical origin.

- Turbulent mixing, discussed in this section, results from turbulent processes in the atmosphere. It causes spatial (3D) heterogeneity in the refractivity during both SAR acquisitions, and affects flat terrain as well as mountainous terrain.

- Vertical stratification, discussed in section 4.8, is the result of different vertical refractivity profiles during the two SAR acquisitions, assuming there are no heterogeneities within the horizontal layers. This affects mountainous terrain only, and is correlated with topography (Massonnet and Feigl, 1998).

As vertical stratification is a function of the topographic height differences, which are assumed to be unknown a priori, this subject is treated separately, see section 4.8. In this section, we will focus on the effects of turbulent mixing.

4.7.1 Introduction

Three important observations characterize atmospheric signal in interferograms. First, the relative character of an interferogram does not enable absolute signal delay measurements. Hence, we can demean the interferogram, which implies that the expectation value of the atmospheric signal for an arbitrary pixel will be zero. As a consequence, there is no dependency on vertical variation in refractivity—different vertical refractivity distributions can result in the same total delay. Second, orbit errors can easily cause a nearly linear trend over the whole interferogram, see section 4.4.6. Such trends are usually hard to distinguish from atmospheric signal delay trends, and are usually eliminated using some kind of residual flattening, for example using tie-points. This strategy is equivalent to a high-pass filter for the atmospheric signal. Finally, we know that the amount of atmospheric signal within an interferogram is highly variable in time: some interferograms seem to have no atmospheric influence at all, whereas others are contaminated significantly. This is also dependent on the climatological characteristics of the location. Nevertheless, for acquisition time intervals of 1 day or more the two states of the atmosphere appear to be practically uncorrelated.

Several studies have shown that the predominant part of the atmospheric signal in interferograms is caused by the water vapor distribution in the lower troposphere (Goldstein, 1995; Tarayre and Massonnet, 1996; Hanssen and Feijt, 1996; Tarayre, 1996; Zebker et al., 1997; Hanssen, 1998; Hanssen et al., 1999, 2001, 2000b). This is a consequence of the fact that for microwave frequencies, the permanent molecular dipole moment of H_2O dominates the variability of the refractive index. Moreover, the water vapor distribution can be considered to be an approximate passive "tracer" of the mechanical turbulence (Ishimaru, 1978). Therefore, the observed statistics of the atmospheric signal will be correlated with those of wind vortices, see, e.g., Hanssen et al. (2000b).

The behavior of atmospheric signal in radar interferograms can be mathematically described using several interrelated measures such as the power spectrum, the covariance function, the structure function, and the fractal dimension. The power spectrum is useful to recognize scaling properties of the data or to distinguish different scaling regimes. Such Fourier-domain representations enable elegant descriptions of, e.g., filter operations and fast processing algorithms. A disadvantage is that Fourier methods require data sampled on a regular grid (Chilès and Delfiner, 1999). The spectrum of the grid is identical to the spectrum of the underlying continuous signal only if frequencies higher than the Nyquist frequency do not occur in the signal, either because they do not exist or because they are removed before sampling. Hence, the bandwidth of the grid spectrum needs to be smaller than the Nyquist frequency. For the SAR data this condition is usually satisfied.

The covariance function is theoretically equivalent to the power spectrum, although its evaluation for irregularly spaced data can be easier and the signal does not have to be bandlimited. It is used, e.g., to construct the variance-covariance matrix Q_φ, which is a prerequisite for data adjustment and filtering. A disadvantage is that it does not directly show the variance of the difference in signal between two points in an image, its use is restricted to second-order stationary signals, and it requires an a priori estimation of the mean value.

The structure function does not have this restriction and can be used for the broader class of intrinsic functions, cf. fig. 4.1. This implies that it has no restrictions concerning bandlimitedness, preliminary estimation of the mean, or stationarity. It provides a useful quantitative expression for the variance of the difference in atmospheric delay between two points with distance ρ. Since it cannot be directly used for data adjustment procedures, it is useful for data quality description if there is only one realization of the data. Finally, the fractal dimension provides an elegant measure for the roughness and scaling characteristics of the signal. In the following we elaborate on the conversions between these measures, focusing on the atmospheric signal in interferometry.

4.7.2   Theory

Turbulent mixing is a result of different tropospheric processes such as solar heating of the earth's surface (causing convection), differences in wind direction or velocity at different layers, frictional drag, and large scale weather systems. Turbulent processes cascade down from such large scales to smaller scales until the energy is dissipated.

The range of spatial scales over which this cascading takes place is known as the inertial subrange. Kolmogorov's turbulence theory assumes that kinetic energy is conserved within the inertial subrange. The turbulent wind vortices act as a carrier for atmospheric constituents such as water vapor, hereby influencing the refractivity distribution. Although the tropospheric refractivity for radio frequencies is mainly dependent on temperature, pressure, and water vapor, it is dominantly water vapor which causes the atmospheric signal within the dimensions of a SAR image, considering that only horizontal variability in refractivity is affecting the observed delay in interferograms.

Since atmospheric delay due to turbulent mixing affects a wide range of scales, governed by strongly nonlinear processes, there is a need to find the simplest and most robust measures of the variability in the delay signal.

Kolmogorov turbulence theory

In the case of isotropic turbulence in three dimensions, Kolmogorov turbulence theory predicts a specific structure function, $D_N(\rho)$ of the spatial variation of the refractivity N (Tatarski, 1961):

$$D_N(\rho) = E\{[N(\vec{\rho} + \vec{r_1}) - N(\vec{r_1})]^2\} = \begin{cases} C_N^2\, \rho^{2/3} & \text{for } l_i \ll \rho \ll l_o \\ C_N^2\, l_i^{2/3}(\rho/l_i)^2 & \text{for } \rho \ll l_i \end{cases}, \quad (4.7.1)$$

where l_i and l_o are the inner and outer scales of turbulence (Tatarski, 1961; Ishimaru, 1978; Treuhaft and Lanyi, 1987; Ruf and Beus, 1997). The first regime in eq. (4.7.1) was developed by Kolmogorov (1941) and Obukov (1941), and is referred to as the inertial subrange where the "two-thirds law" applies. The exponent $2/3$ expresses the rate at which the refractivity decorrelates with increasing distance. The structure coefficient C_N^2 is a measure for the roughness of the spatial heterogeneities.

The inner and outer scales of turbulence have been studied in first instance for optical signals. For those wavelengths, the inner scale is generally considered to be in the order of a few millimeters whereas the outer scale is in the order of 5–10 m, see Coulman and Vernin (1991) or Ruf and Beus (1997). For optical wavelengths, temperature variations are the dominant factor of the refractivity variations (Lay, 1997). For radio wavelengths, the refractivity variations are dominated by water vapor variability, see (Ishimaru, 1978). The outer scale for fully developed 3D Kolmogorov turbulence can be up to several kilometers, and is often indicated as the effective height, scale height, or the tropospheric thickness (Dravskikh and Finkelstein, 1979; Armstrong and Sramek, 1982; Treuhaft and Lanyi, 1987; Coulman and Vernin, 1991; Ruf and Beus, 1997).

The power spectrum $P_N(k)$, where k is the wave number, within the inertial subrange is (Tatarski, 1961)

$$P_N(k) \propto k^{-5/3}, \quad (4.7.2)$$

for observations along a (one-dimensional) line through the three-dimensional volume, which is equivalent to time series at a point. For a two-dimensional cross-section (slice) of the volume, using the radial wave number $k = (k_1^2 + k_2^2)^{1/2}$, we find

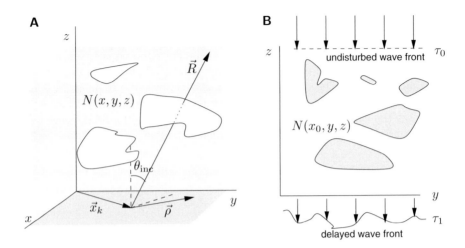

Fig. 4.28. **(A)** The three-dimensional refractivity distribution $N(x, y, z)$ causes the delay along the line of sight $\vec{R}$ at position $\vec{x}_k$. **(B)** In a 2D vertical cross-section, a flat wavefront τ_0 will an inhomogeneously delayed and arrive at the earth's surface as wavefront τ_1.

(Voss, 1988; Turcotte, 1997)

$$P_{N_{2D}}(k) \propto k^{-5/3-1}, \tag{4.7.3}$$

where the addition of minus one is required due to the radial coordinates. Another addition of minus one is required when using spherical wave numbers for the volume. In general, we can write

$$P_N(k) \propto \begin{cases} k^{-5/3} & \text{for a 1D profile through 3D space, with } k = k_1, \\ k^{-8/3} & \text{for a 2D slice through 3D space, with } k^2 = k_1^2 + k_2^2 \\ k^{-11/3} & \text{for a 3D volume, with } k^2 = k_1^2 + k_2^2 + k_3^2, \end{cases} \tag{4.7.4}$$

known as the Kolmogorov Power Spectrum. The wavenumbers k_1, k_2, and k_3 are defined in an orthogonal Cartesian wavenumber frame. The relation between the power spectrum and the structure function is discussed in following paragraphs.

Structure function of the signal delay

In fig. 4.28A the 3D refractivity distribution $N(x, y, z)$ at the time of the first SAR acquisition ($t = t_1$) is sketched. Assuming that the radio wave source is high above the turbulent layer, the 2D vertical cross-section in fig. 4.28B shows a flat, undisturbed wavefront, τ_0, before entering the turbulent refractive layer. After propagation through the medium, arriving at τ_1 at the earth's surface, it is inhomogeneously delayed. The atmospheric phase signal in an interferogram is the superposition of this wavefront for the SAR acquisitions at t_1 and t_2, see section 3.4.

The relation between the refractivity distribution $N(x, y, z, t_i)$ and the one-way signal delay S^{t_i} has been presented in eq. (3.4.2). Consequently, the theoretical structure function of the refractivity, eq. (4.7.1), can be propagated to the structure function of the signal delay.

Assuming the Kolmogorov relation for isotropic 3D turbulence, within the inner and outer scales, we used eq. (4.7.1), with C_N^2 constant over all distances $\vec{r}$. Moreover, in the calculation of the tropospheric delay covariance we will also assume that $E\{N(\vec{r})\}$ is independent of the position $\vec{r}$, i.e., second-order stationarity (Treuhaft and Lanyi, 1987). The tropospheric delay $S_k^{t_i}$ for pixel k at location $\vec{x}_k$ from a satellite at incidence angle (zenith angle) θ_{inc}, see eq. (3.4.2), can be reformulated as

$$S_k^{t_i}(\vec{x}_k) = \frac{1}{\cos\theta_{\text{inc}}} \int_0^h N(\vec{x}_k + \vec{R}(\theta_{\text{inc}}, z)) dz, \qquad (4.7.5)$$

where $\vec{x}_k$ is the 2D location vector on the earth's surface and h is the effective height, see fig. 4.28A. $\vec{R}$ is the vector between position $\vec{x}_k$ and the satellite's position.

For two positions with horizontal distance ρ, the expectation E for the quadratic difference of the delay between the two pixels k and l, at positions $\vec{x}_k$ and $\vec{x}_l = \vec{x}_k + \vec{\rho}$, is the structure function, c.f. eq. (4.1.8):

$$D_S(\rho) = E\{[S^{t_i}(\vec{x} + \vec{\rho}) - S^{t_i}(\vec{x})]^2\}. \qquad (4.7.6)$$

By substituting eq. (4.7.5) in (4.7.6), and exchanging the order of integration and expectation (summation), we can express the structure function of the delay as function of the refractivity (Treuhaft and Lanyi, 1987):

$$D_S(\rho) = \frac{1}{\sin^2\theta_{\text{inc}}} \int_0^h\!\!\int \left[2E\{N^2[\vec{r}(\theta_{\text{inc}}, z_1)]\} \right.$$
$$\left. - 2E\{N[\vec{\rho} + \vec{r}(\theta_{\text{inc}}, z_1)] \ N[\vec{r}(\theta_{\text{inc}}, z_2)]\} \right] dz_1 dz_2. \qquad (4.7.7)$$

This derivation is discussed in appendix B.1. Assuming second-order stationary signals we can write the structure function of N as

$$D_N(R) = 2E\{N^2(\vec{r})\} - 2E\{N(\vec{r} + \vec{R})N(\vec{r})\}, \qquad (4.7.8)$$

and the structure function of the delay $D_S(\rho)$ as a function of the structure function of the refractivity $D_N(R)$:

$$D_S(\rho) = \frac{1}{\sin^2\theta_{\text{inc}}} \int\!\!\int_{0 \to h} [D_N(R_2) - D_N(R_1)] \, dz_1 \, dz_2. \qquad (4.7.9)$$

Using eq. (4.7.1), the structure function of the delay $D_S(\rho)$ behaves as (Tatarski, 1961; Thompson et al., 1986; Treuhaft and Lanyi, 1987; Coulman and Vernin, 1991):

$$D_S(\rho) = C_S^2 \, \rho^{5/3}, \qquad (4.7.10)$$

in comparison with the 2/3-exponent of the refractivity in eq. (4.7.1).

The power spectrum of the atmospheric delay signal, $P_S(k)$, where k is the wavenumber corresponding with the structure function in eq. (4.7.10), is (Tatarski, 1961)

$$P_S(k) \propto \begin{cases} k^{-8/3} & \text{for a 1D profile over the 2D surface, with } k = k_1 \\ k^{-11/3} & \text{for the 2D surface, with } k^2 = k_1^2 + k_2^2 \end{cases}. \qquad (4.7.11)$$

The wavenumbers k_1 and k_2 are orthogonal wavenumbers in the plane.

Power spectrum of the signal delay

The general relation between the structure function and the power spectrum $P_\varphi(f)$ of a random continuous one-dimensional signal with a power-law behavior $\varphi(x)$ is discussed by, e.g., Monin et al. (1975) and Agnew (1992). If the power spectrum is written as

$$P_\varphi(f) = P_0(f/f_0)^{-\beta}, \qquad (4.7.12)$$

where f is some (spatial or temporal) frequency, P_0 and f_0 are normalizing constants, and $-\beta$ is the *spectral index* (often $1 < \beta < 3$). The structure function of a signal $\varphi(x)$ with the power spectral form of eq. (4.7.12) can be written as

$$D_\varphi(\rho) = C_\varphi \frac{P_0}{f_0^{-\beta}} \rho^{\beta-1}, \quad \text{with} \qquad (4.7.13)$$

$$C_\varphi = \frac{-\pi^\beta}{2^{1-\beta}\Gamma(\beta)\cos(-\beta\pi/2)}. \qquad (4.7.14)$$

The derivation of the relation between the power spectrum and the structure function as well as the definition of the structure coefficient is given in appendix B.2.

The power-law model $D_\varphi(\rho) = \rho^\alpha$ is a structure function for $0 < \alpha < 2$ (Chilès and Delfiner, 1999). Since this model doesn't have a "sill" (a range ρ at which the variance of the difference between two points does not increase anymore), it would need to be adjusted for atmospheric signal as observed by radar interferometry. However, since the data ranges in an interferogram are spatially limited, and the observations have a relative character, it is not necessary to expand the model for ranges larger than, say 100–300 km.

Addition and scaling of power-law models

An interferogram contains the summation of the atmospheric signals at the two acquisitions t_1 and t_2. Therefore, the power spectrum and structure function of the signal in an interferogram is different from the signal observed by other space-geodetic techniques, such as GPS or VLBI. Using the power spectrum $P_\varphi(f)$ of (atmospheric) signal $\varphi(x)$, cf. eq. (4.7.12):

$$\varphi(x) \quad \xrightarrow{\mathcal{F}} \quad \Phi(f) \quad \xrightarrow{\Phi(f)\Phi^*(f)} \quad P_\varphi(f) = P_0\left(\frac{f}{f_0}\right)^{-\beta}, \qquad (4.7.15)$$

where $\mathscr{F}$ denotes the 1D Fourier transform, and the operation $\Phi(f)\Phi^*(f)$ results in the power spectrum. For the addition/subtraction of φ_{t_1} and φ_{t_2}, the addition theorem in Fourier transforms states (Bracewell, 1986)

$$\varphi_{t_1}(x) \pm \varphi_{t_2}(x) \xrightarrow{\mathscr{F}} \Phi_{t_1}(f) \pm \Phi_{t_2}(f) = \Psi(f)$$

$$\xrightarrow{\Psi(f)\Psi^*(f)} P_{\varphi_1-\varphi_2}(f) = (P_{0,t_1} + P_{0,t_2})(\frac{f}{f_0})^{-\beta}.$$

(4.7.16)

Therefore, for a model atmosphere with Kolmogorov turbulence characteristics, the observed power spectrum in an interferogram has the same scaling properties as the atmospheric influence during only one SAR acquisition, indicated by the exponent β, although the intensity of the variations is in fact the summation of the two contributing power spectra.

Equivalently, using eqs. (4.7.13) and (4.7.16), the structure function of the summation of the two atmospheric signals is

$$D_{\varphi,\text{ifg}}(\rho) = C_\varphi \frac{P_{0,t_1} + P_{0,t_2}}{f_0^{-\beta}} \rho^{\beta-1}.$$

(4.7.17)

Under the assumption that the atmosphere is statistically similar during the two acquisitions we may write:

$$P_{\varphi,\text{ifg}}(f) = 2P_{\varphi,\text{sar}}(f)$$

(4.7.18)

$$D_{\varphi,\text{ifg}}(\rho) = 2D_{\varphi,\text{sar}}(\rho),$$

(4.7.19)

where the subscript "sar" indicates a single SAR acquisition, and the subscript "ifg" indicates the interferogram.

The *similarity theorem* (Bracewell, 1986) is used when scaling a signal, e.g., when transforming a GPS delay time series to spatial variation using the wind speed v:

$$\varphi(v\,x) \xrightarrow{\mathscr{F}} \frac{1}{|v|}\Phi(\frac{f}{v}) \xrightarrow{power} \frac{1}{v^2}|\Phi(\frac{f}{v})|^2 = \frac{P_0}{v^2}(\frac{f/f_0}{v})^{-\beta}.$$

(4.7.20)

This means that the same $P(f)$ values will end up at a spatial wavenumber $\frac{f/f_0}{v}$, scaled with factor $1/v^2$. Using double-logarithmic axes, this corresponds with a simple shift of the curve, with identical exponent β.

Power spectrum and covariance function

According to the Wiener-Khinchine theorem, the power spectrum $P_\varphi(f)$ and the covariance function $C_\varphi(r)$ of a second-order stationary process are each others cosine Fourier transforms (Chilès and Delfiner, 1999):

$$C_\varphi(r) \xleftrightarrow{\mathscr{F}} P_\varphi(f),$$

with

$$C_\varphi(r) = \int e^{+j2\pi fr} P_\varphi(f)df = \int \cos(2\pi fr)\, P_\varphi(f)\, df$$

(4.7.21)

$$P_\varphi(f) = \int e^{-j2\pi fr} C_\varphi(h)dr = \int \cos(2\pi fr)\, C_\varphi(r)\, dr.$$

(4.7.22)

For a discrete finite signal $\varphi[n] \in \mathbb{R}^1$ with $n = (1, 2, \ldots, N)$, the covariance $C_\varphi[l]$ is derived for $l = (1, 2, \ldots, N/2)$ using

$$C_\varphi[l] = \frac{2}{N} \sum_{k=1}^{N/2} \cos[2\pi(l-1)(k-1)] \, P_\varphi[k]. \tag{4.7.23}$$

The power spectrum $P_\varphi[k]$ is derived using the Discrete Fourier Transform (DFT):

$$\Phi[k] = \mathscr{F}(\varphi[n]) = \sum_{n=1}^{N} \varphi[n] \exp\left(\frac{-j2\pi(k-1)(n-1)}{N}\right), \quad 1 \leq k \leq N \tag{4.7.24}$$

and

$$P_\varphi[k] = \begin{cases} 0, & k = 1 \\ N^{-1}(\Phi[k]\,\Phi^*[k]), & 2 \leq k \leq N/2 \end{cases}, \tag{4.7.25}$$

where we only used the positive wavenumbers. If a function is the Fourier transform of a nonnegative measure it is a positive (semi-)definite covariance function, as proved by Bochner (1959).

Larger scales

Equation (4.7.1) has been defined for homogeneous isotropic 3D turbulence. For horizontal scales commonly associated with, e.g., GPS and radar interferometry, the spatial distances can be larger than the effective height of the wet troposphere (1–3 km), hence, larger than the inertial subrange. This reduces the tropospheric contribution to a relatively thin layer with effectively 2D turbulence. Stotskii (1973) proposed to write the structure function of the signal delay for scales larger than the effective tropospheric height h as

$$D_S(\rho) = h^2 D_N(\rho) = h^2 C_N^2 \rho^{2/3}, \tag{4.7.26}$$

using eq. (4.7.1). For the power-law behavior of the power spectrum we find with eq. (4.7.12) and (4.7.13):

$$P_S(k) \propto \begin{cases} k^{-5/3} & \text{for a 1D profile over the 2D surface, with } k = k_1 \\ k^{-8/3} & \text{for a 2D surface, with } k^2 = k_1^2 + k_2^2 \end{cases}, \tag{4.7.27}$$

where k_1 and k_2 are orthogonal wavenumbers in the plane.

Early experiments investigating atmospheric signal delay were performed using, e.g., VLBI antenna's. Since these have a fixed position, hence fixed spacing, only few observational data were available to verify Stotskii's proposition. The InSAR data analysis discussed hereafter provides independent support of this proposition.

The transition between the two regimes—3D-turbulence, eq. (4.7.11), to 2D turbulence, eq. (4.7.27)—can be described in several ways. For example, Treuhaft and

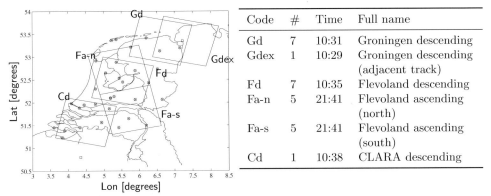

Code	#	Time	Full name
Gd	7	10:31	Groningen descending
Gdex	1	10:29	Groningen descending (adjacent track)
Fd	7	10:35	Flevoland descending
Fa-n	5	21:41	Flevoland ascending (north)
Fa-s	5	21:41	Flevoland ascending (south)
Cd	1	10:38	CLARA descending

Fig. 4.29. Location and coding of the test sites, SAR acquisitions and meteorological stations. The circles indicate synoptic stations, the squares are radiosonde locations.

Lanyi (1987) used a polynomial to describe the transition fluently. Comparison with the InSAR data suggests that such a polynomial form might be restricted to a specific location on earth. In this study, we define a distinct change between the two regimes at scale height h. Since the spectral form is a stochastic representation of the behavior of the signal, such an abrupt change does not introduce artifacts whereas it is a convenient simplification.

4.7.3 Observations

To investigate the stochastic behavior of tropospheric signal in repeat-pass radar interferograms a set of 26 interferograms over the Netherlands—corresponding with 52 weather situations—has been analyzed (Hanssen, 1998). The locations of the interferograms are shown in fig. 4.29. Using a reference elevation model, differential interferograms have been obtained, even though topographic variation is limited to a 5–10 m range for the northern and southern interferograms, and maximally

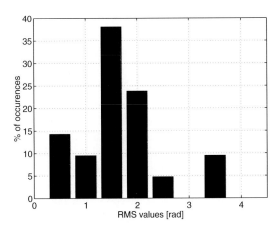

Fig. 4.30. Histogram showing the rms values of the 26 interferograms in the database and the frequencies of occurrence.

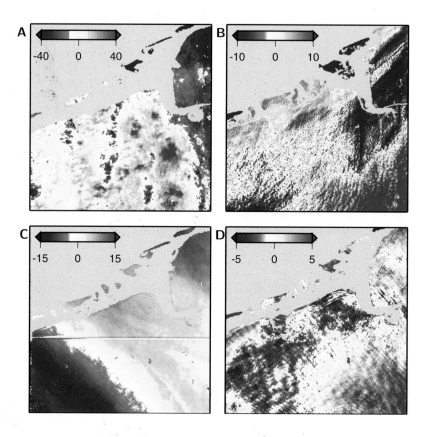

Fig. 4.31. Eight differential tandem interferograms over the northern part of the Netherlands (100 × 100 km), showing only atmospheric signal, expressed in millimeters zenith delay. Note the different ranges in the color scales. **(A)** Jul 1995, **(B)** Aug 1995, **(C)** Dec 1995, **(D)** Mar 1996.

∼100 m for the central part. ERS "tandem" imagery is used to minimize the effects of temporal decorrelation and to exclude possible wide-scale surface deformation. The interferogram data span the whole year, providing a reasonable sampling of the seasonal variation. However, the imaging times are the same for all acquisitions, prohibiting the study of variations between daytime and nighttime acquisitions.

Since orbit errors can introduce phase trends in the interferogram, the data are detrended in azimuth and range using a linear polynomial model. Therefore, the data can be considered second-order stationary. For these data the mean values are zero by definition, but higher order moments vary considerably depending on the weather situation, as shown in fig. 4.30. The observed rms values are shown for an area of ∼50 × 50 km. RMS values of 1.5 rad (6.7 mm slant delay) occurred in most of the situations, almost 40%. Approximately 15% of the situations had a low rms value of 0.5 rad (2.2 mm), whereas 10% of the situations had a significant

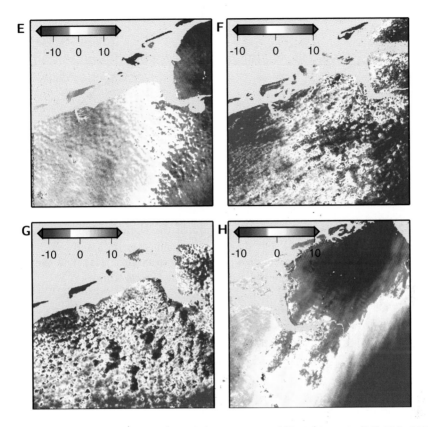

Fig. 4.31.—*continued* **(E)** Apr 1996, **(F)** May 1996, **(G)** Aug 1996, **(H)** Feb 1996.

atmospheric variation of 3.5 rad (15.6 mm) rms. Although these values give an indication of the range of atmospheric contributions, they should always be related to their spatial support. Therefore, other measures as discussed below provide a better parameterization. Figure 4.31 shows an overview of the turbulent signal in eight differential tandem interferograms, acquired over Groningen in the northern part of the Netherlands during 1995/96. The interferometric phase is unwrapped and converted to zenith delay signal in mm.

The state of the atmosphere varied from severe thunderstorms during the both acquisitions contributing to one interferogram (fig. 4.31A) to extremely calm weather without significant convection, e.g. fig. 4.31D. Hanssen (1998) provides a more elaborate meteorological description of these data.

Power spectrum and fractal dimension

The 50 × 50 km windows in the data of fig. 4.31 are used to calculate a set of power spectra. First the mean value of the window is subtracted, and both in azimuth and in range direction a linear trend is removed from the data. For all eight interferograms of fig. 4.31, the one-dimensional power spectra are shown in

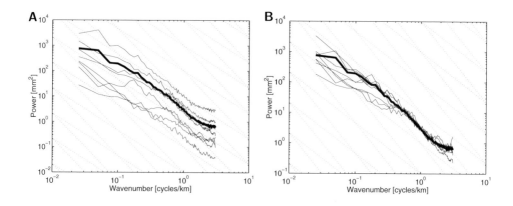

Fig. 4.32. The one-dimensional power spectra of the eight atmospheric situations in fig. 4.31. **(A)** Absolute values. The diagonal lines indicate $-5/3$ and $-8/3$ power-law behavior. The bold line is the average of all eight situations. **(B)** The same spectra multiplied with a factor to obtain the same power at 1 cycle/km.

fig. 4.32A, assuming isotropy. The diagonal lines in fig. 4.32A indicate the slope of $-5/3$ and $-8/3$ power-law functions. The data show that all atmospheric situations exhibit similar power-law behavior, with slopes varying between $-5/3$ and $-8/3$ for the different scaling regimes. There is a range in the absolute power, indicated by the vertical position of the spectra in the graph, of almost two orders of magnitude, indicating more or less severe weather conditions. The upper line is corresponding with the first interferogram in fig. 4.31, which indeed exhibits the most severe zenith delay variability. The bold line is the average of all eight situations. In fig. 4.32B, all power spectra have been multiplied by a factor, yielding the averaged power for 1 cycle/km. Comparison of the shape of the power spectra shows that the data are not entirely scale-invariant, and three dominant regimes can be recognized, of which two are atmospheric while the third one probably reflects noise.

Regime I, with exponent $\beta = 5/3$ and fractal dimension $D_2 \approx 2.7$, see eqs. (4.1.14) and (4.1.15), covers scales larger than the thickness of the turbulent layer. For these scales, between ~ 2 km and the size of the interferogram (~ 100–200 km), the approximation of two-dimensional turbulence can be applied, c.f. eq. (4.7.11). Over these scales the characteristics of the total atmospheric column dominate. For example, convective processes result in significant differences in the overall refractivity between updrafts (warm and moist) and downdrafts (cold and relatively dry). Although there is evidently atmospheric signal at scales larger than the interferogram size, we can safely ignore this since it is outside the measurement capabilities of the system. Stationarity of the physical process is always ensured due to the limitation of the earth's circumference, which enforces the flattening of the power spectrum for large scales. Based on VLBI observations it is proposed that the slope of the power spectrum is zero for scales larger than 3000 km (Treuhaft and Lanyi, 1987).

Regime II covers scales smaller than the thickness of the turbulent layer (the depth

Table 4.5. Three regimes of atmospheric delay as observed by radar interferometry.

regime	scale (km)	D_2 *	β [†]
I	> 2	2.67 (16/6)	5/3
II	0.50–2.0	2.16 (13/6)	8/3
III	0.01–0.5	3.16 (19/6)	2/3

*, fractal dimension;   †Power-law exponent.

of the convective boundary layer, say 2 km), down to the resolution level. The integrated refractivity has a much smoother behavior over these scales, indicated by the steeper power exponent $\beta = 8/3$ and the lower fractal dimension $D_2 \approx 2.2$. The correspondence in the power slopes is clearly visible in fig. 4.32B, where all curves overlap within this regime. The behavior in this regime has been observed in several studies on SAR interferometry, see Goldstein (1995); Hanssen et al. (1998b); Hanssen (1998); Ferretti et al. (1999a). Nevertheless, in contrast to some studies, the combination with regime I indicates that the signal cannot be considered scale-invariant for all spatial scales. In fact, assuming a $-8/3$ power-law for scales larger than 5–10 km will increasingly underestimate the roughness and hence the variability of the atmospheric signal.

Regime III is unlikely to have an atmospheric origin, since there is no physical explanation for increased delay variation at such small scales. Instead, it is likely caused by high wavenumber noise in the data. This can be a result of decorrelation effects or, e.g., by interpolation errors when subtracting the reference DEM. In these examples, regime III noise starts to influence the data for wavelengths less than $\sim$500 m. Although the driving mechanisms differ from regime I and II, it can be advantageous to describe the statistical behavior of regime III the same way, since the transitions between the regimes are gradual and the stochastic model can be constructed uniformly. The power exponent in regime III approaches $\beta = 2/3$ and the lower fractal dimension $D_2 \approx 3.2$, indicating the very rough behavior at these scales.

Table 4.6. P_0, rms, and main cloud type for the eight interferograms in fig. 4.31

Fig. 4.31-	A	B	C	D	E	F	G	H
Code	Gd1	Gd2	Gd3	Gd4	Gd5	Gd6	Gd7	Gdex
Month/Year	7/95	8/95	12/95	3/96	4/96	5/96	8/96	2/96
P_0 * (mm^2)	11.2	4.7	0.6	0.5	0.8	2.5	4.7	0.3
rms [†] (mm)	15.2	6.7	7.1	2.7	6.2	8.0	8.0	8.5
Cloud	Cb	Ci	St/Sc	St/Sc	Ci	Cu	Cu	St

Cb, Cumulonimbus;   Ci, Cirrus;   St, Stratus;   Sc, Stratocumulus;   Cu, Cumulus.   *P_0 is the power initialization parameter at 1 cycle/km †The rms is computed over an area of $\sim$50$\times$50 km.

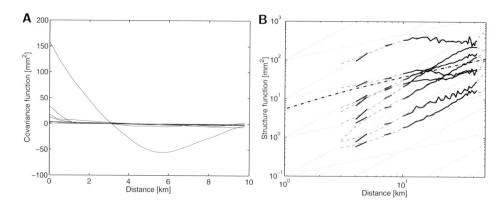

Fig. 4.33. **(A)** Empirical covariance functions corresponding to the atmospheric situations of fig. 4.31, derived from the power spectra of eight 50×50 km segments of the data. **(B)** Derived structure functions for 8 interferogram segments of 50×50 km over Groningen, the Netherlands, using a log-log scale. For comparison we show the Treuhaft/Lanyi model as the dash-dotted line (Treuhaft and Lanyi, 1987). The dotted lines follow a 2/3 and a 5/3 slope for reference. Since the interferograms were significantly subsampled, only long wavelengths are shown.

Summarizing, the important observation from the power spectra is their similar shape, even though the atmospheric circumstances were extremely variable, ranging from calm, cold weather to severe storms with strong convection and humidity gradients. These weather situations covered two orders of magnitude in the power spectrum. Therefore, a reasonable approximation of the power spectra seems to be based on the three regimes, using only the P_0 value (for 1 cycle/km wavenumbers) for initialization, see table 4.5. This observation is used in the atmospheric model proposed in section 4.7.4. The P_0-values for the eight interferograms in fig. 4.31 are listed in table 4.6.

Covariance function

Assuming second-order stationarity and isotropy, the covariance between two observations is a function of only the distance between the observations. Empirical covariance functions $C(r)$, where r reflects the distance between two observations, are derived from the inverse DFT of the power spectra, see eq. (4.7.23), and shown in fig. 4.33A. Clearly, the interferogram with the most severe atmospheric signal shows the most dominant empirical covariance function. The variances ($r = 0$) are dependent of the spatial support of the data, in this case the window size of 50×50 km chosen for evaluation.

The standard approach to modeling the covariance function is by approximating some parametric analytical form for $C(r)$ which ensures positive-definiteness (Rummel, 1990; Daniels and Cressie, 1999). However, since the power spectrum is nonnegative by definition, its Fourier transform is always a positive definite covariance function (Bochner, 1959).

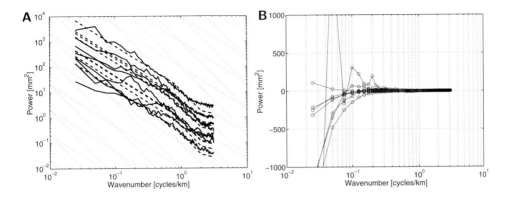

Fig. 4.34. (A) Power spectra for eight independent SAR interferograms, and the models defined by the power spectrum value at $f/f_0 = 1$ cycle/km, see table 4.6. The diagonal lines indicate $-8/3$ and $-5/3$ power laws. **(B)** Difference between model and observations. The positive and negative differences for long wavelengths indicate differences in weather conditions and poor spectral estimation. On the average the fits are reasonable.

Structure functions

The two-dimensional structure functions are derived from the eight interferograms. From these a radial average yields the one-dimensional structure function, as shown in fig. 4.33B. For comparison, the model proposed by Treuhaft and Lanyi (1987) is plotted as the dash-dotted line. This model uses an effective height of the wet troposphere of 1000 m, and an initialization value derived from the standard deviation of the zenith wet delay over a given time interval for sites at mid-latitudes (California, Australia and Spain). It can be observed that these standard parameters provide a reasonable model, but the variability in weather conditions over a coastal zone area such as the Netherlands is too significant to allow for such a generalization. Hence, the initialization parameters need to be adjusted to the specific weather conditions.

4.7.4 Empirical model

Based on the observations in interferograms, an analytical one-parameter model can be constructed to describe the atmospheric signal. The application of atmospheric models for radar interferometry is different from models used for, e.g., VLBI or GPS. First, the interferometric data always reflect the superposition of two independent atmospheric states. Second, the interferogram can be regarded as a snapshot—an instantaneous acquisition of the integrated signal delay over the entire atmospheric column—whereas the other methods measure variability in time at a fixed position. These methods need to apply Taylor's hypothesis of a "frozen atmosphere" to map time series to spatial variation, using the wind speed (Taylor, 1938). Finally, the spatial accuracy of a SAR interferogram is relatively high, say, 40×40 m, whereas the atmospheric zenith delay for a technique such as GPS is a weighted average of all slant delays to satellites at different directions and elevations using a mapping

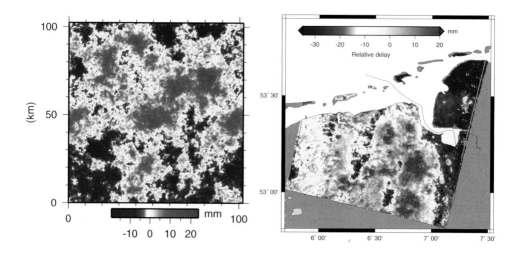

Fig. 4.35. (A) Simulated atmosphere interferogram, based on three fractal regimes. The data are scaled to relative signal delay in mm. **(B)** Example of observed tropospheric signal due to cumulonimbus clouds and convection over Groningen, the Netherlands, cf. fig. 4.31A.

function. As a result, such a measurement cannot be regarded as a point measurement anymore, but as an averaged value over an area with a diameter of several kilometers, introducing aliasing effects. The InSAR data do not suffer from this limitation.

Here we propose a new atmospheric model for radar interferometry, based on the observed scaling regimes in section 4.7.3. The model assumes isotropy and no ionospheric influences. Furthermore, the influence of topography is ignored to obtain a generic model. Topographic influence is discussed separately in section 4.8.

These results can be written as:

$$
P_\varphi(f) = \begin{cases} P_I(f/f_0)^{-5/3} & \text{for } 1.5 \leq (f_0/f) < 50 \text{ km,} \\ P_0(f/f_0)^{-8/3} & \text{for } 0.25 \leq (f_0/f) < 1.5 \text{ km,} \\ P_{III}(f/f_0)^{-2/3} & \text{for } 0.02 \leq (f_0/f) < 0.25 \text{ km.} \end{cases} \tag{4.7.28}
$$

By defining f as the wavenumber in cycles/km ($f_0 = 1$ cycle/km) and enforcing a continuous function, we can initialize this model by one parameter, by measuring the power spectral density at $f = f_0$, which yields P_0. The other two coefficients are easily found to be $P_I = P_0(f_0/f)$ for $(f_0/f) = 1.5$ km and $P_{III} = P_0(f_0/f)^2$ for $(f_0/f) = 0.25$ km.

Figure 4.34A shows the empirical results of this approach, for eight independent interferograms, suffering from atmospheric signal varying from thunderstorms to calm weather. It is clear to see that the basic structure of all situations is similar, following the power-law behavior described above. The diagonal lines indicate $-8/3$

power-law values. Note that it is not possible, as some authors have proposed, to describe the scaling behavior with only one single power-law exponent. For wavelengths larger than 1.5 km none of the signals follow the $-8/3$ lines. Although there is a strong similarity in shape, the difference between the absolute values of these curves varies two orders of magnitude. It is this variation that is important to consider while evaluating the quality of the interferometric data.

The covariance function derived from the model can be used to fill the variance-covariance matrix $\mathbf{Q}_\varphi$, based on the distance between each pair of pixels. The diagonal elements consist of the model values and the phase variance derived from the coherence of the data.

Simulation

Figure 4.35A shows a simulation of the model over a 50×50 km area. The transition of regime I to regime II is chosen at a wavelength of 2 km, the transition of regime II to regime III at a wavelength of 900 m. In fig. 4.35B an example of observed tropospheric signal is shown for comparison. This situation corresponds with storms during both acquisitions.

The isotropic scaling behavior as described above is applicable for many atmospheric situations, from calm to stormy weather, without clouds to cumulonimbus clouds. Although the scaling behavior is similar for all situations, only the absolute values of the variations are causing the differences in the delay of the wavefront. The power initialization parameter P_0 can be observed over a relatively small area of the interferogram, sufficient to estimate the entire behavior over larger scales. By transforming the modeled spectra to the covariance functions, it is possible to estimate the variance of the difference between points in the interferogram. This way, the stochastic model for repeat-pass radar interferometry can be adjusted to account for the atmospheric variability.

4.7.5 Conclusions

The stochastic modeling of atmospheric turbulent signal for radar interferometric measurements is a key element in describing the quality of the data and the derived parameters. Apart from the usual phase noise described by the interferometric coherence, the variances and covariances introduced by the atmosphere need to be included in this model as well. It has been shown that, although the state of the atmosphere can be very different for various interferograms, the shape of the power spectrum shows three distinct scaling regimes which are remarkably similar, independent of the type of atmospheric behavior, as supported by Kolmogorov turbulence theory, extended for ranges exceeding the 3D turbulence characteristics. The absolute values of the observed power spectra differ two orders of magnitude.

Using the atmospheric model discussed in this section, it is possible to quantitatively describe the covariance function of an interferogram using only one parameter. This parameter can be estimated from a small patch of the interferogram, with flat terrain and without deformation. This first order estimate may then be used to describe covariances in the whole interferogram. An alternative approach is the estimation of the initialization parameter using, e.g., GPS time series in combination with wind speed observations, see eq. (4.7.20). In this way, a single GPS receiver in the interferogram area would be sufficient to quantify the covariances in the interferometric data.

4.8 Atmospheric signal: stratification

This section reports on the implications of vertical refractivity profiles in the troposphere on the quality of Digital Elevation Models and surface deformation maps derived from synthetic aperture radar (SAR) interferometry. A representative set of 1460 radiosondes acquired over one year in the Netherlands is used to obtain statistics for the differential delays between the two acquisition dates, and apply these to simulate 1-day and $k \times 35$ day intervals corresponding with ERS-1/2 orbit characteristics. It is shown that differential delays can amount up to more than 1 cm for height intervals of 500 meters or more. For a 2 km height interval and an interferometric baseline of 80 m such delays result in a height error of 180 m. It is not possible to find a generally valid correction scheme for these delays using surface meteorological measurements. Only in situ vertical profile measurements such as radiosondes can be used to correct for these errors. To obtain a first order indication of the extent of these effects on the accuracy of products derived from radar interferometry, the rms of the delay is determined as a function of height. An empirical expression for the rms is presented.

Introduction

Atmospheric stratification only considers variation of the refractivity along the vertical. Assuming an infinite number of thin atmospheric layers, each with constant refractivity, there will be no horizontal delay differences over flat terrain, even for different refractivity profiles during both SAR acquisitions. This is due to the fact that SAR interferograms are not sensitive to image-wide phase biases. However, for hilly or mountainous terrain a difference in the vertical refractivity profile during both acquisitions will affect the phase difference between two arbitrary resolution cells with different topographic height, see fig 4.36A, and may cause an erroneous interpretation. This effect has been recognized during deformation studies of mount Etna by, e.g., Tarayre and Massonnet (1996); Massonnet and Feigl (1998); Delacourt et al. (1998), and Ferretti et al. (1999a).

Since the resulting phase error in the phase difference between two resolution cells has zero-mean (expectation value), and the observed phase gradients are often a combination of topographic residuals, deformation, horizontal atmospheric heterogeneity and differential vertical stratification, it has not been possible yet to obtain reasonable error estimates using interferometric data only. A simple and effective way to address this uncertainty is to study vertical radiosonde profiles, and analyze the statistics of the delay variation for every possible height interval. We analyzed 1460 radiosondes acquired 4 times daily from 1 Jan. 1998 to 31 Dec. 1998, located in a moderate sea-climate at latitude 52.10° and longitude 5.18°. Although climate conditions vary considerably over the globe, it will be shown that it is not the abso-

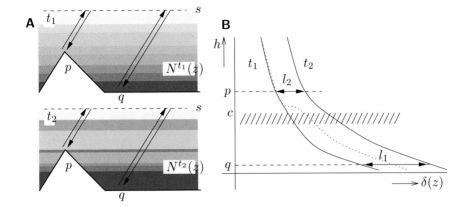

Fig. 4.36. (A) Differential tropospheric delay between point p at the top and point q at the foot of the mountain due to their height difference and a different vertical refractivity profile $N^{t_i}(z)$ during the two SAR acquisitions at t_1 and t_2. **(B)** The cumulative delay curves for t_1 and t_2 indicating the height of points p and q. The effect of the delay on the interferometric phase difference between p and q is determined only by the difference $l_2 - l_1$. The presence of, e.g., an extra cloud layer c at t_1, indicated by the dashed area, will cause a shift of the cumulative delay profile below the cloud, indicated by the dotted line.

lute refractivity value, but the dispersion of the refractivity for a fixed height that determines the amount of error for interferometry. Since water vapor is the main factor influencing the variation of refractivity, we assume that the results reported here can be regarded as representative for a large part of the globe, excluding polar regions, where the atmosphere has a very stable stratification and the cold air cannot hold much water vapor.

Theory

The geometric and delay terms for points p and q, see fig. 4.36, at different topographic heights, during acquisitions t_1 and t_2, assuming zero-baseline, identical incidence angles θ_{inc}, and no horizontal variation in refractivity N, can be written as:

$$\psi_p^{t_i} = \frac{4\pi}{\lambda \cos\theta_{\mathrm{inc}}}(z_{ps} + \delta_{ps}^{t_i}), \tag{4.8.1a}$$

$$\psi_q^{t_i} = \frac{4\pi}{\lambda \cos\theta_{\mathrm{inc}}}(z_{ps} + z_{qp} + \delta_{ps}^{t_i} + \delta_{qp}^{t_i}), \tag{4.8.1b}$$

where z_{ps} is the geometric distance between point p and satellite s, projected onto the vertical. The vertical delay between p and s at t_i is indicated by $\delta_{ps}^{t_i}$.

The interferogram phases at points p and q are

$$\varphi_p = \psi_p^{t_1} - \psi_p^{t_2} = \frac{4\pi}{\lambda \cos\theta_{\text{inc}}}(\delta_{ps}^{t_1} - \delta_{ps}^{t_2}), \qquad (4.8.2a)$$

$$\varphi_q = \psi_q^{t_1} - \psi_q^{t_2} = \frac{4\pi}{\lambda \cos\theta_{\text{inc}}}(\delta_{ps}^{t_1} + \delta_{qp}^{t_1} - \delta_{ps}^{t_2} - \delta_{qp}^{t_2}), \qquad (4.8.2b)$$

and the contribution of the tropospheric delay to the interferometric phase difference between point p and q is

$$\varphi_{pq} = \varphi_p - \varphi_q = \frac{4\pi}{\lambda \cos\theta_{\text{inc}}}(\delta_{qp}^{t_1} - \delta_{qp}^{t_2}). \qquad (4.8.3)$$

From eq. (4.8.3) we conclude that whenever $\delta_{qp}^{t_1} \neq \delta_{qp}^{t_2}$, there will be a contribution of the tropospheric vertical stratification in the interferogram. Note that this effect will only affect points in the interferogram that have a different topographic height. The total integrated effect is proportional to the height difference.

The delay $\delta_{qp}^{t_i}$ is related to the refractivity profile by

$$\delta_{pq}^{t_i} = 10^{-6} \int_q^p N^{t_i}(z)dz. \qquad (4.8.4)$$

Hence, we write the interferometric phase difference in (4.8.3) as

$$\varphi_{pq} = \frac{4\pi}{\lambda \cos\theta_{\text{inc}}} 10^{-6} \int_q^p (N^{t_1}(z) - N^{t_2}(z))dz. \qquad (4.8.5)$$

The difference in the vertical refractivity profile, eq. (4.8.5), can cause a significant contribution in the observed phase difference between point p and q, see the results presented below.

From eq. (4.8.4) it is clear that the dimensionless refractivity, integrated over unit distance, yields the fractional delay in parts per million (ppm). For example, $N = 300$ corresponds with a fractional delay of 0.3 mm/m. Therefore, we can regard integrated refractivity values as cumulative delay. The two curves sketched in fig. 4.36B indicate the cumulative delay at acquisitions t_1 and t_2. The delay differences between points p and q are $(\delta_q^{t_1} - \delta_p^{t_1})$ and $(\delta_q^{t_2} - \delta_p^{t_2})$, respectively. In the interferogram the phase difference, expressed in delay D_{pq}, will be

$$\begin{aligned} D_{pq} &= (\delta_q^{t_1} - \delta_p^{t_1}) - (\delta_q^{t_2} - \delta_p^{t_2}) \\ &= (\delta_q^{t_1} - \delta_q^{t_2}) - (\delta_p^{t_1} - \delta_p^{t_2}) \qquad (4.8.6) \\ &= (l_2 - l_1). \end{aligned}$$

Hence, to be able to correct for the effect of vertical stratification, it is necessary to know the difference in cumulative delay between t_1 and t_2 at point p and point q.

Results and discussion

Daily radiosonde launches at De Bilt, the Netherlands, measured height, pressure, temperature, and relative humidity during their ascent. Using the Clausius-Clapeyron equation the saturation water vapor pressure corresponding to the temperature level is determined, which enables the conversion from relative humidity

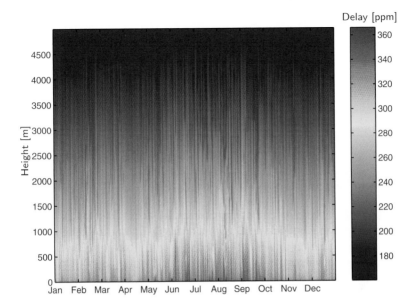

Fig. 4.37. Refractivity or fractional delay in ppm for 365 radiosondes acquired at noon in 1998.

to partial water vapor pressure (Stull, 1995). Using equations presented in (Smith and Weintraub, 1953) we calculated the refractivity for every height level. For all sondes acquired at noon during 1998, the results are shown in fig 4.37. Refractivity values are presented as fractional delay in ppm. Apart from seasonal variation it can be observed that the profiles show a ragged behavior, indicating significant variation in refractivity, especially at lower altitudes. Figure 4.38A shows the rms of this variation as a function of height. After interpolation of the refractivity measurements to a regular height interval we can perform the integration in eq. (4.8.5) for every combination of two profiles during the year. Simulating different ERS-1/2 repeat intervals (1, 3 35, 70, 140 days), statistics of the contribution of atmospheric delay due to vertical stratification can be obtained for all height intervals and time intervals. Figure 4.38B shows the standard deviation of the zenith delay difference between a point with zero altitude and points with altitudes up to 5 km. The size of the standard deviation increases for longer time intervals, indicating that 1-day intervals have better characteristics than intervals spanning seasonal changes.

With a precision better than 2–3 mm we find the following empirical model for the standard deviation of the interferometric phase due to differential tropospheric stratification:

$$\sigma_\varphi = \frac{4\pi}{\lambda \cos\theta_{\mathrm{inc}}} (33.7 + 0.08\Delta t)10^{-3} \sin\frac{h\pi}{2h_s}, \quad 1 \le \Delta t \le 182, \quad 0 \le h \le h_s \quad (4.8.7)$$

where Δt is the time interval in days, h represents height in meters, and $h_s = 5000$ m is a scale height. Above this height the variability of the refractivity is considered

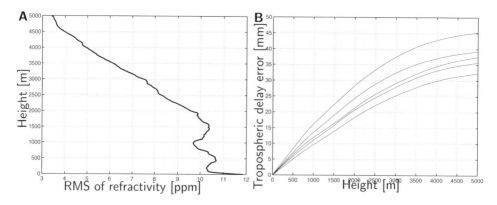

Fig. 4.38. (A) RMS of the refractivity for 365 sondes acquired in De Bilt at noon during 1998, expressed in fractional delay. **(B)** Standard deviation of the observed zenith delay due to vertical stratification as a function of the height interval for the 12 h sonde. The lines represent the 1σ values for 1, 3, 35, 70, and 140 days, counted from the lower to the upper curve.

negligible. For the accuracy of a height difference one can simply use

$$\sigma_\varphi(\Delta h) = \sigma_\varphi(h_2) - \sigma_\varphi(h_1). \tag{4.8.8}$$

Assuming a Gaussian distribution, these results imply that approximately 33% of the interferometric combinations exhibit effects more severe than expressed in eq. (4.8.7). For example, for a time interval of 175 days and a height interval of 2 km, 33% of the interferograms will have more than one phase cycle error due to vertical tropospheric stratification assuming ERS conditions. For a 100 m perpendicular baseline, this translates to a height error of 100 m or worse. For a 1-day interval and 2 km height interval it yields a height error of 76 m or worse.

Possibilities for correction

Since the size of the error described in the previous section is considerable, it is necessary to investigate methodologies to correct for these errors. Three categories of possibilities are investigated: vertical profile measurements, integrated refractivity measurements, and surface measurements in connection with a model. It is obvious that vertical refractivity profiles, acquired at the interferogram location during the SAR acquisitions are the best option. In this case it is possible to insert the refractivity values in eq. (4.8.5) and to determine the interferometric phase difference for every pair of pixels. The second option, integrated refractivity or delay measurements can result from, e.g., GPS observations. The problem with this type of observation is that, in order to determine the integrated quantities in eqs. (4.8.4) and (4.8.5), receivers are necessary in point p as well as in point q. Therefore, to determine the interferometric phase component due to vertical stratification it is necessary to have such a device at many (theoretically all) elevation levels, which is impractical. The third option, surface observations, has been studied first by

Delacourt et al. (1998) on interferometric data of Mt. Etna and assumes that surface observations of pressure, temperature, and relative humidity can be used to approximate the vertical refractivity profile.

The radiosonde data used in this study are ideal to test the hypothesis of retrieving a vertical refractivity profile from surface measurements since an unambiguous comparison between the error signal and the corrected signal can be performed. Therefore, there is no risk of involving other parameters such as errors in the topographic model or due to surface deformation in the results. We tested the Saastamoinen-Baby model, which decomposes the total delay in a hydrostatic component and a wet component (Saastamoinen, 1972; Baby et al., 1988). The hydrostatic component approximates the delay component based on surface pressure assuming an exponential decrease with height. The derivative of this delay versus height yields the hydrostatic component of the refractivity. It is well known that this model is very accurate. The wet component of the model is much harder to determine, since there is much more variability with height. Baby et al. (1988) applied the coarse assumption that the relative humidity at the surface remains constant over a certain height interval, above which it reduces to zero. An expression for the wet component of the delay, using two climate-dependent constants, surface relative humidity, and temperature linearly decreasing with height can be found in Baby et al. (1988). Differentiating this refractivity-height curve yields the wet component of the refractivity.

Summing the hydrostatic and the wet component of the refractivity yields the "model"-refractivity $N_m^{t_i}(z)$, which can be compared with the "true" refractivity $N_m^{t_i}(z)$ as obtained from the radiosonde. To study the feasibility to use this model to correct the atmospheric signal due to vertical stratification in the interferogram, we simulated the interferometric error signal by subtracting the refractivity profiles during two days, both for the true as well as the model refractivity, and subtracting both results:

$$N_{res}^{\Delta t}(z) = (N^{t_i}(z) - N^{t_i + \Delta t}(z)) - (N_m^{t_i}(z) - N_m^{t_i + \Delta t}(z)), \qquad (4.8.9)$$

which yields the residual refractivity $N_{res}^{\Delta t}(z)$ for that simulated interferogram. Comparison of the cumulative sum of $N_{res}^{\Delta t}(z)$ and $N^{\Delta t}(z)$, denoted $\delta_{pq,res}^{\Delta t}(z)$ and $\delta_{pq}^{\Delta t}(z)$, indicates whether the error signal due to vertical stratification is reduced or not. If for every height level

$$|\delta_{pq,res}^{\Delta t}(z)| \leq |\delta_{pq}^{\Delta t}(z)|, \qquad (4.8.10)$$

the correction improves the results. Applying this scheme to all interferometric combinations with $\Delta t = [1, 3, 35, 70]$ yielded, with increasing height, 0–50% significant improvement (correction of errors by 4 mm or better). In 0–35% of all simulations, the error signal increased after correction using the surface observations.

The unpredictable behavior of the wet component of the refractivity is the main reason for these disappointing results. In many cases, surface humidity measures do not reflect the humidity of the total profile. An obvious example of such a situation is fog, where surface relative humidity is 100%, which does not necessarily imply high humidities at higher levels. From the analyzed data, it appears that such misinterpretations are common. The fact that interferometric data reflect a linear combination of two different profiles makes it nearly impossible to model the behavior of vertical stratification based on surface observations.

Conclusions

The accuracy of DEMs obtained from repeat-pass radar interferometry is significantly influenced by atmospheric signal due to turbulent mixing and vertical stratification. An empirical relation between the size of the error due to the latter component has been derived for different height intervals and temporal baselines, using a set of radiosonde observations acquired during 1998. The analyzed dataset is restricted to one location on earth, which limits conclusions on global scales. However, since not the total refractivity but the variability of refractivity influences the signal in radar interferograms, it is expected that the results obtained here give a correct order of magnitude of the error.

Correction of the error due to vertical stratification is only possible using vertical profile measurements. Surface observations combined with a tropospheric model are in general unreliable, whereas integrated refractivity observations such as obtained using GPS are inadequate for correction.

4.9   Error propagation and joint stochastic model

In this chapter the most important error sources for repeat-pass radar interferometry have been analyzed. The joint stochastic model for a specific radar interferogram can be constructed by combining all these models in the variance-covariance matrix $\mathbf{C}_{\varphi}$, see eqs. (4.2.1) and (4.2.2). The error statistics affecting the data can be categorized in two classes: single-point statistics and multiple-point statistics. The single-point statistics describe the quality of a single observation (resolution cell) in the image, and are described by the PDF for that observation. It is assumed that the observations are not correlated. Multiple-point statistics consider possible correlations between the observed values at resolution cells with different locations, and are described using the covariance function, assuming second-order stationarity.

The phase variance σ_{φ}^2 can be (i) predicted a priori or (ii) estimated from the data. For a priori prediction, the characteristics of the interferometric geometry, the sensor, and the processing are used together with assumptions on the terrain to obtain predicted coherence values. The product of the coherence values and the number of effective looks are then converted to phase variance, using eqs. (4.2.24) and (4.2.27), see fig. 4.39. Nevertheless, this approach can be valuable only to estimate an upper bound for coherence and a lower bound for phase variance, since the influence of temporal decorrelation is often spatially variable and difficult to assess. For rocky surfaces, such as lava flows, temporal decorrelation will be minimal even over years, whereas decorrelation due to anthropogenic activity can happen overnight. Usually, knowledge of terrain vegetation in combination with the wavelength of the sensor and some experience can be used to obtain a coarse qualitative indication of the effect of temporal decorrelation.   Regarding the geometric decorrelation it is stressed that these equations hold for distributed scattering mechanisms. Corner reflector

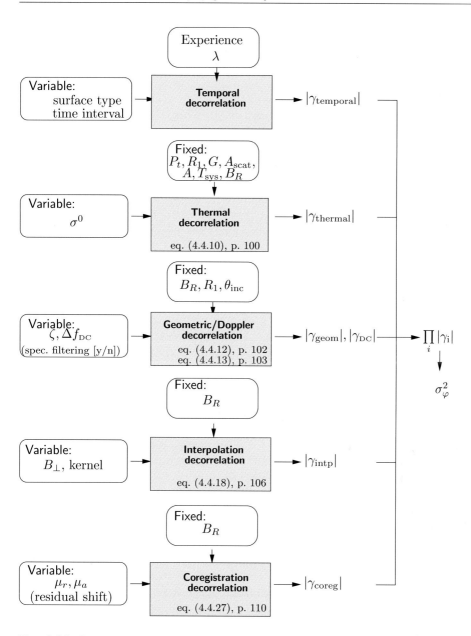

Fig. 4.39. Propagation of decorrelation errors to phase variance. Each source of decorrelation can be modeled using a set of fixed (system) parameters and a set of variable parameters, which depend on surface characteristics and interferometric geometry. The product of all coherence values is used to derive the variance of the interferometric phase $\sigma^2_{\varphi_S}$. (List of symbols at page xvi).

type point scatterers (artificial or natural) can remain coherent over a wide range of baselines. For these targets, the measure of coherence is generally unsuitable due to the ergodicity assumptions. Thermal decorrelation is dependent on the radar cross-section of the reflections. For example, an airport runway will have an extremely low radar cross section, σ^0, whereas mountains with a slope identical to the look angle will have an extremely high value for σ^0. In general, however, this is difficult to predict.

The advantage of a priori prediction of the phase variance is that some information is available even before the data are ordered. This way, suitable interferometric pairs can be chosen from the database. The disadvantage is that usually the quality of an interferogram is spatially variable and temporal decorrelation is often very difficult to predict quantitatively. The alternative is (a posteriori) phase variance estimation from the processed interferogram. In this case, the coherence is simply estimated using an estimation window, see eq. (4.3.2).

The phase variances (estimated or predicted) can be used to fill the diagonal of the variance-covariance matrix $\mathbf{C}_\varphi$ as

$$
\mathbf{C}_\varphi = \begin{bmatrix} \sigma_1^2 & & & \\ & \sigma_2^2 & & \\ & & \ddots & \\ & & & \sigma_n^2 \end{bmatrix} = \sigma^2 \mathbf{Q}_\varphi, \quad (\sigma^2 = 1), \tag{4.9.1}
$$

Spatial correlation in the interferogram data can occur due to orbit errors and atmospheric signal, as described in sections 4.6 and 4.7. Since orbit errors result in a long-wavelength trend over the entire interferogram, second-order stationarity will not be reached. The same holds for phase ramps due to atmospheric influences. Although second-order stationarity is not necessary for quality assessment using the structure function, it is necessary when using the covariance function for data adjustment and filtering. Since the covariance function is needed to fill the variance-covariance matrix, the data need to be detrended, which removes the long wavelength orbit errors.

The remaining spatial correlation, mainly due to atmospheric signal in the data, is described using the atmospheric model proposed in section 4.7.4. The initialization value P_0 can be estimated from (parts of) the data or from other sources such as GPS time series. If no information is available, a conservative value can be chosen for initialization. Using eq. (4.7.23), the positive definite covariance function is obtained.

The variance-covariance matrix $\mathbf{C}_s$ of the atmospheric delay, cf. eq. (3.1.11), can be constructed using the horizontal distance between every pair of observations and the covariance function. The covariance corresponding to this distance is inserted at the appropriate position in $\mathbf{C}_s$, yielding a symmetric variance-covariance matrix, as shown in fig. 4.40. The first column of this matrix corresponds with all combinations of observations with observation φ_1, at the upper left position in the original interferogram. Since the vectorization of the observations is performed as in eq. (3.1.1), see fig. 3.1, the covariance behavior for observations in the same column

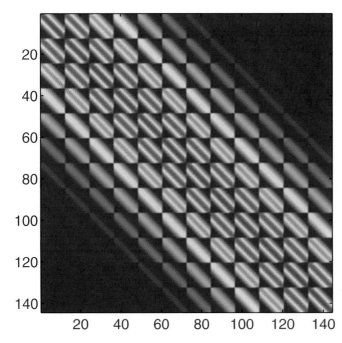

Fig. 4.40. Variance-covariance matrix $\mathbf{C_s}$, obtained for the vectorized observations φ, using the mean covariance function of fig. 4.33A ($\sigma^2 = 30$ mm^2). A 12×12 interferogram is assumed, with a posting of 0.16 km. The matrix exhibits a Block Toeplitz-Toeplitz Block (BTTB) structure.

of the interferogram is smooth, whereas the step to the next column of the interferogram is a discrete transition. Performing this procedure for all combinations with all observations, we find $\mathbf{C_s}$ as in fig. 4.40. Note that for this example, a very small interferogram of 12×12 pixels was simulated, with a posting of approximately 160 m. Notice that the number of elements in the $\mathbf{C_s}$-matrix increases quadratically with the interferogram size.

The structure of $\mathbf{C_s}$ is a Block Toeplitz matrix with Toeplitz blocks, also known as a BTTB matrix. In fig. 4.40, matrix $\mathbf{C_s}$ is a 12×12 block matrix, each block being a 12×12 Toeplitz matrix (Strohmer, 1997). For BTTB matrices, fast inversion techniques exist, see, e.g., Kailath et al. (1978); Bitmead and Anderson (1980); Strang (1986); Ammar and Gragg (1988); Brent (1991), and Strohmer (1997).

The joint stochastic model combines the influence of the single-point statistics (point variances) and multiple-point statistics (covariance functions). By adding $\mathbf{C_\varphi}$ and $\mathbf{C_s}$, we find the joint variance-covariance matrix $\mathbf{C}$:

$$\mathbf{C} \doteq \mathbf{C_\varphi} + \mathbf{C_s}. \qquad (4.9.2)$$

Simply adding the variance in $\mathbf{C_\varphi}$ to the diagonal of $\mathbf{C_s}$ is possible due to the distinct difference in the driving mechanisms of these contributions, which ensures that they are uncorrelated.

It is obvious that the inversion of the variance-covariance matrix $\mathbf{C}$ is a significant numerical challenge, especially when using an entire single-look interferogram (more than 100 million observations). However, with $\mathbf{C}_{\varphi} = \mathbf{U}\mathbf{V}^T$ we can decompose the inversion of $\mathbf{C}$ as (Golub and Van Loan, 1996)

$$\mathbf{C}^{-1} = (\mathbf{C}_{\varphi} + \mathbf{C}_{\mathbf{s}})^{-1} = \mathbf{C}_{\mathbf{s}}^{-1} - \mathbf{C}_{\mathbf{s}}^{-1}\mathbf{U}(\mathbf{I} + \mathbf{V}^T\mathbf{C}_{\mathbf{s}}^{-1}\mathbf{U})^{-1}\mathbf{V}^T\mathbf{C}_{\mathbf{s}}^{-1}. \qquad (4.9.3)$$

Hence, finding an analytic expression for the inverse of BTTB matrix $\mathbf{C}_{\mathbf{s}}$ based on the atmospheric model and the pixel locations could aid a relatively fast inversion of the joint variance-covariance matrix.

Consideration of the influence of vertical atmospheric stratification, as described in section 4.8, is only possible when initial topographic heights are known or derived from the data recursively. If such information is available, the covariance between two points in the image at a different height level can be added to $\mathbf{C}_{\mathbf{s}}$.

For completeness, we remark that we will leave the problem of the integer phase ambiguities outside the scope of this discussion. For now, we assume that phase unwrapping does not add significantly to the total error budget.

4.9.1   Height error

The propagation of an error in the interferometric phase to topographic height differences has been derived by, e.g., Rodriguez and Martin (1992); Zebker and Villasenor (1992); Zebker et al. (1994a). In these approaches the variance of the topographic height h_p at point p follows from eq. (2.4.12)

$$\sigma_{h_p}^2 = (-\frac{\lambda R_{1,p}\sin\theta_p^0}{4\pi B_{\perp,p}^0})^2\sigma_{\varphi_p}^2, \qquad (4.9.4)$$

where the phase variance is found as a function of the coherence and the multilook number using eqs. (4.2.24) and (4.2.27), assuming distributed scattering. This expression, however, suggests an absolute point positioning accuracy, a property which is not justified since the obtained height observations are relative to a reference surface. In practice, the relation with the reference surface can only be obtained by examining spatial gradients between resolution cells. This concept implies that, in order to obtain the variance of the topographic height, an arbitrary reference pixel is needed and the height is in fact a height *difference* between that reference pixel and the pixel under consideration. As a result, the spatial correlations in the data, e.g., as introduced by the atmosphere, need to be included, which yields the variance of the height difference between point p and point q:

$$\sigma_{h_p - h_q}^2 = \sigma_{h_{pq}}^2 = (-\frac{\lambda R_{1,p}\sin\theta_p^0}{4\pi B_{\perp,p}^0})^2\sigma_{\varphi_p}^2 + (-\frac{\lambda R_{1,q}\sin\theta_q^0}{4\pi B_{\perp,q}^0})^2\sigma_{\varphi_q}^2$$
$$- 2\frac{\lambda^2 R_{1,p}R_{1,q}\sin\theta_p^0\sin\theta_q^0}{(4\pi)^2 B_{\perp,p}^0 B_{\perp,q}^0}\sigma_{\varphi_{pq}}, \qquad (4.9.5)$$

where short hand notation $\sigma_{\varphi_{pq}} = \sigma_{\varphi_p\varphi_q}$ is used. In this expression, two neighboring resolution cells will have a small $\sigma_{h_{pq}}^2$ since the covariance function of the phase $\sigma_{\varphi_{pq}}$ has a high value, as the atmospheric contribution for both resolution cells are nearly identical. For two resolution cells with a large spatial separation, the variance of the height difference will be large, since the covariance function of the phase $\sigma_{\varphi_{pq}}$ has a small value, as the atmospheric contribution for both resolution cells will be almost uncorrelated.

4.10   Conclusions

This chapter gave an overview of the error sources in repeat-pass spaceborne radar interferometry. It summarized how the error sources can be described mathematically and how a stochastic model can be devised which includes these influences. It is proposed to include the spatial variation of the atmospheric signal as a covariance function in the stochastic model. In an interferogram, the functional model relates the observations (phase values for every pixel) to the unknown parameters (e.g., the parameters of a deformation model). The stochastic model appoints a variance for every single pixel and covariances between any combination of two pixels. The determination of the variance values has been discussed in sections 4.1–4.4. Covariance functions, mainly considering atmospheric signal, were modeled and reported in section 4.7–4.8. Although the formulation of the stochastic model is a step forward in the analysis and interpretation of interferometric data, there are numerical challenges to overcome, e.g., the inversion of the BTTB variance-covariance matrix.

Chapter 5

Data analysis and interpretation for deformation monitoring

This chapter focuses on the application of repeat-pass (differential) interferometry for deformation monitoring. Some case studies are presented which serve as an example for the feasibility and limitations of the technique.
key words: *Deformation, Differential interferometry.*

Deformations of the earth's surface occur at many different scales and with varying magnitudes. To be observed and quantified by repeat-pass spaceborne radar interferometry the process needs to fulfill certain conditions. In terms of scale, the dominant scales should be less than the size of the interferogram. Generally, a small discrete deformation between two neighboring resolution cells can be more easily reliably quantified than smooth spatial variations which might mimic, e.g., atmospheric signal or trends due to orbit inaccuracies. In terms of magnitudes, the deformation signal needs to be at least, say, two times as much as the noise source related to that specific scale. For example, for neighboring resolution cells, the atmospheric induced noise can be safely ignored, and the deformation signal only needs to be more than the phase noise due to decorrelation. On the other hand, for deformations spanning a large area, atmospheric noise can be significant, forcing larger deformations to take place before obtaining a satisfying signal-to-noise ratio. In practice, e.g., for continuous-type deformation processes which span large time intervals, this might imply that only interferograms with a long temporal baseline can be used. Many of the deformation studies using interferometry have focused on phenomena that fulfilled both the scale and the magnitude (SNR) requirements, and achieved spectacular, uniquely interpretable results.

In this chapter, some case studies will be presented which have three common difficulties compared with the more opportunistic studies mentioned above. First, the problem of temporal decorrelation will result in isolated patches of reasonable coherence, surrounded by areas which are completely decorrelated. Sometimes these coherent patches can be only one or a few resolution cells wide. This poses an identification problem: it needs to be decided whether a patch is reliable (coherent) or not. The conventional solution for this problem is by estimating the coherence of the image and thresholding on some predefined coherence value. This approach, however, is rather pragmatic and completely fails in the case of only one or a few reliable pixels. This is due to the fact that a relatively large coherence estimation window needs to be used to obtain an unbiased coherence estimate, and the ergodicity assumption is not satisfied for a single coherent observation.

Second, after successful identification of the patches (pixels) with a reliable and accurate phase value, the patch needs to be unwrapped. Although this might be possible within the patch, difficulties arise when separated patches need to be connected. Especially when the distance between those patches is too large it appears to be almost impossible to connect them. The only way to solve this problem is to make the problem less under-determined, that is, add more independent observations (other interferograms) or introduce a priori information on the characteristics of the problem.

Finally, the problem of connecting the identified patches is even more complicated by the influence of atmospheric signal, which might result in a bias of an entire coherent patch. The combination of this problem with the general problem of connecting the patches can be another reason to combine many interferograms, aiding interpretation and analysis.

In the following we start with a short discussion on the decomposition of the three-dimensional displacement vector in terms of the radar range-change observation. To obtain some experience with the accuracies of the deformation measurements on a pixel-by-pixel base, we performed an experiment using a set of corner reflectors, which were intentionally displaced in height during the ERS-1/2 "tandem mission." These results are presented in section 5.2. In the same area, with varying land-use due to agricultural activities, it is investigated how conventional coherence estimation windows would need to be applied to obtain a coherent signal over a time span of 3.5 years. This is reported in section 5.3. A more significant occurrence of land subsidence due to the activities of a geothermal plant in Mexico is discussed in section 5.4, and a concise presentation of the extended and complicated deformation pattern associated with the 17 August 1999 earthquake in Izmit, Turkey is given in section 5.5.

5.1   Decomposition of the displacement vector

As a radar is only capable of measuring path length differences in its line of sight or slant-range direction, deformation interferograms imply a non-uniqueness and need to be interpreted with care. A three-dimensional displacement vector $\vec{d}$ with components d_n, d_e, and d_u, in North, East, and Up direction respectively, will be projected to one slant-range component d_r in the radar line-of-sight (LOS). For a satellite orbit with heading (azimuth) α_h, see fig. 5.1A and 5.1B, we find

$$d_r = d_u \cos(\theta_{\text{inc}}) - \sin(\theta_{\text{inc}}) \left[d_n \cos(\alpha_h - 3\pi/2) + d_e \sin(\alpha_h - 3\pi/2) \right], \qquad (5.1.1)$$

where $(\alpha_h - 3\pi/2)$ corresponds with the angle to the azimuth look direction, which is perpendicular to the satellite heading, for a right-looking satellite. The incidence angle is denoted by θ_{inc}. For an incidence angle of $\theta_{\text{inc}} \approx 23°$ and a heading $\alpha_h \approx 190°$, we find a sensitivity decomposition of a slant-range deformation of

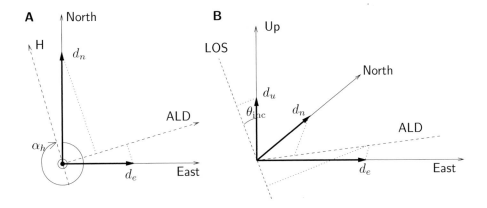

Fig. 5.1. Projection of the three components of the deformation vector $\vec{d} = (d_n, d_e, d_u)$ onto the satellite line-of-sight (LOS). **(A)** Top-view showing the North and East components projection on the azimuth look direction (ALD), which is perpendicular the the satellite's heading (H). The heading angle is indicated by α_h. **(B)** 3D sketch including the projection of the Up-component to the line-of-sight via incidence angle θ_{inc}.

$[0.92, -0.07, 0.38]$ $[d_u, d_n, d_e]^T$. Therefore, assuming only vertical deformation, one fringe corresponds with about 26 mm.

Due to the projection of the three-dimensional displacement vector to the radar line-of-sight, it is not possible to retrieve the full displacement vector. Using a combination of an ascending and a descending interferogram, two of the three components can be retrieved. For the third component, assumptions are necessary on the characteristics of the displacement. For example, for glacier flow it is often assumed that the flow is parallel to the surface, which confines the vector along the gradient of the elevation model (Mohr, 1997).

5.2 Corner reflector experiments

Parts of the northeastern part of the Netherlands (Groningen, see fig. 5.2) have been exploited since 1964 due to their large natural gas reservoirs. Due to this gas extraction, a maximum land subsidence of 23 cm has occurred between 1964 and 1998 in the center of a bowl shaped subsidence pattern (NAM, 2000). The shortest distance from the center of the subsidence bowl to a presumably stable point is approximately 23 km. This implies a maximum subsidence rate of ~7 mm/yr and a subsidence gradient of ~0.3 mm/km/yr. It is expected that the total amount of subsidence at the end of the production period will be around 38 cm in 2050 (NAM, 2000). Assuming a comparable spatial extent of the general subsidence pattern, this results in an expected maximum subsidence rate of ~3 mm/yr and a subsidence gradient of ~0.1 mm/km/yr. Extended leveling campaigns, surveyed every five years, produced a robust network of benchmarks, which is now regularly surveyed using GPS receivers as well. However, based on current experience, the GPS

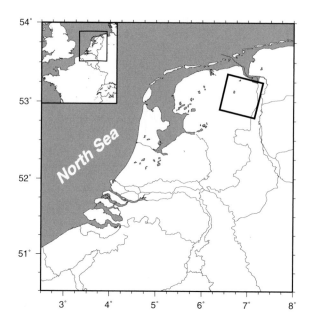

Fig. 5.2. Location of the 50×50 km SAR quarter scene over Groningen, the Netherlands. The area is relatively flat, over large areas less than a few meters of height variation.

network results are not dense and accurate enough to successfully replace leveling (NAM, 2000). Based on the presumed high spatial density of SAR acquisitions it was proposed to investigate the feasibility of radar interferometry for this specific subsidence problem.

An extensive radar interferometry experiment is performed within the bounds of the subsiding area, involving time series with a long temporal baseline, tandem acquisitions for monitoring the magnitudes of the atmospheric signal, corner reflector deployment as stable reference points, and GPS observations to investigate the possibility of estimation or correction of atmospheric signal from the data.

Since the Groningen landscape contains mainly agricultural areas it suffers from severe temporal decorrelation, which makes interferograms decorrelate over more than a couple of months. For some areas the decorrelation can be instantaneous due to anthropogenic activities. The topography of Groningen is very flat, over large areas less than a few meters of height variation.

In this section, we report on an experiment to assess the accuracy of the technique in terms of the vertical movement of the corner reflectors between two SAR acquisitions. The following section will comment on the possibilities for finding stable points for accurate phase estimates. The influence of the atmosphere over this region has been discussed in section 4.7.

5.2.1   Introduction

During the Groningen Interferometric SAR Experiment (GISARE), a test was performed in detecting vertical corner reflector movement in between two ERS tandem passes of 16 and 17 March 1996, similar to an experiment performed by Hartl et al. (1993), Hartl and Xia (1993), and Prati et al. (1993). Eight corner reflectors were installed in the region. To allow accurate monitoring of the heights, all are located within less than 100 m distance of a fixed benchmark of the vertical reference datum (NAP). Of the eight installed corner reflectors, two were moved upward using a screw mechanism. In this section it is analyzed if this movement is detectable in the interferogram, how accurate a height change of a corner reflector can be monitored using differential SAR interferometry, and which parameters are of influence for the statistics of the experiment.

Theory

Following the notation conventions of section 2.4, the received phase φ_{1p}, between satellite 1 and resolution cell P is the summation of two independent components, a *scattering* part $\varphi_{\text{scat},1p}$ and a *path length* part $\varphi_{\text{path},1p}$. Assuming that the scattering part remains relatively unchanged between the two acquisitions, phase differences $\varphi_p = \varphi_{2p} - \varphi_{1p}$ are caused only by path length differences, hence $\varphi_1 = \varphi_{\text{path},2p} - \varphi_{\text{path},1p}$. Path length differences arise due to (i) the oblique viewing geometry in combination with the baseline, (ii) topographic height variation within the scene, (iii) local surface deformation, (iv) spatial atmospheric delay differences, and (v) spatial differences in the moisture or, more general, dielectric constant.

Phase ramps caused by the baseline geometry and earth curvature are reduced using precise orbit information (Scharroo and Visser, 1998). For spatial scales associated with the corner reflector, residual phase ramps can be removed safely. The topographic effect can be estimated using the height ambiguity, see eq. (2.4.14), p. 37. For the interferogram under consideration, the perpendicular baseline $B_\perp = -18$ m, which results in a height ambiguity of approximately 516 m. Since the maximum height differences in the GISARE test area do not exceed 20 meters, topographic phase differences of 1.4 degrees can be expected. These differences are negligible with respect to the thermal noise in the phase measurements. Moreover, the height of the corner reflector with respect to its surroundings is ~ 1 m, which results in a topographic phase difference of $0.7°$, or an error in the vertical deformation measurement of 6×10^{-3} mm. This effect is beyond the measurement accuracy and can be ignored.

Phase differences due to lateral atmospheric delay differences within the scene will not effect the difference between two neighboring resolution cells, see section 4.7. Moreover, examining the total interferogram of this scene, see fig. 5.3, it can be observed that atmospheric inhomogeneities are extremely limited, apart from some slight atmospheric gravity waves. As a result, the only phase variations over small spatial scales around the corner reflectors need to be due to relative deformation between one resolution cell and another, although the unknown behavior of the dielectric constant might have some limited influence as well.

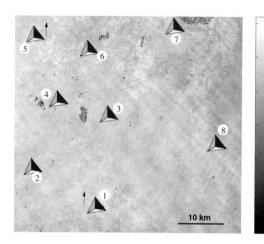

Fig. 5.3. Interferogram of the GISARE test area, acquired by ERS-1 and ERS-2 at 16 and 17 March 1996, with the locations of the eight corner reflectors. Reflectors 1 and 5 have been intentionally elevated by ~1 cm between the two acquisitions.

5.2.2 Ground truth

At March 15, all eight corner reflectors were leveled with respect to a fixed reference point, usually a concrete GPS or NAP marker in the vicinity of the reflector. In table 5.1 this leveling is denoted with **L1**. ERS-1 acquired its radar data on March 16, at 11:31 LT. After this pass, two selected reflectors, 1 and 5, were raised approximately 1 cm. To determine this vertical movement as accurately as possible, before and after the movement the reflector was leveled with respect to a local, temporary benchmark. These surveys are denoted by **L2** and **L3**, respectively. Exactly 24 hours after the pass of ERS-1, ERS-2 acquired its data of this scene from nearly the same orbit as ERS-1. Finally, one day later the first leveling was repeated with respect to the GPS or NAP marker, here denoted by **L4**. In table 5.1 the chronological sequence of events for reflector 1 and 5 is listed. As the corner height adjustment was applied shortly after the pass of ERS-1, the result of leveling **L2** and **L3** can be considered very reliable. However, the time interval between the corner adjustment and the pass of ERS-2 is approximately 24 hours. Therefore, we need to test if

Table 5.1. Sequence of events for the corner reflector experiment

Date	Time (LT)	Code	Event
15–3–96	9:00–13:00	**L1**	All corner reflectors leveled
16–3–96	11:31		SAR acquisition ERS-1
16–3–96	11:45 & 13:30	**L2**	Corner 1&5 leveled (local reference)
16–3–96	11:55 & 13:35		Corner 1&5 moved upward
16–3–96	12:00 & 13:40	**L3**	Corner 1&5 leveled (local reference)
17–3–96	11:31		SAR acquisition ERS-2
18–3–96	9:00–13:00	**L4**	All corner reflectors leveled

Table 5.2. Results of the corner movement leveling surveys

Corner#	(L3-L2)		(L4-L1)		Joint	
	Δh_{lev}	$\sigma_{\Delta h_{\text{lev}}}$	Δh_{lev}	$\sigma_{\Delta h_{\text{lev}}}$	Δh_{lev}	$\sigma_{\Delta h_{\text{lev}}}$
1	1.075	0.020	1.044	0.028	1.060	0.017
5	1.035	0.035	1.102	0.015	1.069	0.019

between these times any additional corner movement took place, perhaps due to an unknown disturbance. To study this possibility we examine the difference

$$(\mathbf{L4} - \mathbf{L1}) - (\mathbf{L3} - \mathbf{L2}). \tag{5.2.1}$$

The left-hand part is referred to as the long epoch, the right-hand part as the short epoch.

The leveling surveys were performed using a closed loop, which yields two height differences h_1 and h_2 between the reference point and the corner reflector. This procedure gives a conventional mean μ and standard deviation σ defined by

$$\sigma = \sqrt{\frac{(h_1 - \mu)^2 + (h_2 - \mu)^2}{N - 1}} = \sqrt{(h_1 - \mu)^2 + (h_2 - \mu)^2} \tag{5.2.2}$$

The mean height difference Δh, between the *before* and *after* measurements is now defined using the difference of the means

$$\Delta h_{\text{lev,L3-L2}} = \mu_{\text{L3}} - \mu_{\text{L2}}, \tag{5.2.3}$$

and standard deviation

$$\sigma_{\Delta h_{\text{lev,L3-L2}}} = \sqrt{\sigma_{\text{L3}}^2 + \sigma_{\text{L2}}^2}. \tag{5.2.4}$$

The results of the measurements for reflectors 1 and 5 are presented in table 5.2, and graphically in fig. 5.4. Based on the curves for **L3-L2** (the short epoch) and **L4-L1** (the long epoch), indicated by the dashed lines in fig. 5.4, two conclusions are possible: (i) Between **L3** and **L4**, which is two days later, reflector 1 has moved downward 0.31 mm and reflector 5 has moved upward 0.67 mm, or, (ii) there has not been any additional movement, apart from the movement we induced ourselves. In the latter case, the differences between the mean values of the two survey combinations is caused by measuring noise. Since the differences are extremely small, it is assumed that the latter possibility is correct. In this scenario the height difference measurements of the short and the long epoch are considered to be uncorrelated, and standard error propagation yields the average reflector movement with associated standard deviations. These values are shown in table 5.2 and indicated by the solid line in fig. 5.4.

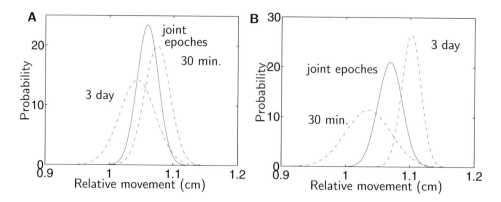

Fig. 5.4. Gaussian PDF's for reflectors 1 and 5. The dashed lines correspond with the **L3-L2** (30 min) surveys and the **L4-L1** (3 day) leveling surveys. The solid line represents the least-squares result of the joint model, assuming no additional motion between the short and the long epoch.

5.2.3   InSAR deformation measurements

To obtain an estimate of the phase difference between the corner reflector and the surrounding area, the phase of the corner reflector is determined at sub-pixel accuracy using complex interpolation. The phase of the surrounding area is called the *reference* phase. The phase difference between corner reflector and reference phase can be mapped to a change in the elevation of the reflector. Although only two of the reflectors are manually adjusted in elevation, any natural movement, which might occur between the two acquisitions, might be detected by the leveling surveys as well. Therefore, we analyze the observed deformations of all eight reflectors.

The corner reflectors can be manually detected in the interferogram amplitude using a topographic map as a reference. An area of 256×256 complex pixels around each reflector is stored for further evaluation. Using the fringe frequency parameters of the interferogram, a flat-earth phase is removed in each of these areas. In the complex image, an area of 32×32 pixels centered around the corner reflector is interpolated to $1/16$ of a pixel using spectral zero-padding at the spectral minimum. The interferometric phase of the corner reflector is calculated at the maximum of the peak power in this interpolated image. Besides the peak power and peak phase also the peak location and 3dB width are stored.

The reference area consists of four sub-distributions of 80×80 m within an area of 1×1 km centered at the corner reflectors position, see fig. 5.5. To avoid contamination by the sidelobes of the corner reflector, a strip of 11 pixels in range and 50 lines in azimuth (approximately 200 meters wide) is excluded from reference phase estimation. Due to these two perpendicular strips, four rectangular areas of approximately 400×400 meters (20×103 pixels) are created around the corner reflector. Within each of these areas, the sub-distributions with minimal phase variance are selected. The four sub-distributions contain 80 phase measurements (4 pixels in

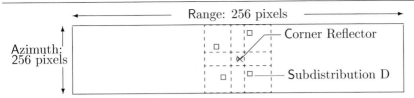

Fig. 5.5. Configuration of analyzed area, reference area (sub-distribution A, B, C, and D) and corner reflector .

range and 20 lines in azimuth), corresponding with an area of 80×80 meters. The mean phase value of the four sub-distributions is chosen to be representative for the background of the corner reflector. The standard deviation of the reference phase is derived using plain error propagation.

Phase statistics of the corner reflector

Since the complex value of a SAR pixel is the coherent superposition of all scatterers in the corresponding resolution cell, the interpolated value of the interferometric phase of the corner reflector is mainly disturbed by the influence of the surrounding scatterers within the resolution cell, as shown in fig. 5.6. Using an estimation of the power of the background reflections P_{bg}, the radius of the corresponding error circle is determined by

$$\sigma_\phi \approx \sqrt{\frac{E\{P_{\mathrm{bg}}\}}{2 P_{\mathrm{cr}}}}. \tag{5.2.5}$$

The expectation of the background power P_{bg} is determined in the master as well as in the slave image, using a mean value of four areas of 2×2 pixels around each corner reflector. The corner reflector power P_{cr} is also determined in master and slave image using complex interpolation, see table 5.3. Average values for the standard deviation of the phase are used in the evaluation of the corner reflector movement.

5.2.4   Results

The results of the experiment are listed in table 5.4. The first part of the table shows the observed phase values for the background φ_{ref} and the corner reflectors φ_{cr}. It is

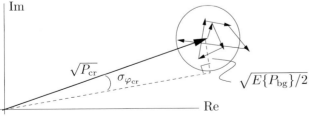

Fig. 5.6. The complex vector (phasor) of the corner reflector and the surrounding scatterers within the same resolution cell and their relation to the standard deviation of the phase of the corner reflector.

Table 5.3. Phase standard deviation corner reflector from intensity observations.

Corner	Master Power (dB) CR	BG	$\sigma_{\varphi_{cr}}$ (deg)	Slave Power (dB) CR	BG	$\sigma_{\varphi_{cr}}$ (deg)
1	66.90	48.49	0.298	59.84	41.85	0.459
2	66.63	47.37	0.289	60.62	42.33	0.431
3	64.66	44.22	0.302	58.74	39.27	0.449
4	64.93	45.91	0.323	58.56	40.60	0.495
5	64.91	46.15	0.328	58.86	40.75	0.482
6	66.37	49.68	0.340	59.91	43.69	0.506
7	65.95	47.16	0.308	59.59	42.14	0.480
8	64.85	46.97	0.346	58.98	41.87	0.507

CR, corner reflector; BG, background; σ_φ, phase standard deviation, derived from power CR and BG measured in master and slave, see eq. (5.2.5).

assumed that the phase difference $\Delta\varphi$ between these values is less than π, although the 2π-ambiguity is important to be considered in the interpretation of the results. Using the local incidence angle θ_{loc} and the wavelength λ the range displacement $\Delta\rho$ and vertical displacement Δh is obtained.

The second part of the table shows the accuracies of the interferometric observations. The standard deviations of the reference phase value $\sigma_{\varphi_{ref}}$ results from the averaging of all reference phase observations. The standard deviation for the reflector phase $\sigma_{\varphi_{cr}}$ was listed in table 5.3. Using the same mapping with the local incidence angles, these values are mapped to the vertical height values $\sigma_{h_{ref}}$ and $\sigma_{h_{cr}}$, and consequently to the standard deviation of the vertical deformation of the corner reflector $\sigma_{\Delta h_{cr}}$.

The third part of the table lists the height changes Δh_{lev} observed by the leveling surveys, and their associated standard deviations $\sigma_{\Delta h_{lev}}$. Finally, the last part of the table shows the differences between the leveling and the InSAR results.

Table 5.4. Results of corner movement evaluation. Number of reference phase observations: 4×80. Phase differences $-\pi < \Delta\varphi \leq \pi$ have been used, acknowledging the possibility of a $\Delta\varphi \pm 2\pi$ solution. Slant-range displacements are converted to height changes, using the local incidence angle θ_{loc}.

		1	2	3	4	Corner # 5	6	7	8
					Interferometry				
$\varphi_{ref}^{\ddagger}$	[deg]	96.0	99.8	5.0	−119.5	119.4	118.5	44.3	−94.6
φ_{cr}	[deg]	−126.9	88.8	−7.1	−149.5	22.9	107.3	9.0	−133.6
$\Delta\varphi$	[deg]	137.0	−11.0	−12.1	−30.0	−96.5	−11.2	−35.4	−39.0
θ_{loc}	[deg]	21.9	22.6	21.9	22.3	22.8	22.1	20.9	20.3
$\Delta\rho$	[cm]	1.077	−0.087	−0.095	−0.236	−0.759	−0.088	−0.278	−0.307
Δh	[cm]	1.161	−0.094	−0.103	−0.255	−0.823	−0.095	−0.297	−0.327
					Accuracies				
$\sigma_{\varphi_{ref}}$	[deg]	10.6	14.4	11.8	15.4	27.0	11.8	13.2	13.7
$\sigma_{h_{ref}}$	[cm]	0.076	0.103	0.085	0.111	0.194	0.085	0.096	0.100
$\sigma_{h_{cr}}$	[cm]	0.003	0.003	0.003	0.003	0.003	0.003	0.003	0.003
$\sigma_{\Delta h_{cr}}$	[cm]	0.076	0.103	0.085	0.111	0.194	0.085	0.096	0.100
					Leveling				
Δh_{lev}	[cm]	1.060	−0.090	−0.130	−0.025	1.069	0.083	0.225	0.070
$\sigma_{\Delta h_{lev}}$	[cm]	0.017	—	0.053	0.038	0.019	0.024	0.051	0.014
					Comparison				
$\Delta h_{lev} - \Delta h_{\varphi}$	[cm]	0.101	−0.004	0.027	−0.230	1.177	−0.178	−0.522	−0.397

φ_{ref} Mean of reference phase. φ_{cr} Corner phase. $\sigma_{\varphi_{ref}}$ Standard deviation reference phase. $\sigma_{\Delta h_{ref}}$ Standard Deviation height (reference area). $\sigma_{\Delta h_{cr}}$ Standard Deviation height (Corner reflector). $\sigma_{\Delta h}$ Standard Deviation height change. $\sigma_{\Delta h_{lev}}$ Lev. Standard Deviation height change. θ_{loc} Local Inc. Angle. $\Delta h_{lev} - \Delta h_{\varphi}$ Difference InSAR-Leveling.

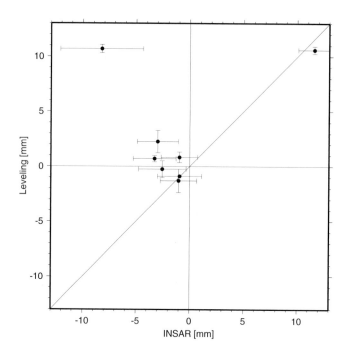

Fig. 5.7. Scatterplot of results corner movement as observed by InSAR and leveling. Error bars reflect the $\pm 2\sigma$ interval.

5.2.5   Discussion

The results listed in table 5.4 are visualized in fig. 5.7. The dots represent the movement as observed by InSAR and the leveling results. Error bars delimit the $\pm 2\sigma$ interval. It can be concluded that the movement of five of the eight reflectors is correct within the 95% probability curve. Within this group, the movement of reflector 1 is captured almost perfectly by the SAR analysis. For two of the reflectors the difference with the leveling results is between 3 and 5 mm, whereas $2\sigma_{\Delta h_{cr}} \approx 2$ mm, and for one of the two intentionally elevated ones the difference is almost 12 mm (with $2\sigma_{\Delta h_{cr}} \approx 4$ mm. It is remarkable that the latter observation has a $\sigma_{\varphi_{ref}}$ which is twice the accuracy of the other corner reflectors. This is an indication that changes in the reference area (the earth's surface surrounding the reflector) between the two SAR acquisitions might be the cause of this bias. Indeed, the surface around reflector 5 was a bare and wet ploughed field. It can be suggested that changes in moisture (dielectric constant) of this field have resulted in a coherent phase change during the one-day time interval. That such changes can occur very locally is shown, e.g., in fig. 5.8, where local coherent phase changes related to agricultural fields are observed. The temperatures during both acquisitions were just above the freezing level, $1°$–$3°$C. Therefore, it is very likely that during the night between the acquisitions the surface of the fields were frozen. If

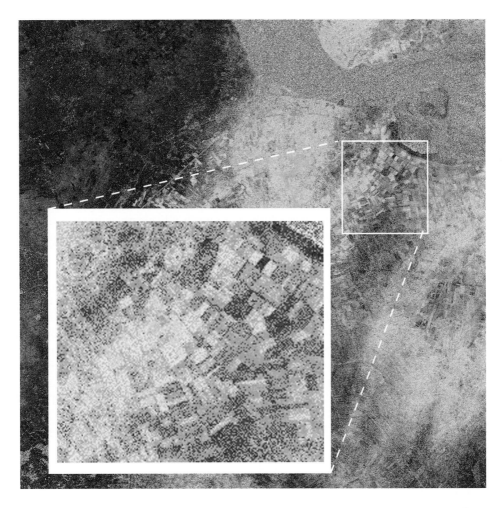

Fig. 5.8. Local differential movements of agricultural fields. Phase values ranging from $-\pi$ to π are related to the gray scale. Especially at the "wraps," where small discrete phase variations are most visible, differential movement of $20°–40°$ can be observed.

the ploughed field surrounding reflector 5 reacts differently to freezing and thawing compared with the smooth grass land surrounding reflector 1, it can be expected that phase changes relate either to the deformation (swelling/deflation) of the field, or to different moisture content.

5.3 Groningen coherence estimation

As discussed in the introduction of this chapter, the identification of coherent pixels forms one of the most severe problems in areas that suffer from significant temporal decorrelation. In fig. 5.9A it is shown how the coherence over the area of Groningen has decreased over a time interval of 3.5 years. From this coherence image it may seem that no significant information can be retrieved in this interval, since the overall tone of the image has a constant and low coherence value. On the other hand one might conclude that this average coherence value ($\sim$0.3) is still high enough to be interpretable in terms of a reliable interferometric phase. After closer analysis of the data, see fig. 5.9B, isolated points with a relatively high coherence estimation can be detected, especially over the urban areas.

As discussed in section 4.3, the discrimination between areas of different coherence is mainly influenced by the number of independent samples in the coherence estimation window. If this number is too small, low coherence areas become biased towards a higher coherence estimation. Figure 5.9A, for which an estimation window of 2×11 samples was used, shows this effect. Although almost all coherence in this (mostly agricultural) area has disappeared, a bias of $\sim$0.3 is apparent in the data. Therefore discrimination between different coherence levels becomes difficult.

In fig. 5.9C it is shown how an increase of the coherence estimation window from 2×11 to 20×110 samples yields an improved discrimination between areas of low coherence, which is due to the corresponding drop in coherence bias. In the figure, this is obvious from the red tone of the decorrelated areas. Note that although a continuous color wheel proved to be the best visualization for these data, coherence levels above 0.95, which are also colored red, can be neglected with this estimation window. On the other hand, areas with a relatively high coherence can be detected as the yellow and green spots. Obviously, the larger estimation window implies a decrease in the absolute coherence levels with respect to those of small window sizes. Therefore, the range of coherence values in the image reduces, whereas the reliability of the estimates increases.

Due to the better possibilities to discriminate between coherence levels, thresholding at a chosen low level becomes possible when applying large coherence estimation windows. Figure 5.9D shows how the interferometric phase values corresponding with pixels with a coherence higher than 0.1 still reveal interpretable phase information. In this masked interferogram, a multilook window equal to the coherence estimation window (20×110) was applied. Although disturbing influences of the satellite orbits and atmospheric heterogeneities still form a major restriction for an unambiguous interpretation, this method of phase analysis reveals promising prospects for the study of interferograms spanning long time intervals and severely decorrelated interferograms. The coherence preserving areas are in this case mainly of anthropogenic origin.

A significant limitation of using large coherence estimation windows is the sensitivity for non-coherent pixels. When decorrelated pixels have a significant magnitude, perhaps due to speckle effects, they will contaminate the phase values of the coherent pixels. This problem can be circumvented by using many coregistered SAR images and forming a temporal instead of a spatial coherence estimation window, as demonstrated by Ferretti et al. (2000).

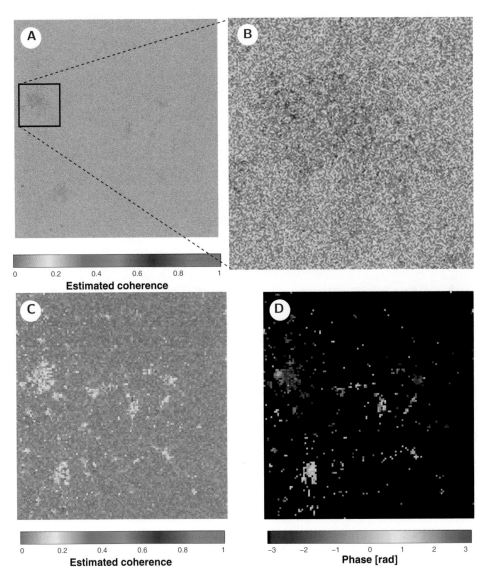

Fig. 5.9. (A) Coherence image corresponding to a 50×50 km interferogram of the area around Groningen, the Netherlands, after a 3.5 year time interval, using a coherence estimation window of 2×11 samples. **(B)** Zoom over the city of Groningen, showing isolated high-coherent points. **(C)** Coherence image using 20×110 samples per pixel. **(D)** Interferogram phase, masked below a coherence level of 0.1 with a multilook window of 20×110 pixels.

5.4 Cerro Prieto geothermal field

The Cerro Prieto Geothermal Field (CPGF) is located 30 km south-east of Mexicali, Baja California, Mexico, see fig. 5.10. Centered between the right-lateral, strike-slip

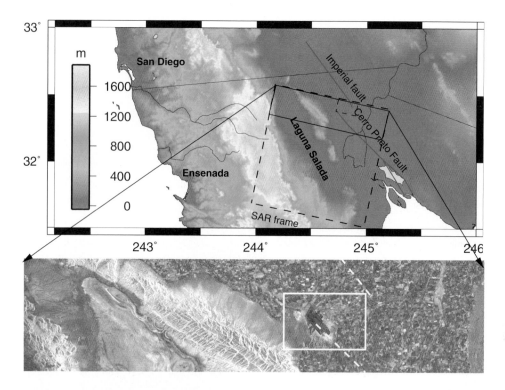

Fig. 5.10. Location of the SAR acquisitions and the Cerro Prieto Geothermal Field (CPGF). The lower image is the amplitude of the upper 24 km of the interferogram. The study area is indicated by the rectangular frame, while the locations of the Cerro Prieto fault and the Imperial fault are indicated by the dashed lines.

Imperial fault and Cerro Prieto fault, the area is a pull-apart basin in the southern part of the Salton Trough. The extraction of water and steam from the geothermal system, at a depth of 1500–3000 m causes a maximum subsidence rate in the order of 8 cm/yr (Fabriol and Glowacka, 1997). Interferometric SAR observations, obtained between May 1995 and August 1997, are used to evaluate the potential of the technique for monitoring the subsidence. The spatial characteristics of the deformation are analyzed using linear combinations of differential interferograms. Information on deformation, production, and recharging by fluid injection is used for a coarse geophysical interpretation of the deformation mechanisms.

5.4.1 Background

The Salton Trough and Gulf of California are the results of tectonic activity that created a series of spreading centers and transform faults, linking the East-Pacific Rise with the San Andreas fault system. (Pacific–North-American plate boundary). As a result, at Cerro Prieto, the earth's crust is being pulled open ("pull-apart basin") an thinning. An upwelling of magma from the asthenosphere, combined with intrusions in the form of dikes and sills is the main heat source for the reservoir. This is causing groundwater in sandstones and gray shales from the Colorado River delta to be heated at depth. Formation temperatures at 2500 m are estimated at 350°C. One of the most notable manifestations of the thermal activity in the area is the Cerro Prieto volcano, with an elevation of about 210 m. Since the Mexicali Valley floor was a part of the Colorado river delta, it is a very flat area with agricultural activity. The area west of CPGF is dry and shows geothermal surface manifestations such as mud volcanoes and fumaroles. The climate in the area is extreme: temperatures vary from -2°C in winter to $+47$°C in summer. Rainfall is very low, with an annual mean of 80 mm.

The CPGF, operated by the Mexican *Comisión Federal de Electricidad* (CFE), is one of the largest producing geothermal fields in the world. It is exploited since 1973 and currently involves three power plants with more than 100 operating wells, covering 12 km^2. Since 1989, the natural fluid recharge of the reservoir has been supplement by (artificial) injection in 7 wells. In October 1996, the three Cerro Prieto power plants, see fig. 5.11B, were generating about 560 MW of energy based on a total production (steam and water) of approximately 12.000 T/h (tons per hour or m^3/h). During that month, some 2.300 T/h of cooled geothermal brine were being reinjected into the reservoir (19% of the fluids produced). Over the years, this substantial net extraction of mass has resulted in systematic land subsidence.

The Cerro Prieto wells produce a mixture of liquid, vapor, and gases. As the fluid from the reservoir rises to the surface it *flashes* in the wellbore. The steam and gases are sent to the power plant turbines. The separated brine is redirected to the evaporation pond, which is clearly visible in the radar images. Part of the brine is reinjected in the injection wells, see fig. 5.11B. The subdivision in the three well groups CP-I, CP-II, and CP-III is caused by the different reservoir and well characteristics. Every well group has a separate power plant. The average distance between the wells is about 400 m.

5.4.2 Leveling results

Ground subsidence has already been observed from first-order leveling between 1977 and 1979. However, due to a strong earthquake in the region in June 1980, it was not clear how much of the subsidence in these and later surveys was due to fluid extraction. In general, leveling and gravimetry show a subsidence rate in the order of 6 cm/yr, averaged over the last 35 years, with a maximum rate of 10 cm/yr.

In order to validate the results of the time period covered by the SAR interferograms, results from leveling campaigns performed by CFE in 1994, 1996, and 1997 have been used. The leveling network of 1996 covered only a limited part of the 1994 and 1997 surveys. The interpolated results for the 1994–1997 epoch are shown in fig. 5.12. The subsidence is imaged using a 10 cm color interval, to provide more sensitivity for gradient differences. The general subsidence bowl is located at the center of CPGF and shows a subsidence rate of $\sim$30 cm over the three years ($\sim$10 cm/yr). The iso-line corresponding with zero-subsidence is indicated by the striped-dotted

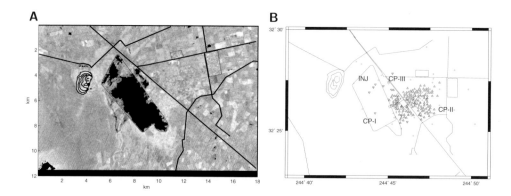

Fig. 5.11. (A) Radar intensity image (size 12×18 km) of the Cerro Prieto Geothermal Field area. The Cerro Prieto volcano is located to the northeast of the evaporation pond, which is masked in this image. The northwest-southeast diagonal line is a railroad, other lines indicate major roads and two villages. The well groups are located to the east of the evaporation pond, see (B). **(B)** Distribution of the three well groups (CP-I, CP-II, CP-III) for extraction and injection.

line in the lower left of the area, at a distance of approximately 9 km of the main subsidence bowl. The iso-line of -25 cm subsidence corresponds approximately with the outer margins of the three well groups.

A small pit with a maximum subsidence of 55.7 cm ($\sim$18 cm/yr) and a diameter of 1.5 km is located between the railroad and the evaporation pond. Although a limited number of benchmarks contributes to this pit, the interpolation seems to be reasonable.

5.4.3 Production and injection

To study the geophysical driving mechanisms of the deformation and to interpret the interferometric data, production and injection data are provided by CFE. Production data are given for the total flowrate (steam and water) as an average value per day per well in tons per hour. The injection data are given in average values per day per injection well in tons per second. Since 1985, the production has been nearly constant. The production figures between April 1994 and March 1998 show a constant total fluid mixture production of $\sim$12 kT/h. Injection of brine started in 1989, using injection wells located in the north of the evaporation pond. Injection is equally divided over all wells, and is nearly constant as well, approximately 2.3 kT/h.

Based on this information, and the knowledge of compaction-type subsidence (hence, the surface subsidence is probably not seismic but largely continuous) it can be expected that the interferometric combinations reflect subsidence rates linearly proportional to the temporal baselines.

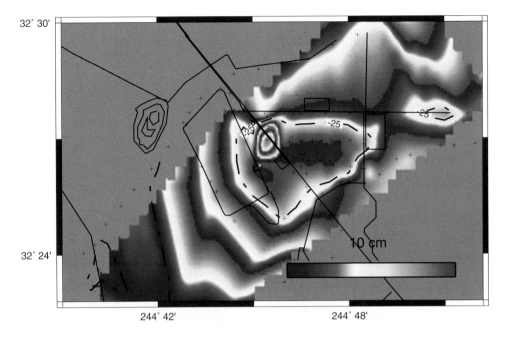

Fig. 5.12. Leveling results 1994–1997, using Delaunay triangulation and linear interpolation. The colors show a 10 cm subsidence interval. Maximum subsidence of 55.7 cm was observed in a localized sub-bowl between the railroad and the evaporation pond. The contour indicating 25 cm subsidence is indicated by the dashed line. Crosses indicate leveling benchmarks.

5.4.4 Interferometric results

Figure 2.12 (p. 44) shows all available SAR acquisitions over the Cerro Prieto Geothermal Field from April 1995 to August 1997. To reduce the effect of geometric decorrelation in the interferograms, ten acquisitions within a baseline band less than 200 m are selected. Two ERS tandem (1 day interval) acquisitions are available, expected to show no deformation signal. The tandem pair with the longest baseline ($B_\perp = 320$ m) is used to obtain an interferogram with significant topographic information, while introducing a minimal amount of atmospheric signal in the differential interferograms. Three-pass differential interferometry is used, scaling the topographic interferogram and subtracting it from every deformation pair. The data are filtered using the Goldstein and Werner (1998) filter to increase signal-to-noise ratio.

Regarding the subsidence at the Cerro Prieto geothermal area, we expect that mainly the vertical displacement is significant. Using eq. (5.1.1), one fringe of slant-range deformation will correspond with approximately 30 mm vertical deformation.

Eight deformation interferograms are processed differentially, using the acquisitions

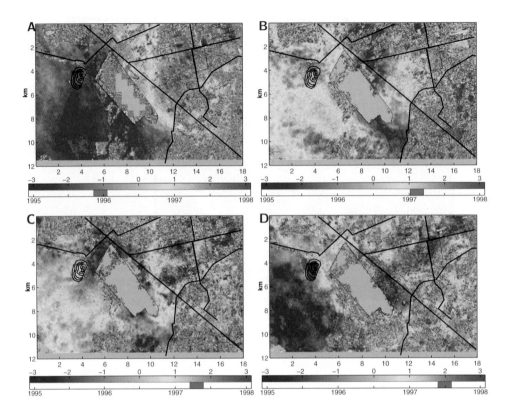

Fig. 5.13. Differential SAR interferograms with a temporal baseline of 70 days, between November 1995 and August 1997. Colors represent the interferometric phase in rad. The time interval is indicated in the colorbar. **(A)** Differential interferogram 12/11/95-21/01/96. **(B)** Differential interferogram 05/01/97-16/03/97. **(C)** Differential interferogram 16/03/97-25/05/97. **(D)** Differential interferogram 25/05/97-03/08/97. The deformation pattern is clearly visible centered over the CPGF area.

indicated by the solid dots in fig. 2.12. All interferograms are processed onto the same reference grid, to enable different combinations and stacking analyses.

Figures 5.13 and 5.14 show all differential interferograms, using the tandem pair as reference. The color bar shows the interferometric phase in radians, and the time difference between the two acquisitions is indicated by the green bar. The contours of the evaporation pond and the Cerro Prieto volcano, as well as the railroad and major roads are superposed on the images for orientation.

Figure 5.13 shows the four 70 day interval differential interferograms. Especially fig. 5.13A is a clear indication of the subsidence pattern, showing 0.5 cycles of deformation, corresponding with 1.5 cm subsidence, or 7.8 cm/year. The smoothness of the interferometric phase suggests minimal atmospheric influence for this specific pair. A relatively strong subsidence gradient is observed to the south of the evaporation pond, where a gradient of 1.2 cm over 1.5 km occurs. Temporal decorrelation is clearly limited to some agricultural fields, whereas the geothermal area remains relatively coherent. The subsidence pattern seems limited to the upper left part of the image, in contrast with leveling data. Unfortunately, temporal decorrelation outside the selected interferogram area is too large, limiting the interpretable area to this crop. Figures 5.13B–D show three consecutive interferograms, each sharing one SAR image with the next one. These three interferograms appear more noisy than interferogram A. The small scale, relatively smooth variability in the interferometric phase is due to atmospheric signal. In first approximation this can be concluded from the phase variation over the flat plain to the west of the evaporation pond. Assuming homogeneous atmospheric conditions, such phase variations can be expected over the subsiding areas as well. In section 5.4.9, the analysis of the atmospheric signal is considered in more detail. Although the atmospheric signal leads to phase variations of almost one cycle, the effects are limited in spatial scale, compared to the subsidence pattern expected based on interferogram A. Therefore, the subsidence pattern is still detectable by visual comparison, although an accurate comparison of the magnitudes is more difficult. In the area containing the well groups very localized "dots" with a different phase can be observed. These peculiar effects may be explained by either very localized subsidence of the well installations, due to their direct connection with the reservoir, or by atmospheric delay difference induced by the very hot and humid steam escaping over the wells.

Figure 5.14 shows four interferograms with longer temporal baselines, 176, 245, 351, and 491 days, respectively. The influence of increasing temporal decorrelation is clear, diminishing the coverage of reasonable coherent data. Nevertheless, although the data consist of isolated patches, it is still possible to follow the main fringes visually. The contours of the fringes show the ellipse-shaped subsidence bowl. The interferograms in fig. 5.14A and B are both combinations with the oldest image (20 May 1995). Since this common image appears as master in fig. 5.14A and as slave in B, the fringe direction is reversed. It is clear to see that both interferograms do not show the localized "sub-bowl," which was observed in the 1994-1997 leveling, cf. fig. 5.12. This suggests that this effect occurred between the 1994 leveling and May 1995, or was an erroneous interpretation of the leveling, perhaps due to benchmark instability. Assuming the deformation rate estimated from fig. 5.13A to be correct, the amount of phase cycles between the center of the subsidence ellipse and the relatively stable area west of the evaporation pond can be extrapolated to these intervals. This results in 1.3, 2.0, 2.6, and 3.5 fringes for fig. 5.14A, B, C, and D, respectively. Especially for interferogram D, this implies that we would "miss" one fringe by visual inspection only. Therefore, phase and patch unwrapping algorithms need to be fine-tuned for this particular situation. Nevertheless, the results of these four interferograms is consistent with those obtained from the 70 day interval pairs.

5.4.5 Stacking experiments

Stacking (adding) interferograms is an integer linear combination, as discussed in section 3.5.3, see also Sandwell and Price (1998); Sandwell and Sichoix (2000). For such combinations, performed using a complex multiplication to prevent wrapping

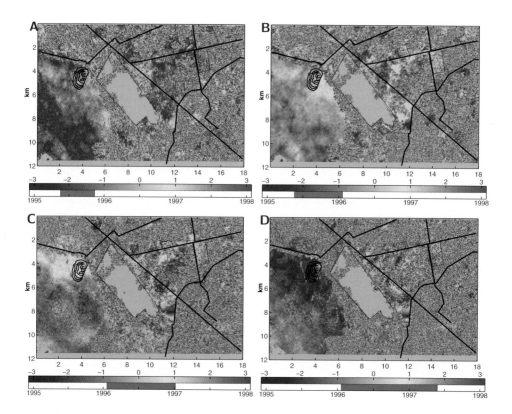

Fig. 5.14. Differential SAR interferograms with increasing temporal baselines between 176 days and 491 days. **(A)** Differential interferogram 20/05/95-12/11/95 (176 days). **(B)** Differential interferogram 20/01/96-20/05/95 (−245 days). **(C)** Differential interferogram 20/01/96-05/01/97 (351 days). **(D)** Differential interferogram 20/01/96-25/05/97 (491 days). The interferograms clearly show the deformation signal, although temporal decorrelation increases considerably.

errors, phase unwrapping is not required. The result of adding deformation interferograms with common images is equivalent to calculating an interferogram with the first and the last image. Here we show two examples of stacking.

Stack 1 (fig. 5.15A) is formed by the linear integer combination of three interferograms 20/05/95-12/11/95, 12/11/95-21/01/96, and tandem interferogram 20/01/96-21/01/96. This stack can be compared with one single interferogram 20/05/95-20/01/96. The stack is shown in fig. 5.15A, the single interferogram is shown in fig. 5.14B. The difference between both is not drawn since it is obvious that it shows 0 rad over the coherent areas, apart from some small differences introduced by the filtering of the contributing interferograms. It is important to note that this example does not imply that it is not necessary to calculate the interferograms over smaller time intervals. Using this simple form of stacking, a decorrelated area in one of

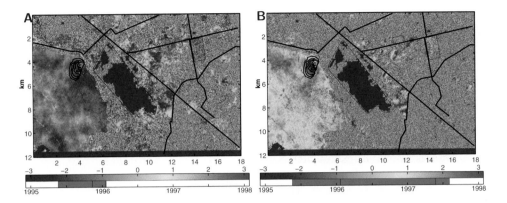

Fig. 5.15. Stacked interferograms. **(A)** Stack 1: 20/05/95-12/11/95 + 12/11/95-21/01/96 − 20/01/96-21/01/96. **(B)** Stack 2, spanning the full period (803 days): 20/05/95–03/08/97.

the contributing interferograms will contaminate the final result in the stack. An improved way would be to use all partial results as well in the interpretation.

Stack 2 (fig. 5.15B) is formed by the linear combination of all interferograms spanning the epoch between 20 May 1995 and 3 August 1997 (803 days). Obviously, such a long temporal baseline results in a loss of coherence over the largest part of the area. Nevertheless, there are some parts where a large phase gradient is still visible. The expected total amount of fringes due to the subsidence is 5.7. Note that the atmospheric influence in a stack is limited to the atmosphere in the first and the last contributing SAR acquisitions, since intermediate atmospheric contributions cancel in the differencing. Therefore, if possible, downscaling a stack covering a long interval reduces the influence of these atmospheric contributions, whereas upscaling of a short time interval exaggerates the atmospheric contribution.

5.4.6 Phase unwrapping strategies

Temporal decorrelation characteristics differ for various types of land use. This often results in relatively coherent "islands" or patches in a decorrelated surrounding, see, e.g., fig. 5.15A and B. As isolated patches are not connected, phase unwrapping can be performed in every patch, but individual patches can have $k2\pi$ offsets. Two (laborious) strategies for resolving phase ambiguities are tested on these data:

- *Johnny Appleseed*[1] phase unwrapping refers to the identification and separate unwrapping of all coherent patches, placing "seeds" for starting the phase unwrapping in a dense grid over the interferogram, followed by patch unwrapping using a priori information.

---

[1] Johnny Appleseed (John Chapman), was a legendary figure, immortalized in American literature by Vachel Lindsay (1879-1931), wandering from coast to coast through the U.S., planting seeds of trees in every town. Also famous character in the Disney film *Melody Times* (1948).

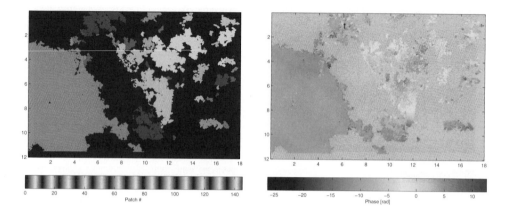

Fig. 5.16. Results of the Johnny Appleseed strategy for the interferogram 20/01/96–05/01/97 (351 days). **(A)** 145 identified patches, color coded. **(B)** Offset unwrapped phase for every patch.

- *Tie-point* unwrapping. This approach is using local phase observations at manually chosen positions, unwrapping them using a priori information or first-order assumptions.

Johnny Appleseed strategy

Using the tree algorithm of Goldstein et al. (1988), branch cuts are defined connecting residues. Phase unwrapping then starts at a (specified or random) seed-position, and progresses within the branch cut limits. The Johnny Appleseed method places a fine grid of seeds over the interferogram, and starts unwrapping for every seed that is not already in an unwrapped patch. Hereby, unwrapped patches are numbered consecutively until every seed position has been processed. This approach results in an unwrapped phase for every individual patch. To unwrap the patches relative to each other, each labeled patch can be offset by $k2\pi$, until a priori defined criteria are satisfied. Often, manual editing of the patches is necessary, such as dividing, joining, or eliminating operations. Figure 5.16 shows the initial results of Johnny Appleseed unwrapping of the 20/01/96–05/01/97 interferogram. Figure 5.16A shows all identified patches, while fig. 5.16B represents the unwrapped results. In this specific case, many patches are identified, making the procedure very laborious. Often, however, interferograms consist of a few large patches, e.g. separated by waterways. In this case, the reference to the water level makes the procedure fast and efficient, albeit not automatic.

Tie-point unwrapping

The *tie-point* method uses manually selected tie-points in the wrapped interferogram. The points are chosen based on coherence considerations, or assumptions about the phase stability of individual or groups of resolution cells. The positions and phase values are stored, and a two-dimensional interpolation is applied using a

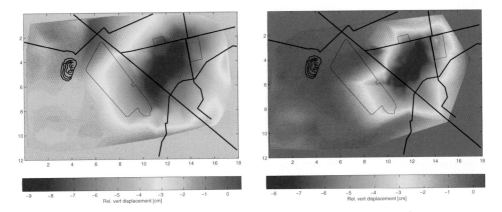

Fig. 5.17. Subsidence maps derived from tie-point unwrapped interferograms. After unwrapping every tie-point, cubic interpolation is applied to obtain the deformation maps. **(A)** Result for interferogram 20/01/96–05/01/97 (351 days). **(B)** Result for interferogram 20/01/96–25/5/97 (491 days). In both results it can be observed that the area of maximum subsidence over long time spans is situated at the position of the well groups.

bi-cubic algebraic polynomial. Two examples of tie-point unwrapping are shown in fig. 5.17. Knowledge about the deformation parameters (such as maximum phase gradient per pixel per time epoch) is used to offset tie-point values by $k2\pi$ until the total pattern satisfies deformation criteria. Since the source of fluid extraction is situated between 1500–4000 m, the deformation signal will have relatively long spatial wavelengths. Maximum deformation gradients are estimated from the leveling results in the order of 0.1 rad/pixel/yr, for a posting of 20 m. By subtracting the re-wrapped interpolated result from the original interferogram, it is possible to analyze the difference image. Local phase ramps indicate erroneous $k2\pi$ offsets, so adjustments to these offsets can be applied.

Integer linear combinations

Since many of the analyzed interferograms show areas with severe temporal decorrelation, an iterative approach for the data analysis can be useful. Additional initialization assumptions can aid post-processing steps such as patch-unwrapping. Assuming constant deformation in time, enables the use of relatively coherent (70 days) interferograms to extrapolate in time and predict deformation rates over longer time spans. The use of integer multiplication factors makes phase unwrapping of the interferogram to be scaled unnecessary. Drawback is the amplification of short wavelength and atmospheric noise, when the scaling is larger than, say, a factor of three. This approach can be used to offset isolated coherent patches by $k2\pi$ to get true absolute phase values.

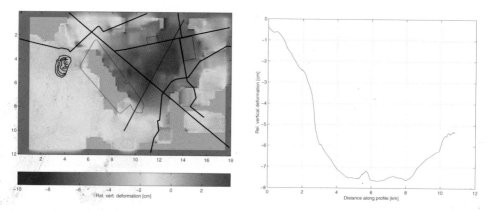

Fig. 5.18. (A) Subsidence rate per year. Result from the Johnny Appleseed strategy, boxcar smoothed (50×50 pixels). **(B)** Cross section A-A'. Note the steep gradient at the boundary of the subsidence area.

5.4.7   Forward modeling

To investigate the geophysical characteristics of the observed subsidence pattern forward models are used (Hanssen et al., 1998a). A reasonable correspondence between the modeled and observed deformation pattern can provide coarse estimates on the driving mechanisms of the subsidence. The subsidence is caused by volume contraction within the reservoir, which can be a result of a poroelastic and thermoelastic mechanism. Point centers of contraction, or Mogi sources, are used assuming that the earth can be modeled as an elastic half space (Mogi, 1958). Using a single Mogi-source model a reasonable first-order fit of the one-year subsidence rates is obtained, see fig. 5.19. The tie-point interpolated analysis of the 20/01/96–05/01/97 interferogram (fig. 5.19A) is compared to the slant-range displacement as inferred from a Mogi-source model at a depth of 5.9 km and a volume change rate of $\Delta V = 3 \times 10^6$ m^3/yr (fig. 5.19B). The difference between the model and the observations (fig. 5.19C) shows residues of maximally 10% in the subsidence region. The cross-section indicated by the blue line is shown in fig. 5.19D. We anticipate that a better fit to the data could be obtained using a multiple Mogi-source model with a shallower source depth and small volume change rate of, say, $\Delta V = 2 \times 10^6$ m^3/yr, as performed by Carnec and Fabriol (1999). On the other hand, a perfect fit using a large number of Mogi sources may overestimate the accuracy and feasibility of the model, especially regarding the influence of the assumptions on which this type of modeling is based and its under-determined nature.

5.4.8   Deformation mechanisms

Fluid extraction from the pores of a volume of rock within the earth decreases the pore pressure, which results in a contraction of the rock (Segall, 1985). As long as this process is reversible, the mechanism is poroelastic. On the other hand, a decrease in the volume can also be caused by the cooling of the reservoir. Although the removal of water from the reservoir does not produce a temperature drop, cooling of the reservoir can be caused by the flashing of the water to steam due to its extraction and the associated fast pressure drop (Mossop et al., 1997). As such, a thermoelastic mechanism could also contribute to the observed deformation.

The reservoir dimensions of $\sim$20 km^2 spatially and a 2 km thickness give an upper bound for the volume of the reservoir of $V = 40$ km^3. Hence the volume strain rate, ϵ_{kk}, is about $\Delta V/V = 0.5 \times 10^{-4}$. The volumetric coefficient of thermal expansion, α_v, relates an increase in temperature, ΔT, to volume strain, as $\alpha_v = \epsilon_{kk}/\delta T$. The volumetric coefficient of thermal expansion for greywacke at a similar geothermal field (The Geysers, California) is assumed to be similar to the greywacke here. Using this value, $\alpha_v = 3 \times 10^{-5}$ K^{-1}, the strain requires a temperature decrease of more than 1 K/yr to be thermoelastic. Compared with studies at The Geysers, by Mossop et al. (1997), this value is far too large. At The Geysers (with very similar production and injection rates, 80×10^6 m^3/yr, and 40% of the production, respectively) a temperature decrease of 0.25 K/yr was inferred (Mossop et al., 1997). Since the Geysers geothermal field is more efficient (higher steam production) and cooler than Cerro Prieto (higher temperature decrease due to injection), the Geysers temperature decrease can be taken as an upper bound for the expected temperature decrease at Cerro Prieto.

This indicates that the observed deformation cannot be explained by thermoelastic volume changes only. At most 25% of the deformation is thermoelastic, so the larger part will be of poroelastic nature.

5.4.9 Identification of atmospheric signal using linear combinations

The identification of significant localized atmospheric signal is possible using linear combinations, or *pair-wise logic* (Massonnet and Feigl, 1995a). Figure 5.20 is a demonstration of this concept. Two interferograms (05/01/97–16/03/97 and 16/03/97–26/05/97, fig. 5.20A and B) share a common image acquired at 16/03/97. Atmospheric signal during that acquisition will contaminate the subsidence signal in both interferograms. Such anomalies are visible, e.g., in the center of the interferograms, west of the evaporation pond, indicated by the arrows. Since the 16/03/97 acquisition is used as "slave" in the first interferogram and as "master" in the second interferogram, the sign of the anomalies is reversed in both interferograms. Therefore, the sum of both interferograms results in complete removal of these atmospheric anomalies (fig. 5.20C), whereas subtraction of the interferograms results in an amplification of the atmospheric signal (fig. 5.20D) with a factor 2. For this particular situation both interferograms cover 70 days, and the deformation rate can be considered approximately constant in time according to the production and injection data. Therefore, addition of the interferograms results in a doubled (140 day) deformation signal, whereas subtraction removes the deformation signal entirely.

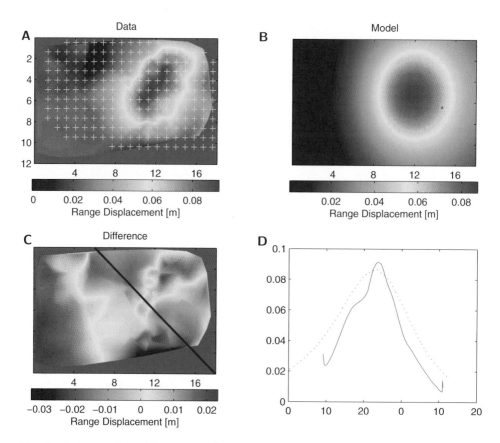

Fig. 5.19. Forward modeling results (Hanssen et al., 1998a). **(A)** Tie-point interpolated analysis of interferogram 960120–970105 (fig. 5.14C). **(B)** Grid search result for a single Mogi source. **(C)** Difference between data and model. **(D)** Profile indicated by the blue line in (C).

5.4.10   Conclusions

Land subsidence over the Cerro Prieto geothermal area can be clearly detected by using radar interferometry. Systematic monitoring of the area using radar acquisitions acquired every 35–70 days can provide a significant improvement of the knowledge on the spatial distribution of subsidence as well as the subsidence rate and its temporal fluctuations. Especially the monitoring of strong localized gradients in the subsidence rates, see fig. 5.18, can be important for engineering purposes, e.g., to avoid damage to infrastructural works. Furthermore, the combination of the radar interferometric data with other geophysical parameters may lead to improved understanding of the thermoelastic or poroelastic characteristics of the subsurface. Based on the interferograms analyzed, the comparison with the leveling data shows that the subsidence rate of the main bowl is approximately equal. However, a localized

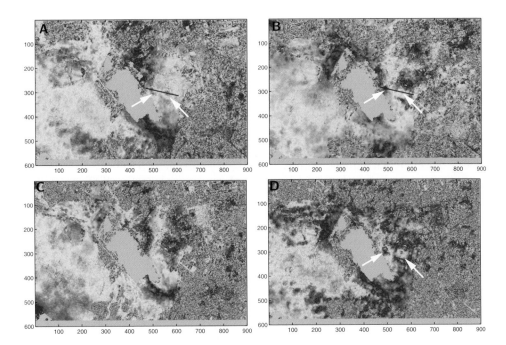

Fig. 5.20. Atmospheric effects, for example as indicated by the arrows, identified by addition and subtraction of interferograms. **(A)** Interferogram 05/01/97–16/03/97. **(B)** Interferogram 16/03/97–26/05/97. **(C)** Addition of (A) and (B), the atmospheric effects of 16/03/97 are removed. The deformation spans 140 days. **(D)** Subtraction of (A) and (B), the atmospheric effects of 16/03/97 are doubled, whereas the deformation fringes are removed.

sub-bowl, identified in the 1994-1997 leveling (see fig. 5.12), is not observed in the interferometric data. This implies that it is either a phenomenon which occurred between the 1994 leveling and the oldest SAR image (May 1995), or an artifact in the leveling, perhaps due to benchmark instability.

Problems in the interferogram analysis mainly occur due to temporal decorrelation and atmospheric signal in the data. The temporal decorrelation leads to isolated patches of coherent phase information, surrounded by decorrelated areas. Phase unwrapping and patch unwrapping may yield ambiguous results. Atmospheric signal may deteriorate single interferograms, making a sound interpretation of a single interferogram cumbersome. The use of a priori information, as well as an increase in the number of available images can help reducing these problems.

5.5   Izmit earthquake

On 17 August 1999, 3 AM local time, the area around Izmit, Turkey suffered a strong earthquake: 7.5 on the Richter scale (Reilinger et al., 2000). The earthquake was caused by an accumulation of stress during many decades around the North Anatolian Fault (NAF) zone, where the Anatolian plate and the Eurasian plate meet. It was the biggest earthquake since the 1967 Mudurnu Valley earthquake of Ms=7.1, and had catastrophic consequences—killing more than 17.000 people, leaving thousands homeless, and causing $10–$25 billion in damage (Parsons et al., 2000). Since the earthquake sequence of August–November 1999 can be regarded as part of a largely westward propagation of ruptures along the NAF since 1939 (King et al., 1994; Stein et al., 1997; Stein, 1999), it is likely that it has increased the stress on faults beneath Marmara Sea. This, in combination with an earthquake history that indicates a seismic gap beneath Marmara Sea and tectonic loading poses an increased earthquake risk for the city of Istanbul (Hubert-Ferrari et al., 2000; Parsons et al., 2000), just a few hundred kilometers northwest of the Izmit region.

This section contains a concise presentation of the coseismic interferometric data (acquired before and after the earthquake), using the optimal interferometric combination. It is not intended to give an in-depth geophysical analysis, but merely as an example of an interferometric deformation signal that serves as a comparison for the error sources discussed in the previous chapters.

5.5.1   Background

The North-Anatolian Fault forms the boundary between the Eurasian and the Anatolian plates. Right-lateral strike-slip occurs along a nearly vertical fault plane. Plate motion amounts up to 2.5 cm/yr (Hubert-Ferrari et al., 2000). The North-Anatolian Fault branches of in a northern and a southern branch. The 17 August 1999 Izmit earthquake occurred on the northern branch, and ruptured a 110 km long section.

5.5.2   Interferometric data

Izmit earthquake stimulated ESA to perform dedicated SAR observations over the region, optimized for obtaining coseismic interferograms. ERS-1, which was in hibernation mode at the time, had coincidentally been used one month before the earthquake, and was switched on again the first possibility after the earthquake. An orbit maintenance manoeuvre for ERS-2 was scheduled ahead, to enable near-zero baseline conditions for the coseismic ERS-2/ERS-2 interferogram used in this study. The perpendicular baseline was ∼18 m, decreasing to nearly zero from near-range to far-range, which corresponds with a height ambiguity of ∼566 m. Especially at the north of the Gulf of Izmit, topographic height variation is less than 600 m (Fielding et al., 1999), which results in less than one-fringe topographic contribution. To the south of the Gulf, topography is higher, up to 1000 m, resulting in maximally 2 fringes topography.

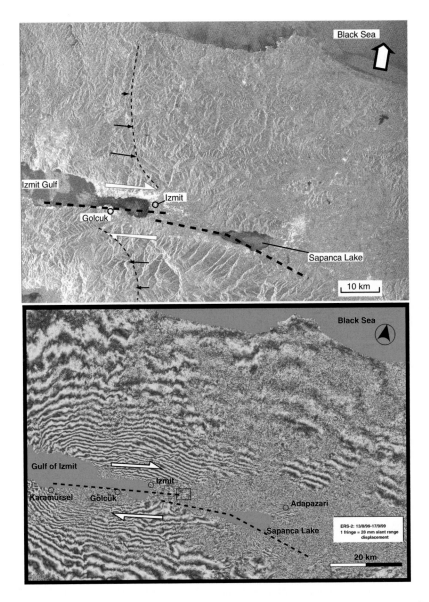

Fig. 5.21. (A) Radar intensity image, with the Black Sea in the northeast, the Gulf of Izmit in the west, and Sapanca lake in the center. Topography is visible in the south. Urban areas such as Izmit can be recognized by the bright intensity. The North Anatolian Fault is indicated by the fat dashed line. **(B)** Coseismic interferogram of 13 August and 17 September 1999, showing the deformation pattern due to the Izmit earthquake at 17 August. The color cycles correspond with 28 mm in the satellite line of sight.

Figure 5.21A is the radar intensity image of ERS-2, acquired from 789 km height. Brightness variation in the image shows topography and urban areas. The dotted horizontal line indicates the approximate location of the North-Anatolian Fault. Figure 5.21B is the coseismic (ERS-2/ERS-2) interferogram between 13 August and 17 September 1999. Since the topographic contribution is less than 2 fringes, the interferogram shows primarily slant-range displacements due to the earthquake and possibly postseismic deformation between 19 August and 17 September 1999.

5.5.3   Interpretation

Every fringe in fig. 5.21B corresponds with 28 mm of deformation, directed to the radar or away from it. In this case a yellow-blue-red (YBR) color cycle means relative deformation away from the radar, while yellow-red-blue (YRB) suggests relative deformation towards the radar.

Approximately 27 fringes span the range between the Black Sea coast and the northern shores of the Gulf of Izmit, corresponding with $d_r = {\sim}75$ cm in slant range. Traveling this range from North to South, the order of the colors is YBR, which implies that the northern shores of the Gulf of Izmit have moved 75 cm away from the radar, relative to a point at the Black Sea coast.

Here, the measured $\sim$75 cm is the deformation in the radar line-of-sight. Since the dominant part of the strike-slip deformation along this fault system is horizontal (Barka, 1999), we find with eq. (5.1.1) a horizontal component in the azimuth look (ALD) direction of $\sim$7.2 cm/fringe. The decomposition of this horizontal displacement vector in a north and east component is not possible without any a priori information. Nevertheless, since the fault is oriented nearly perpendicular to the satellite track, the ALD component will be of the same order of magnitude as the absolute horizontal deformation vector. As a result, the northern shore of the Gulf of Izmit has moved $\sim$1.95 m in ALD direction, relative to a reference point at the Black Sea coast.

From the interferogram it can be concluded that it is very difficult to count the fringes near the fault. This is caused by two mechanisms. First, near to the fault the deformation rate is probably to high, so that undersampling occurs. This is effectively the upper deformation gradient of 10^{-3}, see section 2.4.3 and fig. 2.11. Second, temporal decorrelation in the rupture zone area is extreme, resulting in a loss of coherence over many of the damaged area. Therefore, extrapolation of the fringes, or use of additional observations is necessary to obtain deformation rates near the fault. From field observations it is known that horizontal displacement between 2 and 3 m has occurred near the fault. Near the Gulf of Izmit observed deformations amounted up to 4–4.5 m.

5.5.4   Analysis of the observed pattern

In fig. 5.22 a first order, coarse interpretation of the fringe pattern is presented. The fault plane represents the boundary between the Eurasian plate in the north and the Anatolian plate at the south side. The dotted line indicates the shortest distance

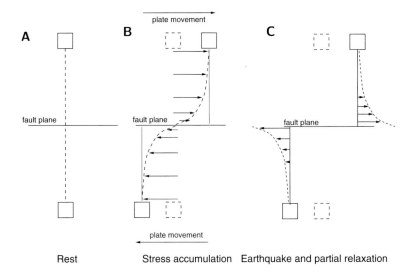

Fig. 5.22. Sketch of the deformation as a result of plate movement. In **(A)** both plates are static, and a straight line can be drawn between the two squares, perpendicular to the fault plane. After a specific time span, both squares have moved with respect to each other, as in **(B)**. Since the fault plane might have sufficient friction both plates will deform in an elastic way. Therefore, strain will accumulate, also on the fault plane. The earthquake, see **(C)**, is a response of the strain accumulation. After the earthquake, both plates will be deformed along the fault plane. The dotted line, which indicated the strain in (B), is now relieved and will be (more or less) back in the original position. The straight line, which indicated the perpendicular projection on the fault plane just before the earthquake, is now (C) curved, indicating the amount of surface deformation. A significant curvature corresponds with a high density of phase cycles in the interferogram.

between the two squares. The general plate movement causes the two squares to move away from each other, parallel to the fault plane (fig. 5.22A and B). Without direct slip or creep along the fault, the dotted line will bent significantly with time. Just before the earthquake, the solid line indicates the shortest distance to the fault plane. Immediately after the earthquake, when both plates have slipped, the situation is as in fig. 5.22C. As the accumulated strain has been (partially) released, the red line has moved to its new position. The blue line now indicates the amount of deformation between shortly before the earthquake and shortly after. In fact, the observed deformation pattern can be regarded as a mirrored version of the strain pattern before the earthquake; cf. the red line in fig. 5.22B with the blue line in fig. 5.22C.

As long as the fringes in the interferogram are parallel at both sides of the fault, the two sides have moved parallel to each other. At those areas the strain which was accumulated over the years is relieved at least partially. In the interferogram it can be observed that the color patterns are not parallel the entire range of the fault: in the western part they bend towards the Gulf of Izmit, both on the northern and the southern side of it. Assuming that this behavior continues through the crust under the Gulf of Izmit, this implies that both sides of the fault have not moved with respect to each other. Since the fringes cannot be followed across the water, there is can be an inaccuracy of several fringes in this assumption, but still the deformation is significantly less than closer to the main epicenter. Since the end of the rupture zone was under water, it could not be clearly detected. The direction of the fringes, however, indicates that the end of the rupture zone was near the left side of the interferogram Assuming that at this position the rupture zone ended, a large part of the strain will still be apparent, which could be an indication for an increased probability for future earthquakes, which is in agreement with general understanding of stress induced triggering (King et al., 1994; Stein et al., 1997; Stein, 1999). Proper ways to localize areas where strain accumulation is still in progress are essential in hazard monitoring and risk assessment.

5.5.5 Conclusions

This brief presentation of the coseismic interferogram of the Izmit earthquake shows some of the pros and cons of the technique. The advantages are clearly the increased spatial coverage with respect to conventional geodetic techniques, see Reilinger et al. (2000) for an overview of these data for the Izmit earthquake. Furthermore, the possibility to combine an archived SAR acquisition prior to the earthquake with a new acquisition after the event (including orbit optimization) results in a reduced demand on monumentation and in situ survey networks. Nevertheless, the InSAR data only yield line-of-sight deformations, which demand either reliable assumptions on the physics of the deformation, or a complementary combination with other geodetic or geophysical techniques. Temporal decorrelation can eliminate the possibility of a coherent interferometric combination. For example, if the earthquake would have happened a few months later, in the snow season, the chances to obtain a coherent interferogram would have been reduced dramatically.

It needs to be emphasized that the study of the geophysical interpretation of the Izmit SAR data is still in progress, aiming at the development of a best-fit model to describe this earthquake, see Riva et al. (2000).

5.6 Conclusions

Before commenting on the general feasibility of radar interferometry for deformation studies, it is necessary to make a, perhaps more philosophical, comment. In the case studies presented here it was already known that a deformation occurred, and the problem was approached from a survey-strategy point of view. It needs to be stressed that the technique has an unsurpassed quality to serve as a monitoring tool, even if the quantitative results are not uniquely interpretable. In contemporary geodesy, the goal of obtaining more accurate measurements is slowly shifting—in terms of user requirements—towards simply *more* spatially distributed observations, acquired *more often.* In many cases, areas which suffer from deformation (earthquakes, subsidence, landslides, volcano dynamics) are not monitored at all, or observations are

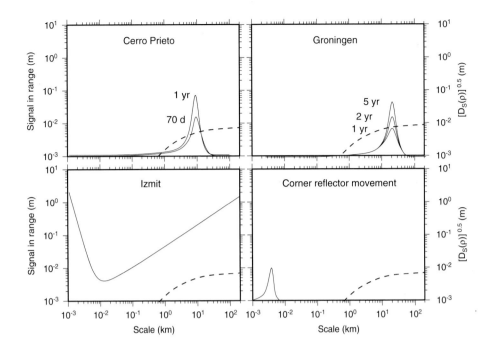

Fig. 5.23. Characteristics of the discussed deformation studies compared with the atmospheric signal, expressed as the square root of the atmospheric structure function and indicated by the dashed line. The sketches serve as an indication of the spatial scales and magnitudes of the deformation events. The vertical axis indicates the deformation signal in slant range, as well as the atmospheric standard deviation as a function of scale.

only available *after* a catastrophic event. A flexible method for acquiring new radar images combined with an archive of historical data may be the only means to analyze this type of situations, see for example volcano studies in Alaska (Lu et al., 1997) or at the Galapagos (Amelung et al., 2000a,b).

To evaluate the feasibility of radar interferometry for a specific application, especially regarding the influence of atmospheric signal in the data, it is necessary to compare the scale characteristics of the deformation phenomena with those of the atmospheric signal. Figure 5.23 shows a sketch of the expected scales for the examples discussed in this chapter. It can be regarded as complementary to the instrumental evaluation shown in fig. 2.11, cf. p. 41. In fig. 5.23, spatial scale is expressed on the horizontal axis, whereas the vertical axis shows (i) the estimated magnitude of the slant-range deformation signal in range, and (ii) the square-root of the atmospheric structure function, or the standard deviation of the atmospheric signal between two points with a certain spatial separation. The dashed line represents an approximation of the square-root of the structure function based on the experimental results in chapter 4. The solid lines in the four plots represent the characteristic scales of the discussed deformation processes. For Cerro Prieto, the main subsidence bowl covers scales ranging from 1–15 km. The magnitude depends on the time interval between the SAR acquisitions, as this is a nearly linear phenomenon. From the analyzed data in section 5.4 it is clear that the subsidence signal is detectable in a 70 day interferogram, although atmospheric anomalies may contaminate scales less than a few kilometers.

For the subsidence problems in Groningen, the spatial scales are larger, say 5–30 km. Since the subsidence rate is slow, less than 1 cm/yr, it takes a considerable amount of time before the subsidence signal dominates over the atmospheric noise level. Obviously, this time period strongly influences the amount of temporal decorrelation. The earthquake in Turkey shows deformation signal over a wide range of scales, up to ∼1.5 m can be observed in slant range. Over smaller scales, the amount of deformation decreases, apart from the rupture zone, where very strong signal is apparent over just a few meters. Note that the latter is not measurable by interferometry due to the steep phase gradient, see fig. 2.11. Over all scales, the deformation signal is stronger than the atmospheric signal. Although this implicates that the deformation signal is detectable, it needs to be stressed that the atmospheric signal still contaminates the data. Finally, the corner reflector experiment (∼1 cm deformation) only concerns a localized group of a few pixels, indicating a negligible atmospheric contribution.

Simple analyses as sketched above may serve as an important tool to guide the decision whether to apply spaceborne repeat-pass radar interferometry for a specific deformation problem.

Chapter 6
Atmospheric monitoring

This chapter presents first results of Interferometric Radar Meteorology (IRM), showing how radar interferometric delay measurements can be used to infer high resolution maps of integrated atmospheric water vapor. Maps of the water vapor distribution can be readily related to meteorological phenomena, and suggest that radar observations can be used to study atmospheric dynamics and may contribute to forecasting. The theory and sensitivity analysis of the various atmospheric refractivity contributions is discussed. IRM results are validated using conventional meteorological techniques and GPS observations.
key words: *Atmosphere, Troposphere, Ionosphere, Water vapor, Meteorology*

The spatial distribution of water vapor in the earth's atmosphere is important for climate studies, mesoscale meteorology, and numerical forecasting (Rind et al., 1991; Emanuel et al., 1995; Crook, 1996). Local concentrations of water vapor vary between 0–4%, which makes it one of the most varying atmospheric constituents (Huschke, 1959). Although extensive ground-based and upper-air sounding networks are routinely used, these measure the water vapor distribution only at coarse scales. Even spaceborne radiometer observations, such as Meteosat or GOES, have spatial resolutions that are currently too coarse for, e.g., quantitative precipitation forecasting or cloud formation studies. Moreover, radiometric water vapor observations from these satellites usually originate from atmospheric layers above 3 km, due to the strong absorption by water vapor (Weldon and Holmes, 1991). This often restricts quantitative interpretation to upper-tropospheric moisture distribution (Schmetz et al., 1995).

These limitations form the main source of error in short-term (0–24 hr) precipitation forecasts (Emanuel et al., 1995). For example, thunderstorm initiation and strength are sensitive to spatial and temporal variations in moisture of the order of 1 g/kg (1.2 hPa) and temperature of 1–3°C (Crook, 1996; Mueller et al., 1993). Such variations are common on a 1-km spatial scale (Weckwerth et al., 1997). SAR, however, provides horizontal resolution as fine as 10–20 m over a swath typically 100 km wide.

It is widely accepted that spaceborne geodetic observations such as GPS and VLBI provide an independent source of meteorological information by analyzing the propagation delay of the electromagnetic signals. Availability of these data on a routine basis is improving numerical forecasting and mesoscale meteorology. In this chapter, we will show that radar interferometry can be used for meteorological purposes as well, even though the configuration is currently not suited for operational meteorology. Nevertheless, the combination of high accuracy and fine spatial resolution with the huge archive of interferometric datasets can be readily used to provided unsurpassed insights in atmospheric dynamics.

The atmosphere below about eighty kilometers naturally divides into two layers, the troposphere and the middle atmosphere, where the latter consists of the stratosphere and the mesosphere (NorthWest Research Associates, 2000). The troposphere is destabilized by radiative heating and by vertical wind shear near the surface. While radiative heating leads to convection, the vertical wind shear leads to baroclinic instability that manifests itself in synoptic-scale storms and frontogenesis. Unlike the troposphere, the overlying middle atmosphere is relatively weather-free. More precisely, the meteorological phenomena of the middle atmosphere do not resemble anything commonly experienced in the troposphere. The stratosphere is surprisingly dry, so cloud formation is almost nonexistent apart from noctilucent and polar stratospheric clouds (NorthWest Research Associates, 2000). Therefore, signal delay in the neutral part of the atmosphere is mainly originating from the troposphere.

Mesoscale Shallow Convection (MSC) represents a specific class of atmospheric systems that could benefit from interferometric radar observations. *Mesoscale* systems have horizontal length scales between ten and a few hundred km. *Shallow* systems occupy the lower 1 to 2 km. This layer is referred to as the *planetary boundary layer*; the lowest part of the atmosphere, where the earth's surface has a profound effect on the properties of the overlying air. In this framework, *convection* represents mass motions of air resulting in vertical transport and mixing of its properties, which is a principal means of transporting heat. Atkinson and Zhang (1996) state three reasons why MSC are important. First, the transport of moisture, heat, and momentum to the free atmosphere are crucial for general circulation. Second, the net radiation budget of the earth, important for understanding the earth's climate system, is significantly influenced by stratocumulus clouds. MSC is related to the occurrence and forming of such clouds. Finally, as long as the process of MSC is not fully understood, there is a gap in the understanding of the dynamics of the planetary boundary layer.

This chapter is intended to be self-contained, allowing for a direct reference by, e.g., meteorologists and for atmospheric studies. It starts with a brief description of the theory necessary for the meteorological interpretation of radar interferometric data. It will be argued that for short spatial scales, say less than ~50 km, the dominant atmospheric signal is due to the water vapor distribution. In section 6.2, the influence of water vapor, temperature, and liquid water (droplets) is discussed. Section 6.3 elaborates on the effects of the ionospheric electron density. In a previous study (Hanssen, 1998), 26 interferograms have been analyzed and interpreted from a meteorological point of view. In sections 6.4–6.7, a number of these case studies are discussed in more depth, focusing on the interpretation and validation of the observations. Section 6.8 concludes this chapter summarizing the potential and limitations of water vapor mapping with radar interferometry.

6.1 Theory

A SAR image contains information on the path length between the radar antenna and the resolution cells on earth, see chapter 3 for a more elaborate discussion. The interferometric combination of two radar images with a temporal separation of 1 day provides a sensitive tool to measure these path length differences as a fraction of the radar wavelength, which is 5.66 cm for C-band radar used here. Conventionally, path length differences can be attributed either to topographic height differences, depending on the relative positions of the satellites (Zebker and Goldstein, 1986), or to surface deformation, depending on the time interval between the two observations (Gabriel et al., 1989; Massonnet et al., 1993, 1995). However, effective path length variations are also caused by radar signal delay variability within the imaged area, due to the heterogeneous refractivity distribution in the atmosphere. Atmosphere induced distortion has been observed in radar interferograms (Goldstein, 1995; Massonnet and Feigl, 1995a; Tarayre and Massonnet, 1996; Hanssen and Feijt, 1996; Zebker et al., 1997) but has typically been treated as noise.

Signal delay, in seconds, is equivalent to an excess path length by multiplication with the speed of light in vacuum. The excess path length can be directly obtained by integrating over the (dimensionless) refractivity along the line of sight. Over small spatial scales, the variation in the integrated refractivity is mainly due to the spatial variation of water vapor during the two image acquisitions. To a lesser degree temperature, liquid water, and pressure gradients influence the delay variation.

Delay measurements observed by space-geodetic techniques can be used to derive precipitable water vapor in the atmosphere (Saastamoinen, 1972; Hogg et al., 1981). Precipitable water vapor is the amount of vertically integrated water vapor and can be expressed in kg/m^2 or as the height of an equivalent column of liquid water in meters. GPS measurements provide temporal variations in precipitable water vapor at one position, see e.g., Bevis et al. (1992). Using InSAR, the data reflect spatial variations in precipitable water vapor during the two image acquisitions. Figure 6.1 shows the geometric configuration of the radar acquisition and the GPS zenith measurement.

6.1.1 Interferometric phase review

The SAR of the ERS-1 and ERS-2 satellites provides an amplitude and a phase value for every resolution cell of approximately 4×20 m. Information on the path length between the radar antenna and a ground resolution cell is contained in the phase measurement. Unfortunately, the phase observation $\psi_p^{t_i}$ of resolution cell p in a SAR acquisition at t_i is a superposition of a number of contributions:

$$\psi_p^{t_i} = \psi_{p,\text{geom}}^{t_i} + \psi_{p,\text{prop}}^{t_i} + \psi_{p,\text{scat}}^{t_i}, \quad -\pi \leq \psi < \pi, \tag{6.1.1}$$

where $\psi_{p,\text{geom}}^{t_i}$ is related to the geometric distance and $\psi_{p,\text{prop}}^{t_i}$ to the signal propagation velocity variations. Most important, the scattering component, $\psi_{p,\text{scat}}^{t_i}$, is the contribution of many arbitrary scatterers in the resolution cell, which add up to produce a uniform probability density function (PDF) for $\psi_{p,\text{scat}}^{t_i}$. As a consequence,

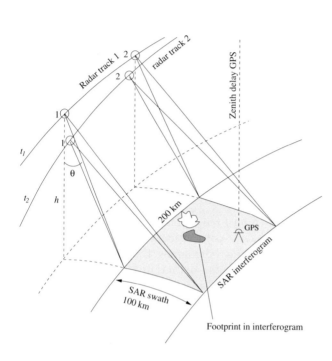

Fig. 6.1. Geometric configuration of radar acquisitions. The SAR image is observed from satellite observations between point 1 and 2 of track 1 at t_1. The same area is viewed from track 2 at t_2. The mean look angle θ is approximately 21°, and the satellite height $h = 785$ km for ERS. A localized heterogeneity in the refractivity during one of the two observations will result in an increased signal delay for the corresponding footprint in the SAR image. A GPS receiver can be used to derive the zenith delay at that point during the SAR acquisition, based on slant delays to a number of GPS satellites.

the sum of the PDF's of all components in eq. (6.1.1) will have a uniform distribution as well, and no useful phase information can be obtained.

In the repeat-pass interferometric combination, two SAR images, acquired at different times, are accurately aligned and differenced, which yields the interferometric phase:

$$\phi_p = \psi_p^{t_1} - \psi_p^{t_2}, \quad -\pi \leq \phi < \pi, \quad (6.1.2)$$

consisting of a differenced geometric component $\phi_{p,\text{geom}}$, propagation component $\phi_{p,\text{prop}}$, and scattering component $\phi_{p,\text{scat}}$. If the second satellite orbit is sufficiently close ($< \sim 600$ m) to the first and the physical scattering characteristics remain constant ($\psi_{p,\text{scat}}^{t_1} \approx \psi_{p,\text{scat}}^{t_2}$), the scattering component in each interferometric phase observation will be eliminated in the differencing. In that case, useful information may be retrieved from the resulting difference phase image or interferogram. The assumption of a stationary scattering component is the limiting factor for the application of SAR interferometry. For example, over water or rapidly changing surfaces it is not possible to obtain coherent phase observations. Over many agricultural areas, as in the test site in this study, phase noise increases with increasing time intervals or due to anthropogenic activity. During the ERS-1 and ERS-2 "tandem mission," which lasted from August 1995 to April 1996, SAR images of the same area were acquired with a sampling interval of 24 hours. This short time interval ensures a sufficiently high correlation between consecutive acquisitions over most land surfaces.

In order to analyze the propagation component of the interferometric phase, the influence of the geometric component needs to be eliminated. Geometric phase differences are caused by either a change in satellite position or a coherent change in the position of the scatterers on earth, between the two acquisitions. A difference in satellite positions will measure topographic height variation in the SAR image. Using a reference elevation model, a synthetic topographic interferogram can be constructed that can be subtracted from the observed interferogram, resulting in a so-called "differential," topographic-free, interferogram (Massonnet et al., 1993). Other variations in the geometric component, for example, due to surface deformation, can be safely ignored for these short time intervals. Therefore, observed phase gradients in the differential interferogram can only be attributed to propagation delay variability and residual trends due to the inaccuracy of the satellite position during the acquisitions. Finally, the interferometric phase, which is originally "wrapped" to the interval $[-\pi, \pi)$, is unwrapped using dedicated phase-unwrapping algorithms, see, e.g., Goldstein et al. (1988); Ghiglia and Pritt (1998).

6.1.2 Interferometric delay analysis

After obtaining the differential interferogram, the observed phase differences can be interpreted as (i) the spatial delay variation between the radar antenna and millions of pixels in the interferogram and (ii) the difference between the two generally uncorrelated states of the atmosphere during the two SAR acquisitions. Due to satellite orbit errors and the wrapped nature of the phase observations, it is only possible to measure the lateral variation of the delay, rather than the total delay. The delay variation $\delta_{p,q} = \delta_p - \delta_q$ between pixel p and q is directly related to the interferometric phase difference $\phi_{p,q} = \phi_p - \phi_q$ by

$$\delta_{p,q} = \frac{\lambda}{4\pi}\phi_{p,q} = 28\frac{\phi_{p,q}}{2\pi} \text{ [mm]}, \qquad (6.1.3)$$

for C-band. Mapping the incident delay variation to zenith values can be achieved by using a mapping function, e.g., a simple cosine function:

$$\delta^z_{p,q} = \delta_{p,q}\cos\theta_{\text{inc}}, \qquad (6.1.4)$$

with incidence angle θ_{inc} (look angle θ plus the earth's curvature, see fig. 2.9) varying between 19° and 23°, see fig. 6.1. Such a simple mapping function is sufficiently accurate for steep incidence angles (Bean and Dutton, 1968).

As discussed in section 4.2, the standard deviation of the phase, σ_ϕ, is derived from the coherence γ, that is, the amount of correlation between the two SAR images, with $0 \leq \gamma \leq 1$. For $\gamma \geq 0.8$ we find $\sigma_\phi \leq 52°$. Using eq. (6.1.3) and (6.1.4), and averaging over 5 pixels to obtain 20×20 m resolution cells yields a formal accuracy of the zenith delay (vertically integrated refractivity) observations of $\sigma_{\delta^z} \approx 2$ mm.

Additional spatial averaging to a ground resolution of approximately 160×160 m yields a phase standard deviation $\sigma_\phi \leq 5°$ and consequently $\sigma_{\delta^z} \leq 0.2$ mm.

Possible systematic errors in these delay observations are manifested as long-wavelength gradients in the interferogram, caused by inaccuracies in the satellite's position during image acquisition. These additional tilts are removed from the interferogram before analyzing the delay differences. This procedure also removes long-wavelength delay gradients caused by pressure and temperature gradients or gradients in the ionospheric electron density.

The technique now reveals slant delay differences or integrated refractivity along every path between the antenna position, a, and the resolution cells on earth. The relation between zenith delay observed at resolution cells p and q and the refractivity distribution during SAR acquisition t_i can be written as

$$\delta_{p,q}^{z,t_i} = 10^{-6} \cos\theta_{\text{inc}} \left(\int_p^a N dz - \int_q^a N dz \right), \qquad (6.1.5)$$

where $N(x,y,z,t)$ is the dimensionless refractivity. The zenith delay variation $\delta_{p,q}^z$ observed in the interferogram can be defined as

$$\delta_{p,q}^z = \delta_{p,q}^{z,t_1} - \delta_{p,q}^{z,t_2}. \qquad (6.1.6)$$

The refractivity N can be written using two equivalent expressions (Smith and Weintraub, 1953; Kursinski, 1997; Hanssen, 1998). The first, original, expression is

$$N = k_1 \frac{P_d}{T} + \left(k_2 \frac{e}{T} + k_3 \frac{e}{T^2} \right) - 4.028 \times 10^7 \frac{n_e}{f^2} + 1.4W, \qquad (6.1.7)$$

where P_d is the partial pressure of dry air in hPa, e is the partial pressure of water vapor in hPa, and T is the absolute temperature in Kelvin. The constants k_1, k_2, and k_3 have been first determined by Smith and Weintraub (1953), but results from Thayer (1974) are also commonly used. We use $k_1 = 77.6$ K hPa^{-1}, $k_2 = 71.6$ K hPa^{-1}, and $k_3 = 3.75 \times 10^5$ K^2 hPa^{-1}. These constants are considered to be accurate to 0.5% of N (Resch, 1984). The electron density per cubic meter is expressed by n_e, f is the radar frequency (5.3 GHz), and W is the liquid water content in g/m^3. The first term at the right-hand side of eq. (6.1.7) is often labeled as the *dry* term, the following two terms (in brackets) as the *wet* terms. The last two terms are the ionospheric term and the liquid term respectively.

Assuming that the total atmospheric pressure $P = P_d + e$ and using the equation of state[1] a second, equivalent, expression for N can be obtained (Davis et al., 1985):

$$N = k_1 \frac{P}{T} + \left(k_2' \frac{e}{T} + k_3 \frac{e}{T^2} \right) - 4.028 \times 10^7 \frac{n_e}{f^2} + 1.45W, \qquad (6.1.8)$$

---

[1] Note that in this case, the equation of state should involve the virtual temperature T_v, see Haltiner and Martin (1957).

where P is the total atmospheric pressure in hPa, and

$$k_2' = k_2 - \frac{R_d}{R_v} k_1 = 23.3 \quad [\text{K hPa}^{-1}], \tag{6.1.9}$$

with $R_d = 287.053$ J K^{-1} kg^{-1}, and $R_v = 461.524$ J K^{-1}. In eq. (6.1.8), the first term at the right-hand side is referred to as the *hydrostatic* term, instead of the dry term (Davis et al., 1985).

If we consider the contribution of the hydrostatic term in eq. (6.1.8) on the zenith delay in eq. (6.1.5), we may write for point p at time t_i, using the ideal gas law $P = \rho R_d T$,

$$\delta_p^{z,t_i} = 10^{-6} \cos\theta_{\text{inc}} \, k_1 \int_p^a \frac{P}{T} dz = 10^{-6} \cos\theta_{\text{inc}} \, k_1 R_d \int_p^a \rho(z)dz \tag{6.1.10}$$

Since the measured total surface pressure can be written as

$$P_s = g_m \int_0^\infty \rho(z)dz, \tag{6.1.11}$$

where g_m is the local gravity at the center of the atmospheric column (Saastamoinen, 1972), which is approximated by

$$g_m = 9.784 \, (1 - 0.0026 \cos 2\Phi - 0.00028z_0) \quad [\text{m/s}^2], \tag{6.1.12}$$

dependent on the surface height z_0 and latitude Φ, the hydrostatic delay is obtained from a simple barometric measurement P_s:

$$\delta_{p,\text{hydrostatic}}^{z,t_i} = k_1 \times 10^{-6} \frac{R_d}{g_m} P_s. \tag{6.1.13}$$

Using these parameters and the surface pressure P_s measured with an accuracy of 0.4 hPa or better, this delay can be predicted with an accuracy of 1 mm or better (Bevis et al., 1996). For the test sites used in this study, $\delta_{p,\text{hydrostatic}}^{z,t_i} = 2.275 \times 10^{-3} P_s$, hence in the order of 2.3 m. For comparison, the wet delay is less than $\sim$0.3 m (Elgered, 1982).

6.1.3 Discussion

Standard observations of surface pressure within a typical interferogram area of 100×100 km show (i) minimal spatial variation, usually less than 1 hPa, and (ii) a smooth behavior. This is due to the large scales of high and low pressure zones. Therefore, the hydrostatic delay observed in interferograms is smooth as well and manifests itself usually as a phase trend of maximally a few millimeters over the entire interferogram. Such phase trends are likely to occur due to orbit errors as well, see fig. 4.23, and are often corrected for using tie-points. As a result, the hydrostatic delay has a very limited influence on interferograms, and can be safely ignored for spatial scales of $\sim$50 km and smaller. If spatially distributed surface pressure measurements are available, these can be used to correct for the hydrostatic delay component.

The wet part of the delay or refractivity, parameterized by temperature T and the partial pressure of water vapor e, is much more spatially variable then the hydrostatic delay, while the variations have a considerable effect on the observed phase delay. The sensitivity of the observed delays to the wet part of the refractivity is discussed in the following section.

6.2 Refractivity sensitivity analysis

This section elaborates on the interpretation of lateral delay variations, emphasizing on the possibilities to distinguish the contribution of temperature and moisture (partial water vapor pressure) to the total refractivity. To select feasible atmospheric parameters, the partial pressure of water vapor is ranging from 0 hPa to 30 hPa, and the temperature is ranging from $-40°$C to $40°$C. We can express the *relative humidity*, RH, by

$$RH = \frac{e}{e_s}, \tag{6.2.1}$$

where the saturation pressure for a certain temperature, e_s, can be found using the Clausius-Clapeyron equation (Stull, 1995)

$$e_s = e_0 \exp\left(\frac{L}{R_v}\left[\frac{1}{T_o} - \frac{1}{T}\right]\right), \tag{6.2.2}$$

where $e_0 = 6.11$ hPa, $T_0 = 273.16$K, T is the temperature in Kelvin, and L is the latent heat. Over a flat water surface, $L = 2.5 \times 10^6$ J kg^{-1}, which is the latent heat of vaporization. In this case, cloud droplets are being formed. The relative humidities for these ranges of temperature and partial water vapor pressure are represented by the iso-lines in fig. 6.2.

6.2.1 Clear air refractivity decomposition

Under the assumption that we can ignore or derive the hydrostatic component of the delay as discussed in the previous section, we can concentrate on the wet component. Using the clear-air wet tropospheric part of eq. (6.1.8), the partial derivatives can be derived to determine the sensitivities:

$$\partial N_{\text{wet}}/\partial T = -\frac{k_2' e}{T^2} - \frac{2k_3 e}{T^3}, \tag{6.2.3a}$$

$$\partial N_{\text{wet}}/\partial e = \frac{k_2'}{T} + \frac{k_3}{T^2}. \tag{6.2.3b}$$

Note that a temperature increase yields a decrease in refractivity. Hence a localized area with a higher temperature will show up as a relative increase in propagation velocity, whereas a localized area with higher water vapor pressure will show up as a relative decrease in propagation velocity. Figure 6.2 shows the ratio $R =$

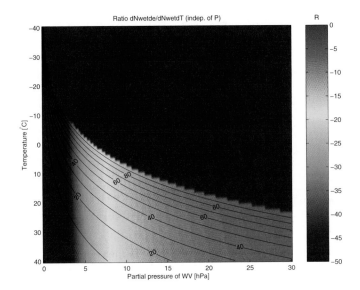

Fig. 6.2. Sensitivity ratio $R = \frac{\partial N_{\text{wet}}}{\partial e} \Big/ \frac{\partial N_{\text{wet}}}{\partial T}$ of the wet part of the refractivity, N_{wet}, dependent of the water vapor pressure e and of the temperature T. Relative humidity isolines are superposed.

$\frac{\partial N_{\text{wet}}}{\partial e} \Big/ \frac{\partial N_{\text{wet}}}{\partial T}$. In other words, we compare (i) the sensitivity of N_{wet} for a 1 hPa change in the partial pressure of water vapor e with (ii) the sensitivity of N_{wet} for a 1°C change in temperature T. If this ratio $|R| > 1$ the refractivity is R times more sensitive for a 1 hPa change in e than for a 1°C change in T. On the x-axis, the partial pressure of water vapor is plotted, on the y-axis the temperature. By combining a specific e and T value it is possible to evaluate the sensitivity ratio. Relative humidity isolines are superposed to show the likelihood of a specific e–T combination. From fig. 6.2, we may conclude that the sensitivity ratio R has wide range of values. Although some very high ratio's are apparent for very dry air (less than 5 hPa water vapor pressure), we can safely state that $4 < |R| < 20$. The ratio is lowest for high temperatures combined with high partial water vapor pressures. These conditions are likely in hot and humid regions, especially near the earth's surface. For cloud development, reaching higher altitudes, temperatures drop with $\sim$6.5 K/km, which increases the ratio.

As a result, the *wet* part of the refractivity is in any scenario at least 4–20 times as sensitive to a 1 hPa change in the partial pressure of water vapor than to a 1°C change in temperature. Moisture variations of 1 g kg^{-1} (1.2 hPa) are common even on a 1-km spatial scale (Weckwerth et al., 1997). At 0°C, such a variation already produces 6 mm delay per vertical km.

6.2.2 Propagation through clouds and droplets

From the signal delay obtained from an interferogram, precipitable water vapor differences can be obtained, under the assumption that the atmosphere is a clear gas. However, the occurrence of clouds (liquid water droplets in a saturated environment) and precipitation demonstrates that this assumption is obviously often not correct.

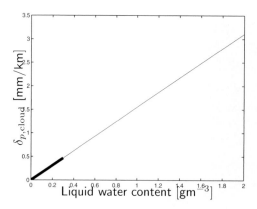

Fig. 6.3. Delay caused by liquid water in clouds, assuming a homogeneous cloud layer of 1 km thickness. An incidence angle of 21° has been assumed. The bold part of the line indicates typical values for non-precipitating clouds (Russchenberg, 2000).

In many space geodetic techniques where the *total* wet delay is measured, the contribution of the droplets is often neglected, since it is only a small part (estimated 1–5% by Kursinski et al. (1997)) of the total wet delay, hence 1–15 mm. InSAR, however, is effectively blind for the total delay, but observes lateral differences in delay instead. Although maximum lateral differences of 10–12 cm have been observed, see Hanssen (1998), the sensitivity of the radar measurements allows for detecting mm-scale fluctuations. Therefore, it might not be sufficient to restrict to the clear gas hypothesis. In the following, we investigate the influence of droplets and comment on the sensitivity of the precipitable water vapor estimate.

Phase delay induced by droplets

The phase shift of spaceborne C-band radar pulses caused by liquid water (droplets) can cause limited additional signal delay in the interferometric observations. The interaction of radar waves with droplets is a *forward scattering* problem: the wave induces a dipole moment in the droplet, which will act as a secondary wave front. After passing the droplet, the principal (undisturbed) wave front will interfere with the secondary wave front and hereby cause a phase shift.

Hall et al. (1996) have listed the liquid water content, W, of clouds, see table 6.1. The liquid water content is the particle number density times the volume per particle times the density of liquid water. Its maximum is usually found at $\sim$2 km above the cloud base, and can be related to the dielectric refractivity using the Clausius-Mossotti equation, (Solheim et al., 1997):

$$N_{\text{cloud}} = \frac{3}{2} \frac{W}{\rho_w} \frac{\varepsilon_0 - 1}{\varepsilon_0 + 2}, \qquad (6.2.4)$$

where ε_0 is the permittivity of water and ρ_w is the density of liquid water. This relation is independent of the shape of the cloud droplet (Born and Wolf, 1980). Although the permittivity of water is a weak function of temperature, it is possible to approximate eq. (6.2.4) to within 1% by (Solheim et al., 1997):

$$N_{\text{cloud}} = 1.45\,W. \qquad (6.2.5)$$

Table 6.1. Liquid water content in clouds, after Hall et al. (1996) and Bean and Dutton (1968)

Type of cloud	Liquid water content [g/m³]	Slant delay [mm/km]
Stratiform clouds	0.05–0.25	0.1–0.4
Small cumulus clouds	0.5	0.7
Cumulus congestus and cumulonimbus	0.5–2.0	0.7–3.1
Ice clouds	< 0.1	< 0.1

Since the value of N_{cloud} reflects the difference with the vacuum refractive index, eq. (6.1.8) can be used to determine the additional delay $\delta^{z,t_i}_{p,\text{cloud}}$ caused by the liquid water in clouds:

$$\delta^{z,t_i}_{p,\text{cloud}} \;\; [\text{in mm}] = \frac{1.45}{\cos\theta_{\text{inc}}} W\,L, \tag{6.2.6}$$

where L is the thickness of the cloud layer in km. Note that cloud-droplet refractivity is in fact *dispersive*. However, since the dispersive part of the refractivity is much smaller than the non-dispersive part it can be ignored for C-band frequencies. Table 6.1 shows the zenith delay for four cloud groups. For repeat pass SAR interferometry, stratiform clouds and ice clouds do not cause large phase disturbances, due to their large horizontal extent and small additional delay. However, especially the cumulus type of clouds can result in a significant additional phase delay, as they have a relatively limited horizontal size combined with a large vertical height and liquid water content.

6.2.3   Comparison of water vapor and liquid water delay

Neglecting the hydrostatic and the ionospheric term in eq. (6.1.8) we find the refractivity expressed in the wet (vapor) and liquid terms:

$$N = k'_2 \frac{e}{T} + k_3 \frac{e}{T^2} + 1.4W. \tag{6.2.7}$$

If precipitable water vapor is defined as

$$PWV = \frac{1}{\rho_{\text{l}}} \int \rho_{\text{v}} dh, \tag{6.2.8}$$

where ρ_{l} is the density of liquid water (10^6 g/m³), and ρ_{v} is the density of water vapor, we can relate the precipitable water vapor to the delay observed in the interferogram, eq. (6.1.8), using the equation of state $e = \rho R_v T$, as

$$\delta^{z,t_i}_{p,\text{v}} = 10^{-6} \rho_{\text{l}} R_v (k'_2 + k_3/T_m) \frac{1}{\rho_{\text{l}}} \int \rho_{\text{v}} dh, \tag{6.2.9}$$

where $R_\mathrm{v} = 461.524$ [J K^{-1}kg^{-1}], and T_m is the mean temperature of the column containing the water vapor, see section 6.6. This can be written as

$$\delta_{p,\mathrm{v}}^{z,t_i} = \Pi^{-1}\, PWV. \tag{6.2.10}$$

If the amount of precipitable liquid water is given by

$$PLW = \frac{1}{\rho_\mathrm{l}} \int W\, dh = 10^{-6} \int W\, dh \quad [m], \tag{6.2.11}$$

this can be related to the delay induced by the droplets $\delta_{p,\mathrm{l}}^{z,t_i}$ by

$$\delta_{p,\mathrm{l}}^{z,t_i} = 1.4\, PLW. \tag{6.2.12}$$

Using a typical value for $\Pi \approx 0.15$, (Bevis et al., 1996), we obtain the following approximations

$$\delta_{p,\mathrm{v}}^{z,t_i} = 6.5\, PWV, \quad \text{and} \tag{6.2.13}$$

$$\delta_{p,\mathrm{l}}^{z,t_i} = 1.4\, PLW. \tag{6.2.14}$$

For a cumulus congestus, cf. table 6.1, $W \approx 1$ g/m^3. Using a cloud depth of 4 km, and using eq. (6.2.11), we find $PLV = 4$ mm. Following eq. (6.2.14), this yields 5.6 mm of zenith signal delay, or 0.2 phase cycles in the interferogram, which is very well detectable. For other cloud types, such as stratiform clouds, ice clouds, and small cumulus clouds, the effects are less than 1 mm, under usual atmospheric circumstances. It is important to note that cloud droplets will only occur in a saturated air mass, and therefore the water vapor concentration will be relatively high. Hence, for the delays observed in the interferogram water vapor will be the main driving force.

6.2.4 Discussion

These theoretical studies show that the sensitivity of the tropospheric refractivity is highest for spatial variation in water vapor content, between 4 and 20 times greater than for temperature variation. This sensitivity, in combination with the fact that water vapor exhibits a strong spatial variability, supports the interpretation of small scale (< 50 km) phase delay variations in terms of water vapor. Nevertheless, the results show that temperature effects and liquid water in the troposphere cannot be ignored, and result in deviations in the quantitative analysis of the delays. Regarding liquid water, especially cumulus congestus clouds, cumulonimbus clouds, and precipitation will modulate the observed water vapor quantities, even though the delay differences due to the strong water vapor gradients will be dominant. Computing precipitable water vapor without taking the liquid effects into account may result in an overestimation of PWV of, say, less than 10%.

For the sake of completeness it needs to be stressed that, applying radar interferometry over a relatively flat area, there is no sensitivity for the vertical profiles of e, P, T, and W as long as changes in these profiles are homogeneous for the image.

There is only sensitivity for lateral inhomogeneities in the refractivity N, hence, lateral inhomogeneities of P, T, e, and W. Note that a different vertical layering during the two SAR acquisitions will, however, influence the interferogram if significant topography is present, see section 4.8. The signatures of these effects will have strong correlation with the topography, which provides a possibility to identify them. For the test sites analyzed here, with height variation within a range of 100 m, no topographic induced effects are expected nor observed.

6.3 Ionospheric influence

The ionosphere can be physically characterized by the ability of external sources to "knock-off" electrons from atoms, hereby creating *free* electrons that are not bound to their remaining ions. These external sources are mainly the solar ultraviolet radiation and energetic electrons of solar and magnetospheric origin. The number of (free) electrons is represented by the *electron density*, in electrons per m^3.

Although the electron density is temporally and spatially variable, the ionosphere is often treated as a spherical shell between 60 and 600 km height with a *constant* electron density in height, homogeneously varying in time. Radio signals traversing through the ionosphere are dispersively delayed along their paths by interactions with the free electrons. The dispersive (frequency dependent) nature of the delay is the key for e.g. dual-frequency GPS to estimate the ionospheric (or plasma) delay. Physically, the ionospheric delay is a path integral through the ionospheric electron density. This integral therefore has the dimension electrons per m^2, and is commonly known as the Total Electron Content or TEC. If the path integral is taken in the zenith direction it is known as the Vertical TEC (VTEC). To achieve comprehensible quantities, one TEC unit (TECU) is equivalent to 10^{16} electrons/m^2. Using dual frequency GPS observations collected by the IGS network (International GPS Service for Geodynamics), maps of the Vertical TEC can be produced, see fig. 6.4, showing two-hourly vertical TEC over Europe, provided by the DLR Remote Sensing Ground Station Neustrelitz.

The delay of a radio wave, propagating through the ionosphere, is different for the phase and the group velocity. Phase delay depends on electron content and affects carrier signals. Group delay depends on dispersion in the ionosphere as well, and affects signal modulation, i.e. the codes on GPS signals. The phase and group delays are of the same magnitude but have an opposite sign. The group signal delay $\delta_{p,\text{iono}}$ through the ionosphere can be accurately represented by a series in the reciprocal of the carrier frequency f:

$$\delta_{p,\text{iono}} \approx A/f^2 + B/f^3 + C/f^4 + \dots \tag{6.3.1}$$

Dual-frequency GPS receivers observe a differential delay $\Delta\delta_{p,\text{iono}}$ between the two frequencies L_1 and L_2, which enables estimation of the unknown A, B, and C coefficients.

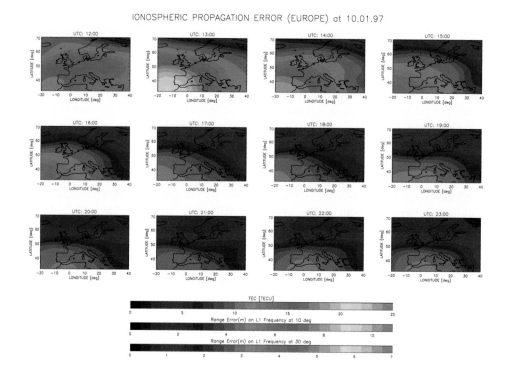

IONOSPHERIC PROPAGATION ERROR (EUROPE) at 10.01.97

Fig. 6.4. Ionospheric map for 10 Jan 1997 (DLR Neustrelitz). The subimages show the hourly change in TEC from 12:00 UTC to 23:00 UTC. The daily variation can be observed as well as an ionospheric storm over south Spain and Morocco. The lower-two color bars give the range error for the L_1 frequency, one of the two GPS frequencies, at different elevation angles.

6.3.1   The mapping function

Radio signals that are not incident from zenith direction are generally delayed (or advanced) as well as bended. To derive the mapping function, M, that accounts for these effects, an idealized single-layer approximation is often applied. The altitude at which a ray between satellite and the earth's surface pierces the layer with maximal electron density is defined as the *effective ionospheric height* h_{sp}. The projection of this point on the ground is labeled the *sub-ionospheric point*, indicated by the subscript 'sp'. In the single-layer approximation, it is assumed that the whole electron density profile, which varies with altitude, is reduced to a very narrow spherical layer at this effective ionospheric height. In general, an effective ionospheric height $h_{sp} = 400$ km is used.

For calculating the mapping function we need the zenith angle χ of the rays at the sub-ionospheric point at height h_{sp}. The mapping function is then defined as

$$M = 1/\cos\chi, \tag{6.3.2}$$

where the cosine of the zenith angle can be approximated by

$$\cos\chi \approx \sqrt{1 - \left(\frac{R_e\cos\varepsilon}{R_e + h_{sp}}\right)^2}, \tag{6.3.3}$$

where R_e is the earth's radius and ε is the elevation angle relative to the horizon. Using the ERS SAR incidence angle $\theta_{inc} = 90° - \varepsilon = 23°$, the earth radius R_e of 6371 km and effective ionospheric height $h_{sp} = 400$ km, the value of the mapping function is $M(23°) = 1.07531$.

6.3.2   Ionospheric delay

The zenith ionospheric range error for point p at $t = t_i$ is in first approximation given by the formula:

$$\delta_{p,iono}^{z,t_i} = \frac{K}{f^2}\,\text{TEC}, \tag{6.3.4}$$

where $\delta_{p,iono}^{z,t_i}$ is the zenith range error in meters, f is the frequency of the signal and TEC is the Total Electron Content. The factor $K = -40.28$ m^{3}s^{-2} (Jakowski et al., 1992). This is the ionospheric delay in the zenith direction. TEC values may vary between 0 m^{-2} at night to 20×10^{16} m^{-2} at the minimum of the solar cycle to 100×10^{16} m^{-2} at the solar maximum. Note that the sign of the factor is negative, i.e., an increase in TEC results in a phase advance.

For a SAR acquisition, under an incidence angle of approximately $23°$ and a radar frequency of 5.3 GHz, this yields

$$\delta_{p,iono}^{t_i} = \frac{K}{f_{SAR}^2}M(23°)\,\text{TEC} \approx -1.54 \times 10^{-18}\,\text{TEC}. \tag{6.3.5}$$

Taking the derivative of this function and using TEC-units (TECU), we obtain for C-band SAR

$$\partial\delta_{p,iono}^{t_i}/\partial\text{TECU} = -0.015. \tag{6.3.6}$$

In other words, for a repeat-pass interferometric cycle of 28 mm, a change in 1 TECU over the scene yields a phase ramp of approximately half a cycle. Of course, we have to take into account that an interferogram is the difference of two SAR acquisitions during different days, but at the same local time. The status of the ionosphere at both instances needs to be accounted for. Therefore, an identical TEC ramp in both images, e.g., due to the diurnal variation, will cancel in the interferogram. Examples of the quantity of delay are given in table 6.2.

Following eq. (6.3.6), for ionospheric effects to have an observable influence in a SAR interferogram, the dominant wavelength should be less than 100 km, to distinguish

Table 6.2. Ionospheric delay for X-band, C-band, and L-band wavelengths.

band	f_0 [GHz]	θ_{inc}	$\frac{\partial \delta^{t_i}_{p,\text{iono}}}{\partial \text{TECU}}$ [m]	$\delta^{t_i}_{p,\text{iono}}$ [m] Solar min	Solar max
X	9.6	30°	−0.005	−0.10	-0.50
C	5.3	23°	−0.015	−0.31	-1.54
L	1.25	39°	−1.241	−6.40	-31.99

$\delta^{t_i}_{p,\text{iono}}$, one-way ionospheric delay. The factor $\partial \delta^{t_i}_{p,\text{iono}}/\partial\text{TECU}$ shows the sensitivity of the SAR data for ionospheric disturbances. Solar minimum and solar maximum conditions correspond with 20 and 100 TECU, respectively.

them from orbit errors, while the amplitude should be more than 0.36 TECU, to get a signal of 0.2 phase cycles. Moreover, the anomaly should appear in only one of the two acquisitions.

The only natural phenomena that could possibly cause such effects are small scale Traveling Ionospheric Disturbances (TIDs), with a wavelength of tens of km (Spoelstra and Yi-Pei, 1995; Spoelstra, 1997). These TIDs are wave effects, mostly propagating from polar regions to regions with lower latitudes.

6.3.3 Spatial scales

The ionosphere can be divided in a number of layers, which have different characteristics. The D-layer (80–100 km) receives only a minor part of the solar radiation, since most of the energy has been absorbed by higher ionospheric layers, and is therefore only weakly ionized. After sunset, this layer disappears completely. The E-layer (100–140 km) has little more ionization, but mostly fades into the F1-layer at night. The F1 and F2 layers, are located at 140–200, and 200–400 km, respectively (Afraimovich et al., 1992). TIDs occur mainly in the F-layers.

Medium-scale TIDs have scale lengths of 100–200 km, time scales of 10–20 min, and result in a 0.5–5% variation in the total electron content. Large-scale TIDs are relatively uncommon. They have scale lengths of 1000 km, time scales of hours, and can cause up to 8% variations in the total electron content (Thompson et al., 1986, p.449). Smith et al. (1950) found that irregularities in the ionization (blobs) occur at scale sizes of a few kilometers or less. It is not reported, however, how large these disturbances are as a fraction of the total electron content (Thompson et al., 1986).

Ionospheric gravity waves occur, e.g., when a sudden disturbance in electron density (e.g., a solar eclipse, waves over a mountain edge, heavy thunderstorms, earthquakes or rocket launches) triggers an expanding wave front (Calais and Minster, 1995, 1996; Ho et al., 1998; Calais and Minster, 1998). These waves normally have enormous wavelengths (hundreds of kilometers), and have gravity as restoring force. Cheng and Huang (1992) reported waves with a wavelength between 160 and 435 km after the eruption of Mount Pinatubo in 1991.

6.3.4   Comparing tropospheric and ionospheric signal

The difference between ionospheric and tropospheric effects on SAR interferometry is based on two characteristics: the sign of the delay and the shape of the feature. From eq. (6.3.5), it follows that an increase in the ionospheric electron content results in a decrease of the observed range, or a *phase advance*. An area of increased partial water vapor pressure, see eq. (6.1.7), results in an increase in the observed range, or a *phase delay*. However, there are three complications. First, one has to know during which of the two SAR images the effect occurred: a phase *advance* in the first image gives the same result as a phase *delay* in the second one. Second, one has to distinguish which area in the interferogram can be regarded as reference area, where the phase is relatively undisturbed: a localized area with *reduced* water vapor density (e.g., a "hole" in a cloud layer) yields a localized relative *advance* of the phase. Third, a localized increase in temperature may also result in a relative phase advance, although this effect usually has a limited magnitude.

The first complication can be resolved by using more than two SAR images. The different interferometric combinations of images enables the unambiguous identification of the image containing anomalies, as long as other error sources such as temporal decorrelation allow for such an evaluation. As long as single interferometric pairs are considered, it is not possible to distinguish between phase delay and advance based on only interferometric data.

The second complication, however, is much harder to circumvent. Just as localized areas with an increased water vapor content occur frequently, in a relative sense also areas with a lesser amount of water vapor occur. This can be clearly observed in, e.g., Meteosat water vapor channel imagery. Therefore, without any additional information it is not possible to uniquely identify ionospheric effects in interferograms based on the phase sign of the feature.

The remaining possibility for identification is based on the shape and characteristics of the feature. The main difficulty in this respect is the limited knowledge about the spatial characteristics of the ionosphere within ranges of less than 100 km. Although different devices using dual frequencies have observed the ionosphere, their results have either been only one-dimensional integrated profiles in time (as is, e.g., possible with GPS), or large-scale interpolated ionospheric maps that do not have the spatial resolution needed for InSAR.

Based on these evaluations, it is difficult to uniquely identify ionospheric effects in SAR interferograms at present, unless strong support from additional sources is available. It should be noted that further research in this field, using, e.g., dual-frequency SAR interferometry, could be extremely valuable for its ionospheric imaging capacities.

6.3.5   Reported results

Saito et al. (1998) used GEONET, the permanent GPS array of Japan with a mean distance between the receivers of 25 km, to map the two-dimensional TEC perturbations. TIDs, with a wavelength of 300–400 km, traveling with a speed of ~150 m/s, were observed. The amount of TEC variation was maximally 0.8 TECU. This would result in 1.2 cm phase variation in a C-band SAR interferogram. However, for these wavelengths such variations would result only in a slight curvature of the interferometric phase.

Gray et al. (2000) observed ionospheric wave effects by examining the azimuth offset vectors during the coregistration of two SAR images over the Antarctic (81.3°S). This resulted in a wave-pattern with wavelengths of ∼5 km which could be explained by a variation of 0.12 TECU (less than 2 mm zenith delay difference) during one of the SAR acquisitions. The variable electron content caused an erroneous Doppler frequency, resulting in an offset of the coverage of resolution elements in azimuth. In the interferometric phase these effects were not visible, likely due to their limited magnitude and the phase contribution of topography, deformation, and tropospheric signal.

Massonnet and Feigl (1995a) reported ionospheric effects in interferograms over Landers, California (34.5°N). They identified a kidney-shaped anomaly by pair-wise logic, to occur in a single SAR acquisition. This anomaly covered an area of 10×30 km, oriented approximately north-south, with a maximum amplitude of ∼3 cm (approximately one fringe). The main reason for assuming an ionospheric driving mechanism is that the anomaly shows a phase advance, instead of a phase delay. Massonnet et al. (1995) show another example of presumed ionospheric origin over Etna volcano, Italy (37.7°N). Here a triangle shaped anomaly with sides of ∼4 km and a phase advance of approximately one fringe is observed.

A number of arguments can be listed that oppose this argumentation. First, the magnitude of the effects is probably too large. From eq. (6.3.6) we find that a rather large local increase in electron content of 2 TECU is needed to explain 3 cm magnitudes. For example, TIDs observed by Saito et al. (1998) with a wavelength of 300–400 km show a maximum effect of 0.8 TECU. It seems reasonable to assume that amplitudes decrease for decreasing wavelength, following some sort of power-law behavior. This argument is supported by the observations by Gray et al. (2000), which show that wave-like features with a wavelength of ∼5 km in auroral zones have a magnitude of 0.12 TECU. Second, the size, shape, and isolated nature of the two observed anomalies is peculiar. Known ionospheric anomalies behave like wave effects. Even ionospheric perturbations observed after a Space Shuttle ascent—a very discrete phenomenon—show wave effects, with a magnitude of 0.04–0.2 TECU (Calais and Minster, 1996). The wave effects observed by Gray et al. (2000) and Joughin et al. (1996) had wavelengths of several km's, but extending over hundreds of km's. The isolated nature of the anomalies observed by Massonnet and Feigl (1995a) and Massonnet et al. (1995) does not correspond with any form of anomaly found in literature. Finally, although this does not necessarily prove that such anomalies cannot exist, the lack of a possible driving mechanism for the presumed ionospheric perturbations makes this hypothesis less likely.

An alternative hypothesis to explain the two examples can be easily found by considering a tropospheric explanation. For example, a relative phase advance can result from a cloud, or (better) water vapor, cover with "holes" in it. A layer of homogeneous stratus or stratocumulus cloud could easily result in a fringe of phase delay. In fact, if such a layer would cover the entire interferogram, it would not even be observed over a flat area. A localized area with less water vapor and/or cloud droplets would result in a relative phase advance as well. Finally it needs to be noted that a localized increase in temperature will also result in a phase advance, see section. 6.2.1, although the magnitude of these effects is usually limited to a couple of millimeters.

6.3.6   Discussion

Based on (i) the theoretical considerations discussed in this section and (ii) the experimental results based on GPS and InSAR observations, we conclude that until today the influence of ionospheric disturbances on repeat-pass spaceborne C-band radar interferograms is at the least controversial, especially for mid-latitudes. The results of Gray et al. (2000) give a first proof of ionospheric effects in auroral zones, even though the derived magnitudes were not clearly identified and validated by analyzing the interferometric phase. Therefore, we currently adapt the hypothesis that ionospheric effects may result in long wavelength gradients/curves over a single SAR image, but will not noticeably affect phase variations at scales less than ~50 km.

6.4   Water vapor mapping

Three SAR interferograms, obtained over the Netherlands (fig. 6.5), show several prominent and representative features. The SAR data were acquired by the ERS-1 and ERS-2 satellites, operating in a one-day interval mode, ensuring sufficiently coherent images over land areas. The principle ground resolution of 4×20 m is spatially averaged to 160×160 m, reducing the delay standard error from a couple of millimeters to below 1 mm. Maps of the water vapor distribution associated with a precipitating cloud, a partly precipitating cold front, and horizontal convective rolls reveal quantitative measures that are not observed with conventional methods, and suggest that such radar observations can be used for forecasting and for studying atmospheric dynamics. These result have been reported by Hanssen et al. (1999).

6.4.1   Precipitating cumulonimbus

The signature of a precipitating cumulonimbus cloud is shown as a localized delay difference in the SAR line of sight, caused by the cloud and the saturated sub-cloud layer see fig. 6.6A. Using the method described in Davis et al. (1985); Bevis et al. (1992) and a simple cosine mapping function, see eq. (6.1.4), the delay differences are mapped to differences in zenith integrated precipitable water (ΔIPW), the vertically integrated water vapor liquid equivalent. The positive sign of the delay indicates that the feature appeared in the first of the two combined SAR images. This inference is verified by the weather radar reflectivity signal, expressed as rain rate, on the first day (fig. 6.6B). During the second SAR acquisition, no precipitation was observed in the weather radar data. Surface observations and radiosonde data estimated that the cloud base height was 800 m; weather radar data indicated that cloud tops were as high as 4.5 km.

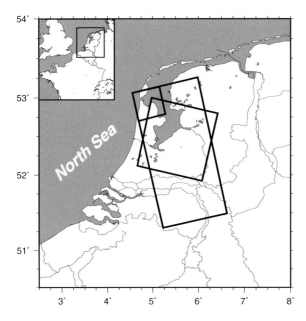

Fig. 6.5. The location of the discussed radar interferograms in section 6.4. Figure 6.6 is located at the small square in the upper-left corner of the rectangle. Figure 6.7 is indicated by the large rectangle, and fig. 6.8 by the big square.

The shape and magnitude of the anomaly in the interferogram correspond closely to the weather radar echo. Phase variations over areas with no precipitation suggest that water vapor gradients are seen in the interferogram that are not observable by the weather radar. The ΔIPW signal over the cloud has a mean value of 1.3 mm, which is reasonably close to the 0.9 mm value inferred from a radiosonde profile considering it was launched about 90 km away and 4 hours prior to the SAR overpass. The interferometric technique can measure only relative changes in water vapor, nonetheless the data reveal that this storm system has approximately 2.6×10^8 liters more water compared to the surrounding air. For this system to reach equilibrium, this excess water would have to be redistributed over a larger area or precipitate out. Such numerical constraints form important information for forecasting purposes.

6.4.2 Partly precipitating cold front

A cold front, propagating from the northwest, is shown in fig. 6.7A. The interferogram reveals a narrow diagonal band of enhanced water vapor, suggesting the presence of a narrow cold-frontal rain band (Parsons, 1992). The weather radar data (fig. 6.7B), however, show that only parts of the band are precipitating. Surface winds (fig. 6.7B) indicate some convergence at the cold front. The frontal circulation may have produced a localized maximum in magnitude or depth of water vapor at the convergence zone, visible as the yellow band in the interferogram. Studies using in situ aircraft and radiosonde observations suggest that regions of enhanced moisture may occur at low-level convergence zones (Simpson, 1994). The sickle-shaped area visible at the southern end of IJssel Lake likely reflects an evaporatively

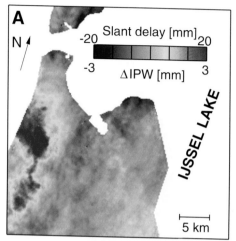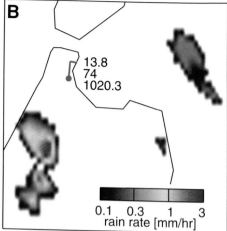

Fig. 6.6. Images of precipitation. **(A)** The SAR interferogram (29 and 30 Aug 1995, 21:41 UTC) shows slant delay variation, mapped to zenith integrated precipitable water differences. The average of the undisturbed area is set to zero, yielding relative precipitable water estimates. **(B)** shows the weather radar rain rate (29 Aug 1995, 21:45 UTC). The surface wind velocity is 4.1 m/s, from 350 degrees, indicated by the wind barb. Temperature (°C), the percentage of relative humidity, and pressure (hPa) are plotted beside the station.

cooled surface outflow (gust front) generated by the rain showers.

The post-frontal relatively cold and humid air produces a delay of about 28 mm compared to the pre-frontal, relatively warm air. This increase is partly caused by an overall temperature difference of approximately 3°C associated with the two air masses. The frontal band is 3–5 km wide and has an average increase in delay between 11 and 64 mm. Using a cold front depth of 3 km, this delay could either be produced by a 3 hPa increase in water vapor or by an unreasonably high temperature decrease (10°C) at the frontal boundary. Additionally, the contribution of liquid water in a cumulus cloud line would produce a maximum delay difference of less than ~2 mm (Kursinski et al., 1997). Thus water vapor, rather than temperature or liquid water, is likely the primary mechanism causing the increased delay at the frontal boundary.

The areas with delay differences of more than 30 mm correspond to precipitation regions. The sudden increase of water vapor content in the sub-cloud layer, caused by partial evaporation of the precipitation, accounts for this increase. The liquid cloud and rain particles cannot produce the enhanced delay, as the precipitating clouds would have to be over 21 km deep to account for the maximum observed delay of 64 mm. The observed features in the interferogram illustrate how localized areas with an increased water vapor content may give an indication for the occurrence and development of precipitation.

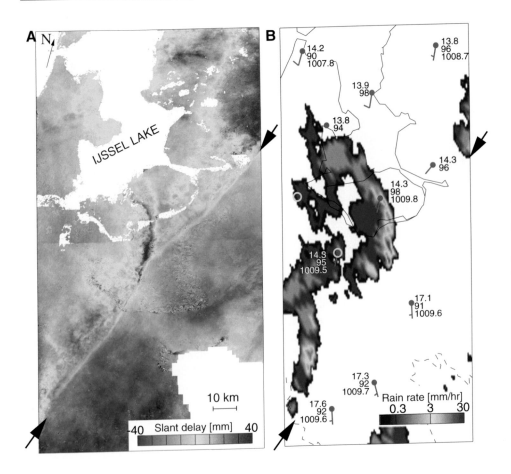

Fig. 6.7. The effect of a cold front. **(A)** SAR interferogram of 3 and 4 Oct 1995, 21:41 UTC. The boundary of the cold front is visible as the line between the arrows. Water areas and areas where no elevation data were available are shown in white. **(B)** Weather radar image from 4 Oct 1995, 21:45 UTC. The position of the two weather radars is indicated by the yellow circles. Superposed are the surface observations (22:00 UTC) as in fig. 6.6B. A half wind barb corresponds to 2.5 m/s and a full barb to 5 m/s.

6.4.3 Boundary layer rolls

The interferogram in fig. 6.8 illustrates the frequently-occurring phenomenon of horizontal convective rolls, a form of boundary-layer convection in which there are counter-rotating horizontal vortices with axes aligned along the mean boundary-layer wind direction (LeMone, 1973; Brown, 1980). Some evidence that the linear features of the interferogram depict rolls rather than atmospheric waves is that the bands are oriented along the wind direction, as shown by the surface station data overlaid on the interferogram (fig. 6.7A). Moreover, radiometer satellite imagery

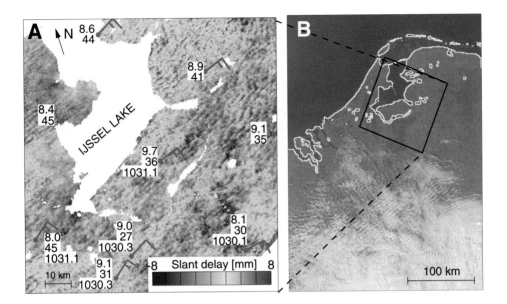

Fig. 6.8. (A) SAR interferogram of 4 and 5 Apr 1996, 10:35 UTC, showing "water vapor streets," caused by horizontal convective rolls. Surface station observations of 4 Apr, 11:00 UTC are superposed as in fig. 6.7B. **(B)** Radiometer (NOAA-AVHRR, channel 1, 2, and 4) color composite image acquired at 4 Apr 1996, 12:01 UTC.

(fig. 6.7B) to the south of the interferogram location shows some evidence of cloud streets oriented with the interferogram bands. Sufficient moisture in the presence of rolls often produces cloud streets atop roll updraft branches (LeMone and Pennell, 1976). Furthermore, a local radiosonde on 4 April 1996 at 12:00 UTC indicated a boundary layer depth of 800 m. This result, combined with the band spacing of 2–3 km, gives an aspect ratio of 2.5–2.9, consistent with the presence of rolls (Kuettner, 1971).

Roll updrafts are warmer and moister than roll downdraft branches (Weckwerth et al., 1997). Moisture variability is likely dominant in the observed 6–10 mm delay differences, because vapor pressure variations associated with rolls (3 hPa) would produce a delay of 13 mm, while temperature variations of 0.5°C would produce less than 1 mm delay. The observations show how streets of water vapor can be identified even when cloud streets are not visible. In fact, no clouds were observed from the weather stations in the interferogram area. Knowledge of these systematic patterns of water vapor can aid in forecasting weather behavior.

6.4.4   Discussion

Radar interferometry can provide worldwide coverage of mesoscale atmospheric phenomena over land and ice, and easily monitor more exotic features such as gravity waves accompanying fronts. Readily attainable integrated precipitable water amounts over large areas may make pinpoint weather forecasting a possibility. The delay maps could eventually be used by bench forecasters or serve as an additional constraint in variational data assimilation models. Contemporary satellites, however, have an orbital repeat period that is far from ideal for operational applications, resulting in loss of coherent phase information over many areas. Ideally, SAR data should be acquired with multiple short temporal spacings in order to maintain high coherence and permit a large number of interferogram pairs to be generated. Multiple observations give not only better precision in time, but keep coherence and hence data quality high. Proper use of airborne and spaceborne SAR systems with short repeat periods could lead to much greater accuracy in meteorological understanding and forecasting.

6.5   Combining amplitude and phase information

In the previous section it was shown how observations from radar interferometry can be used to extract information on the moisture distribution in the boundary layer. Here we demonstrate how the same radar observations can also be used for wind field measurements, which enables an integrated analysis of wind and moisture parameters from a single sensor. Since mesoscale surface winds modulate fluxes of momentum, heat, and moisture—the driving forces of atmospheric circulation—there is close correspondence between the wind field and the transport and distribution of moisture. Examples include, e.g., cumulus convection and atmospheric dynamics near fronts. Improved understanding of both spatial and temporal behavior of air masses is important for parameterizations in weather forecasting and climate models.

Generally, basic meteorological parameters such as wind velocity and water vapor distribution are not measured on km-scales (see, e.g., Stoffelen (1998) for wind measurements and Susskind et al. (1984) for temperature and humidity information). A related problem is that wind and moisture observations are usually acquired by different sensors, that don't necessarily coincide in time. For the analysis of causal connections between both types of parameters on scales between 100 m and 10 km, temporal coincidence is imperative due to their spatial and temporal variability.

Here we elaborate on two case studies exploiting both the radar backscatter intensity and phase information. This approach provides a more thorough examination of previous findings, which were based on phase information only. Using two pairs of interferometric SAR observations the moisture distribution over land areas is retrieved from the interferometric phase, whereas wind information is retrieved from the backscattered radar energy over water areas. The case studies constitute a comprehensive quantitative examination of vortices associated with a system of boundary layer rolls and with a rain band in relation to a cold front. This study has been reported by Hanssen et al. (2000b).

6.5.1 Methodology and data analysis

Spaceborne SAR provides fine-resolution radar images over broad areas. The resolution of the ERS images analyzed here is ~20 m over a swath of 100 km in width. The radar data comprise a grid of complex vectors, one for every pixel. The length of the vector is a measure for the backscatter intensity and yields information on the roughness of the imaged surface. Over water surfaces, the backscatter intensity is dominated by Bragg scattering (Curlander and McDonough, 1991). Contrasts in surface roughness, quantified in the normalized radar cross section (σ^0), can be interpreted as differences in mainly wind speed and direction (Stoffelen, 1998). Sea state or heavy rain are further atmospheric phenomena that may influence the intensity. SAR images can detail small-scale structures over seas such as boundary layer rolls (Alpers and Brümmer, 1994). Here we used a C-band model, CMOD4, to compute the wind speed from the normalized radar cross section and the incidence angle of the radar waves (Stoffelen, 1998). This model correlates backscatter intensity with the wind at 10 m height. Note that other models correlate to a so-called neutral equivalent wind, but in that case accurate ancillary information is needed on atmospheric stratification (Stoffelen, 1998).

Here we eliminate phase signal due to geometric path length differences using an elevation model (Massonnet et al., 1993). As the interferometric pairs are acquired with a time interval of only 24 hours, coherent phase information is obtained in which no surface deformation occurred. This methodology ensures that the interferometric phase is only due to atmospheric propagation delay.

The delay of the radar signal is caused by an integration over the refractive index of the propagation medium, along the line of sight. Horizontal and vertical heterogeneities in refractive index are influenced by the spatial distribution of water vapor, pressure, temperature, liquid water, and electron content. Yet, over distances less than ~50 km, the main signal in the interferogram is due to water vapor, albeit temperature and liquid water can add some additional millimeters of delay. Using surface temperature observations, the integrated water vapor signal can be converted to precipitable water—its liquid equivalent, when assuming a fixed vertical temperature profile (Bevis et al., 1992).

6.5.2 Results

Case study 1: Precipitation and convection

An ERS-2 SAR image (orbit 2388, frame 1053) was acquired during a cold front passage over the Netherlands, on 4 October 1995. The interferometric combination with an ERS-1 SAR image (orbit 22061), acquired 24 hours earlier, in the absence of significant weather in our target area, enabled the construction of a radar interferogram, see fig. 6.9. The figure shows coherent phase information over all land areas. Over the water area the intensity image of the SAR acquisition of 4 October is shown. The diagonal line in the lower-right corner of the image is associated with the location of the cold front, and the curved anomaly in the center of the image, south of the lake, can be associated with a gust front. Weather radar observations

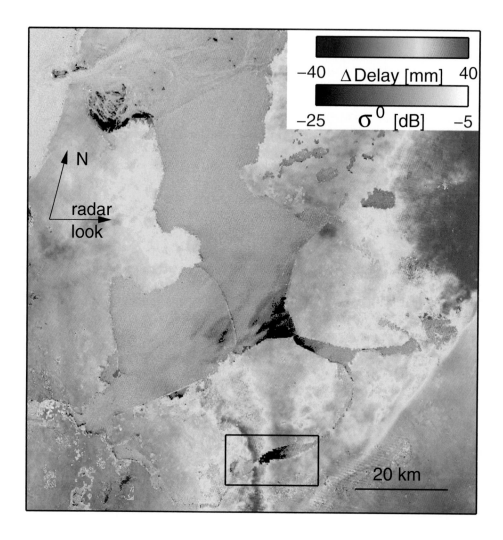

Fig. 6.9. Differential SAR interferogram of a cold front, indicated by the diagonal line in the lower-right corner. The area left of this line is the cold air mass (14°C), the area right is the warm air mass (17°C). Precipitation occurs only over the red areas in the interferogram. The curved feature in the lower part of the interferogram (indicated by the box) is a region with strong precipitation.

show strong precipitation in this region, see fig. 6.7B.

Figure 6.10 shows a small section of this interferogram, centered over Veluwe Lake. The footprint of the storm is moving over the area from left to right. Over the land area, the increased humidity associated with the convective clouds and rain is measured as an increase in signal delay of approximately 30 mm. At this position, P4, weather radar observations indicated rain rates up to 10 mm/hr. In front of the storm (P5) the interferogram shows approximately uniform color, indicating not much variation in moisture. Behind the storm front (P3), variations in moisture are in the order of 8–12 mm, which can be associated with moisture variations due to, e.g., convective cells. The small circular anomalies near P5 are related to processing errors over single pixels.

The roughness of the water surface is expressed by the normalized radar cross section, σ^0, and is well correlated with the near-surface wind conditions (Stoffelen, 1998). It is likely that the signatures on the water surface are mostly due to wind effects, as the signature of the rain would cause fine-scale structures. Moreover, the rain rate is probably too low and the wind speed likely too high for there to be a rain effect by impact on the water surface. A cross-section of fig. 6.10A is shown in fig. 6.10B. Collocated with the rain band, an increase in σ^0 of approximately 18 dB can be observed, at position P4. To the east of the storm front (P5) the water surface is relatively smooth. Moving towards the west, from P4 to P2, the roughness first decreases approximately 8 dB (P3), to increase approximately 4 dB at P2. This nearly symmetrical wind shear appears to be closely associated with the storm front. Assuming a wind direction somewhere between upwind and crosswind (the horizontal wind components in the radar look direction and perpendicular to it, respectively, see fig. 6.10A) the wind velocities following from CMOD4 are ~ 0 m s^{-1} in front of the storm (P5), increasing up to ~ 15 m s^{-1} in the storm (P4), within a spatial distance of 300 m.

At P4, there is a surplus of water vapor with respect to the surroundings, resulting in an excess delay of 3.6 cm, which corresponds to 5.3 ± 0.5 mm integrated precipitable water. The width of the main rain band is 2 km.

From wind observations and common circulation patterns in high-latitude convective systems, one can infer that the radar look direction is crosswind to the flow at location P5 in fig. 6.10A, and parallel to the rain band. Behind the rain band, P1–P3, the wind is more along the radar look direction and probably veering. More to the northeast of the rain band surface measurements indicate that the wind has also a component along the radar look direction. This means that besides wind speed changes, there are also wind direction changes that may be relevant for the signature in the radar image (P1–P5). The radar is most sensitive to wind direction changes for wind directions at 45, 135, 225, and 315 with respect to the radar look direction, whereas there is no sensitivity at 0, 90, 180, and 270 degrees.

From these observations, we find a circulation pattern connected to this system termed "optimal shear" by Parsons (1992), depicted in fig. 6.10C. In our case the complete system moves to the east, such that the vertical shear in front of the storm consists of an area with no wind at the surface and probably SW-winds aloft.

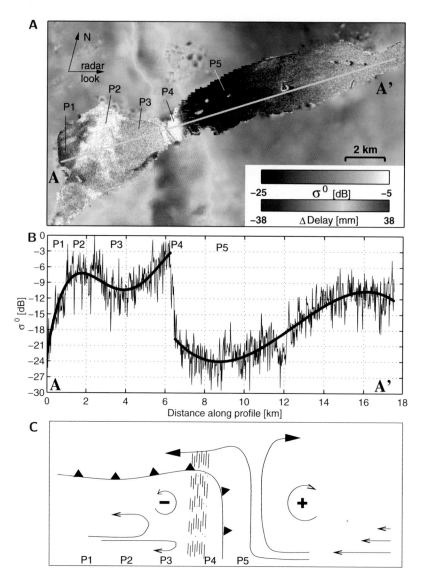

Fig. 6.10. (A) Phase-intensity observations over Veluwe Lake, the Netherlands. The colored section shows interferometric phase observations, converted to slant signal delay variations in mm. The gray-scale part of the figure is derived from the backscatter intensity during the ERS-2 acquisition, and is scaled to logarithmic σ^0 values. The band with increased delay indicates strong rain, and is clearly correlated to the increased surface roughness over the water. **(B)** Cross section AA' of the radar cross section values in (A). The σ^0 values increase approximately 18 dB due to wind patterns associated with the storm. Behind the first rain band, indicated by label P4, roughness decreases and increases again, as observed in the first 6 km of the cross section. **(C)** Sketch of the circulation patterns associated with the cold front, after Parsons (1992). The gust front circulation patterns are likely similar. The vertical shear in front of the storm consists of a windless area at the surface and probably a southwesterly wind aloft.

Case study 2: Boundary layer rolls

In section 6.4.3 it was discussed how during an ERS-1 acquisition on 4 April 1996, 10:34 UTC (orbit 24688, frame 2547), convective rolls occurred in the boundary layer. In the interferogram with an ERS-2 acquisition on 5 April 1996 (orbit 5015), see fig. 6.11, these rolls are clearly visible over the land areas. Since no clouds were observed at the surface stations in the area, the delay variation is a clear air effect.

Several studies have indicated that the ascending parts of the rolls transport warm, moist air at the surface upward, while the descending parts transport dry air originating within the inversion downward, see, e.g., LeMone (1973). Weckwerth et al. (1997) observed variations in the water vapor mixing ratio of 1.5–2.5 g/kg associated with rolls, whereas temperature variations were only 0.5 K. Mapped to water vapor pressure, these moisture variations are 2.4–4 hPa. The sensitivity of the delay to a 1 hPa change in water vapor pressure is in this situation at least 3.5 times as large as to a 1 K change in temperature. The delay variations of up to 10 mm are therefore primarily caused by water vapor differences in the updraft and downdraft branches of the rolls. Not more than 5% of the observed delay can be explained by the temperature variations in the rolls. The rolls are approximately parallel to the wind direction, and the wind speed observed at several surface stations is 6.2–7.7 m s^{-1}.

Over the water areas in IJssel lake, wind streaks are clearly visible, oriented in the same direction as the rolls over land areas. Although streaks caused by rolls are frequently observed in SAR images, see Alpers and Brümmer (1994), Mourad and Walter (1996), the location of the streaks with respect to the downdraft and updraft branches has been uncertain. Spectral analysis reveals a streak spacing of approximately 2 km. Over land, the spacing between the bands is also approximately 2–3 km. A radiosonde in the area, launched 90 minutes after the SAR acquisition, indicated a boundary layer depth of 800 m, which gives an aspect ratio of 2.5–2.9, characteristic for rolls (Atkinson and Zhang, 1996).

For wind speed estimation with a ±1–2 m s^{-1} accuracy, an averaging area of 25 km^2 is needed (Lehner et al., 1998). Using square areas of 10 × 10 km, average wind speeds of 5.3–5.9 m s^{-1} were found. For the analysis of the wind speed differences between the streaks, square areas of this size cannot be used. Instead, we applied a normalized Radon transform to two areas of 11 × 11 km and used a profile perpendicular to the streaks to estimate the average differences in σ^0 (Hanssen, 1994). This method reduces speckle effects significantly and enables direction dependent amplitude estimation. Variations over a range of 0.4–0.5 dB are observed, corresponding with wind speed variations of 1 m s^{-1} or wind direction variations of up to 45 degrees (i.e., the sensitivity to lateral wind velocity changes is about four times the sensitivity to transverse velocity variations).

We note the change in characteristics of the boundary layer rolls as they are advected over the cold water surface, where the fluxes of momentum and heat are expectedly quite different from those over land areas that are heated by the sun. The combination of measurements shown in this paper is ideally suited to show the transformation of air flow after advection over a land-water boundary.

The location of the streaks aligns approximately with the part of the rolls that have less delay, which correspond to the relatively dry and descending air. This is in correspondence with classical theory, which expects the maximum wind speed in a roll in the downdraft area (Alpers and Brümmer, 1994; Atkinson and Zhang, 1996). Mid-latitude cases, without cloud and with important dynamical and thermal instability, such as presented here, are uniquely documented by the SAR methodology presented here.

6.5.3 Discussion

The possibility to simultaneously combine both radar intensity observations over water and phase observations over land surfaces enables specific studies on the interrelation between moisture distribution and wind fields. The fine spatial resolution is particularly suitable for the study of mesoscale shallow convection in the planetary boundary layer in coastal areas or areas with large lakes to describe for example the transformation of air across the land-water interface. Although the phase delay measurements are always a summation of the delay during two acquisitions, specific patterns can often be recognized and analyzed.

6.6 Validation radar interferometry using Meteosat radiometry

Validation of the radar interferometric results can be difficult since conventional imaging radiometers do not provide quantitative measures for water vapor content over the entire tropospheric column and have poor spatial resolution (Weldon and Holmes, 1991). Moreover, comparable quantitative data such as signal delay observed by Global Positioning System (GPS) receivers are only available as time series at a fixed location. In this section, the technique of InSAR integrated refractivity mapping is validated for a specific atmospheric situation with relatively transparent air. It is investigated whether it is possible, for this specific situation, to obtain a parameterization of radiometer brightness temperatures as a function of precipitable water vapor, using Meteosat 6.7 μm [water vapor (WV) channel] observations and GPS time series. After establishing such a relationship for a single point, it is extended spatially to derive precipitable water vapor that can be used for validating the SAR interferogram.

In section 6.6.1 the basic characteristics of the Meteosat observations are described. Section 6.6.2 describes the methodology to derive precipitable water vapor estimates, and presents the main results of this study. Discussion and conclusions make up section 6.6.3. These results are reported by Hanssen et al. (2001).

6.6.1 Meteosat water vapor channel

The Meteosat WV channel, centered around 6.7 μm, is used in operational meteorology to observe the development of structures of upper-tropospheric water vapor, which carry the signatures of atmospheric conditions. Especially subsidence inversions, that is, downward vertical transport of dry air behind a frontal zone, are clearly visible in the WV images. Due to the strong absorption by water vapor at this wavelength, the observed brightness temperatures usually originate from tro-

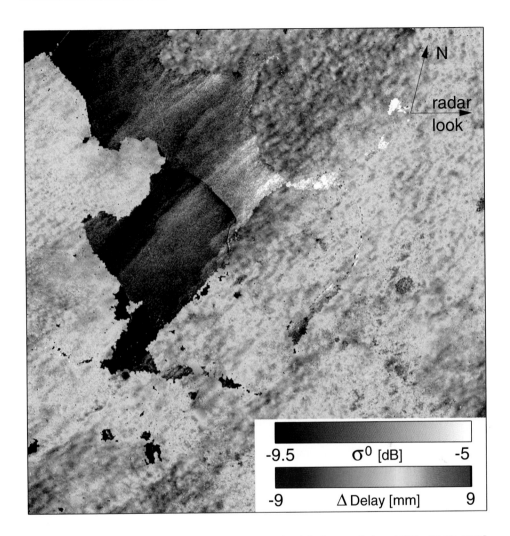

Fig. 6.11. Radar intensity observations over IJssel Lake on 4 Apr 1996, 10:35 UTC, combined with signal delay differences derived from the interferometric phase between 4 and 5 Apr 1996. The signal in the interferometric phase is mainly due to water vapor differences between the moist updraft and the drier downdraft regions (Weckwerth et al., 1996). The signal in the intensity observations over water is due to variations in the wind speed, and to a lesser extent due to the wind direction associated with the rolls.

pospheric layers above 3 km (Weldon and Holmes, 1991). Therefore, quantitative analysis is restricted to upper-tropospheric water vapor as described by Schmetz et al. (1995). Unfortunately, however, the concentration of water vapor is highest near the earth's surface, where relatively high pressure and temperature allow the air to contain more water vapor. Therefore, in general it is not possible to make one unique parameterization of WV channel brightness temperatures into precipitable water vapor. In section 6.6.2 it is discussed whether GPS tropospheric delay time series are suitable for establishing a tailor-made parameterization to obtain Meteosat spatial precipitable water vapor values for the special case of a subsidence inversion, at a local resolution of 5×9 km.

6.6.2 Methodology and results

SAR interferogram analysis

An interferogram, covering a strip of 100×200 km, has been formed using ERS SAR acquisitions on 26 and 27 March 1996, 21:41:05 UTC (22:41:05 local time). The orbital separation parallel to the look direction, see fig. 6.1, is approximately 32 m, which makes the configuration moderately sensitive to topographic height differences. An a priori reference elevation model is used to correct for the topographic phase in the interferogram (TDN/MD, 1997). Some additional corrections for tilts in the length and width direction are applied, and water surfaces are masked. The resulting differential interferogram is shown in fig. 6.12. The phase values are converted to zenith wet delays, using eqs. (6.1.3) and (6.1.4), and consecutively to differential precipitable water vapor using eq. (6.6.3), where we use the term "differential" to indicate the difference between the two states of the atmosphere during the SAR acquisitions.

A strong, large-scale, gradient is clearly visible in the south of the interferogram, aligned approximately perpendicular to the radar flight direction. The interpretation of this phenomenon is a key topic of this study. Previous studies have shown that phase variation perpendicular to the flight direction might be caused by oscillator drift errors in the onboard reference clock (Massonnet and Vadon, 1995). A possible way to examine this possibility would be the calculation of a long swath of connected SAR images, but this is outside the scope of this study. Here it is assumed that the phase variation is caused by atmospheric water vapor only. The likelihood of this assumption can be tested using the combined analysis with Meteosat and GPS.

Note that, apart from the gradient in the south of the interferogram, wave phenomena also can be observed in the interferogram, see the lower left-hand part of fig. 6.12. These are possibly due to low-level moisture variations of which the structure seems to indicate the presence of boundary layer rolls.

Wet delay differences δ_{pq}^z can be related to integrated precipitable water (I) values, the liquid equivalent of the integrated water vapor:

$$\delta_{pq}^z = \Pi_{T_s}^{t_1}(I_p^{t_1} - I_q^{t_1}) - \Pi_{T_s}^{t_2}(I_p^{t_2} - I_q^{t_2}). \qquad (6.6.1)$$

Using the density of liquid water, we find that 1 mm integrated precipitable water is equal to 1 kg m^{-2} integrated water vapor. The conversion factor $\Pi_{T_s}^{t_i}$ is approximated using surface temperatures T_s for the image acquired at t_i, defined by Askne and Nordius (1987). After first models for $\Pi_{T_s}^{t_i}$ were derived by Davis et al. (1985) and Bevis et al. (1994), a polynomial model was developed to approximate $\Pi_{T_s}^{t_i}$ for De Bilt, the reference site in the Netherlands, using 1461 radiosonde profiles (Emardson

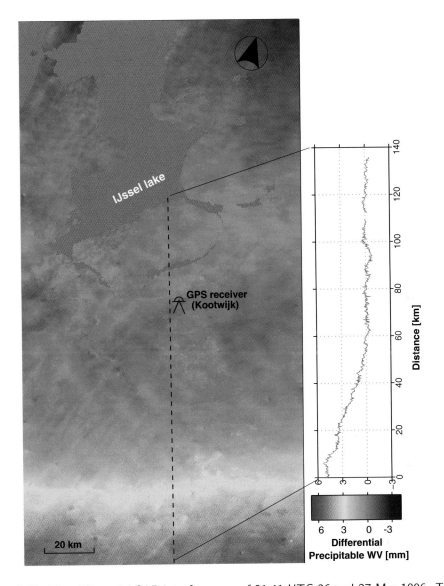

Fig. 6.12. The differential SAR interferogram of 21:41 UTC 26 and 27 Mar 1996. The location of the interferogram in the region is indicated in fig. 6.13. Interferometric phase observations are unwrapped and converted to differential precipitable water vapor—the difference between the precipitable water vapor distributions during the two acquisitions. Topographic information has been removed using a reference elevation model, and water areas are masked. The dashed line indicates the location of the analyzed profile, see fig. 6.17. The location of the GPS receiver at KOSG is sketched in the figure. Diagonal waves are observed in the lower left-hand corner of the interferogram.

and Derks, 2000):

$$\Pi^{t_i}_{T_s,t_D} = a_0 + a_1(T_s - T_0) + a_2(T_s - T_0)^2$$
$$+ a_3 \sin(2\pi\frac{t_D}{365}) + a_4 \cos(2\pi\frac{t_D}{365}), \qquad (6.6.2)$$

where T_0 is the mean annual surface temperature for the location, and t_D is the day number of the year. The used coefficients are $a_0 = 6.443$, $a_1 = -1.33 \times 10^{-2}$, $a_2 = 0.18 \times 10^{-4}$, $a_3 = 3.6 \times 10^{-2}$, $a_4 = 3.0 \times 10^{-2}$, and $T_0 = 283.80$ K (Emardson and Derks, 2000). Using surface temperatures provided by 18 stations in the area, we find that $1/\Pi^{t_1} = 0.1513 \pm 0.0004$ and $1/\Pi^{t_2} = 0.1507 \pm 0.0005$, for 26 and 27 March respectively. Applying one factor for both days, $1/\Pi_m = 0.1510 \pm 0.0005$ is justified, since the fractional error in Π is one order of magnitude smaller than the fractional error in the delay measurement. We can rewrite eq. (6.6.1) as

$$\nabla I^{t_1,t_2}_{pq} = (I^{t_1}_p - I^{t_1}_q) - (I^{t_2}_p - I^{t_2}_q) \doteq \Pi^{-1}_m \delta^z_{pq}, \qquad (6.6.3)$$

where we define $\nabla I^{t_1,t_2}_{pq}$ as the double difference of integrated precipitable water in space and time. Note that a single, isolated anomaly will have a different sign in the interferogram, depending on whether it appeared in the first or the second acquisition. In that case, one of the single spatial $\Delta I^{t_i}_{pq}$ differences can be regarded as nearly zero, in which case we can interpret the double difference as a single spatial difference during the other acquisition.

Meteosat WV channel analysis

On 27 March 1996 a strong subsidence inversion passed the Netherlands from north to south, which was clearly visible in the WV channel images (fig. 6.13). The subsidence inversion can be identified by the dark band in the image, corresponding to relatively high brightness temperatures. Because the descending air in the subsidence inversion is rather dry, the absorption (and emission) of radiation is low and, therefore, the air is relatively transparent. This enables radiation from lower (warmer) layers to contribute to the signal, which results in high apparent brightness temperatures. The center of the subsidence inversion moves in one day from about 53°N to 48°N, as apparent from fig. 6.13.

Assuming that the brightness temperature variations in the Meteosat WV images reflect the variations of the total water vapor column, we use GPS-derived precipitable water vapor observations to parameterize the brightness temperature. The reference to the WV channel are collocated measurements from GPS ground station Kootwijk, at 52.17°N, 5.80°E, see fig. 6.12. From the GPS measurements reliable absolute values for precipitable water vapor are derived, using the standard methodology described in Bevis et al. (1992, 1994). The accuracy of the GPS-derived precipitable water vapor is approximately 2 mm. For 27 hourly observations of GPS precipitable water vapor and Meteosat signal intensity, shown as scatterplot in fig. 6.14, the correlation can be parameterized as

$$I = 36.7 - 0.56\zeta + 0.0022\zeta^2, \qquad (6.6.4)$$

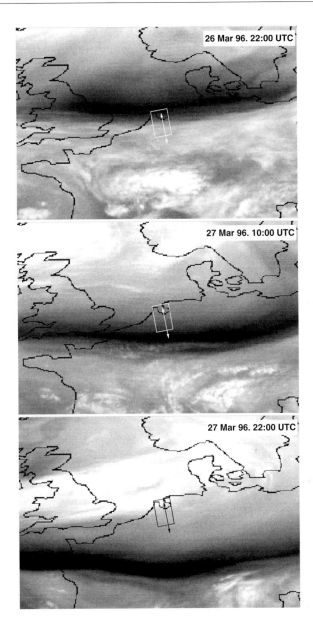

Fig. 6.13. Meteosat 6.7 μm water vapor images at 22:00 UTC 26 Mar, and 10:00 and 22:00 UTC 27 Mar. Brightness temperatures range from $-50°$C (white) to $-10°$C (black). The interferogram area is indicated in the center of the image. The arrow indicates the range over which brightness temperatures were used for comparison with the GPS observations.

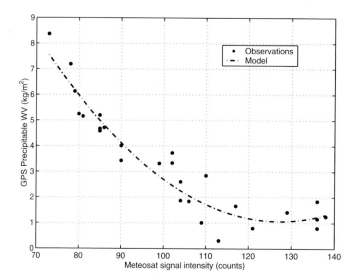

Fig. 6.14. Scatterplot of observed GPS precipitable water vapor and Meteosat water vapor channel signal intensity, for hourly observations between 18:00 UTC 26 Mar and 20:00 UTC 27 Mar. The derived polynomial is indicated by the dashed-dotted line. The rms with respect to the derived polynomial is 0.6 kg m^{-2}.

with I in kg m^{-2} and ζ the signal intensity of the WV channel in number of counts. The rms value of the difference between model and observations is 0.6 kg m^{-2}.

In fig. 6.15 the change in I over Kootwijk is shown derived from both GPS and Meteosat. We find that the rms of 0.6 kg m^{-2} is accurate enough for our analysis. The passage of the subsidence inversion from north to south is clearly visible. Values range from 1 to 7 kg m^{-2}.

Using the parameterization in eq. (6.6.4), precipitable water vapor can be derived from the Meteosat WV channel image. The parameterization yields accurate values for the area near Kootwijk (location indicated in fig.6.12), but with increasing distance the accuracy is likely to decrease.

For the comparison with the SAR interferogram an elongated area is selected (indicated by the arrow in fig. 6.13), which ranges from 51.38°N, 5.93°E to 52.72°N, 5.43°E. The area includes the GPS station Kootwijk Observatory for Satellite Geodesy (KOSG), to optimize the validity of our parameterization, and is parallel to the

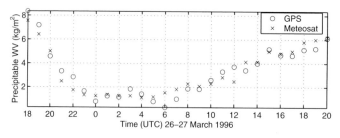

Fig. 6.15. Comparison of hourly precipitable water vapor over Kootwijk derived from Meteosat brightness temperature, using eq. (6.6.4), and derived from GPS zenith delays.

ground track of ERS. The subsidence inversion band is nearly perpendicular to the profile and moves approximately parallel to it. As a result the analyzed precipitable water vapor signal is dominated by only one atmospheric process—the subsidence inversion. Therefore, the parameterization is assumed to be sufficiently accurate for the selected profile.

All pixels that contain high-altitude clouds (temperature below $-20°C$) are excluded from the analysis, to increase the reliability of the parameterization. In a cloud, the fraction of total water that is clustered in particles is generally small, but the absorption of radiation by these particles is much larger than the absorption by the water vapor in the cloud. Therefore, the correlation between the Meteosat WV channel observations and the correct amount of precipitable water vapor will deteriorate if clouds are present.

To obtain similar quantities as in the interferogram the extracted profiles corresponding with 22:00 UTC 26 March and 22:00 UTC 27 March are differenced. The resulting values are indicated by the triangles in fig. 6.17, where the position of the pixels along the profile is expressed by their latitude. From the Meteosat WV images in fig. 6.13 it appears that at 26 March, 22:00 UTC, the selected area was in the front part of the subsidence inversion, which results in a decreasing amount of precipitable water vapor with latitude. At 27 March, 22:00 UTC, the selected area was in the back part, resulting in an increasing amount of precipitable water vapor with latitude. As a consequence of these opposite trends, the quantities in fig. 6.17—which are formed by subtracting the values at 27 March from the values at 26 March—have an amplified north-south gradient. In the following these results will be compared with the results from SAR interferometry.

Intercomparison

The evaluation of the water vapor observations from the Meteosat WV channel and the radar interferogram is subject to 5 degrees of freedom.

1. The interferogram shows relative delay differences. Therefore the analyzed profile has an arbitrary bias when compared with the absolute values of the Meteosat profile.

2. Due to the oblique viewing geometry of the radar ($23°$ from zenith) the profile will be shifted some kilometers parallel to the west, see fig. 6.16. Since the atmospheric situation during the two acquisitions was nearly symmetric (i.e., perpendicular to the profile), the effect of this lateral shift is negligible.

3. The inaccuracy in the satellite orbits might lead to a small tilt in the profile. Here we assume that this tilt is sufficiently eliminated by using a number of reference points during preprocessing.

4. The Meteosat positioning accuracy is approximately 0.5 pixels, corresponding to a 4.5-km uncertainty in north-south direction.

5. Due to the geostationary position over the equator, there is an additional shift D in north-south direction when the majority of the water vapor is at

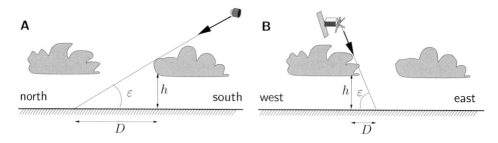

Fig. 6.16. Sketch of the horizontal shift of the position of water vapor due to the oblique viewing geometry of **(A)** Meteosat and **(B)** ERS. For latitude 52°N, the elevation angle ε is 30.5° for Meteosat and 69° for ERS.

a height h, see fig. 6.16. In that case, the information will appear to be shifted northward in the image. For the location analyzed here, this shift is approximately $D = 1.7 \times h$. Note that the time difference between the SAR acquisition and the Meteosat scan of the latitudes of the Netherlands (13 min) might also account for a slight additional shift.

Estimates of the a priori standard deviation of both the Meteosat and InSAR precipitable water vapor estimates are needed to evaluate the correlation between the two profiles. For the Meteosat observations, $\sigma_{I,\mathrm{msat}}$ is assumed to be uncorrelated and equal for every observation. In section 6.6.2, an rms value of 0.6 kg m^{-2} was derived from a comparison with GPS observations. The analyzed profile values are differences between corresponding observations for two consecutive days. Assuming zero covariance between the two days, standard error propagation yields the a priori value for $\sigma_{\Delta I,\mathrm{msat}} = 0.85$ kg m^{-2}.

The expected variance for the InSAR observations is assumed to be uncorrelated between adjacent values and equal for every observation. In section 6.1.2, the delay standard deviation $\sigma_{\delta z} = 0.2$ mm (or kg m^{-2}) was derived. Using eq. (6.6.3), simple error propagation yields an a priori value for the integrated water vapor observations of $\sigma_{I,\mathrm{sar}} = 0.08$ kg m^{-2}.

A goodness-of-fit parameter between the InSAR and the Meteosat observations can be optimized by adjusting the 5 degrees of freedom mentioned above, or by adjusting the a priori variances of both observations. Here, a combination of both approaches is suggested. Parameters 2–4 are assumed to be sufficiently well approximated. A northward shift of the Meteosat observations (parameter 5) of 5 km is used as a first-order approximation, derived from an average height of the dominant water vapor signal in the WV channel. The additional Meteosat profile shift, the bias of the InSAR profile (parameter 1), and the scaling factor for the variances are now estimated by deriving the minimal reduced chi-square value, χ_ν^2, of the two profiles as a function of the bias of the InSAR profile (Bevington and Robinson, 1992):

$$\chi_\nu^2 = \frac{1}{n-1} \sum_{i=1}^{n} (\frac{r_i}{\sigma_i})^2. \tag{6.6.5}$$

The number of measurements is denoted by n, r_i are the differences between the Meteosat and the InSAR profile values, and σ_i are the a priori standard deviations of the differences, that is, $\sigma_i^2 = \sigma_{I,\text{sar},i}^2 + \sigma_{\Delta I,\text{msat},i}^2$. Here we find that $\sigma_i = 0.85$ kg m^{-2} for all $i \in [1, \dots, n]$. Note that the contribution of $\sigma_{\Delta I,\text{msat},i}$ is one order of magnitude larger than $\sigma_{I,\text{sar},i}$.

The χ_ν^2 values are expressed as a function of the InSAR profile bias and the horizontal shift of the Meteosat profile. The search window of the bias has been confined between -3 and 3 kg m^{-2}, whereas the horizontal shift is confined between -1.1 and 1.1 km. The minimum value, $\chi_\nu^2 = 2.1$, is found for a bias of 1 kg m^{-2} and a horizontal shift of -1.1 km. Figure 6.17 shows the original position of the InSAR profile, indicated by the thin line, and the new position, indicated by the bold line. The original position of the Meteosat profile values is indicated by the triangles, while the new positions are indicated by the squares. A 95% confidence interval for the difference of the two data sources is indicated by the inner error bars. Note that the error bars are drawn at the Meteosat positions, although they include a small component of the InSAR profile as well.

From this evaluation it is clear that the adjustment of the free parameters is not sufficient to reach a satisfying comparison. The estimated variance $\hat{\sigma}^2$ can be approximated by (Bevington and Robinson, 1992)

$$\hat{\sigma}^2 = \chi_\nu^2 \sigma_i^2, \tag{6.6.6}$$

if all variances σ_i^2 are equal. If both profiles describe the same physical process, the estimated variance should agree well with the a priori variance and the value of the reduced chi-squared should be approximately unity. In order to reach this situation, the a priori standard deviations of the difference need to be scaled by a factor $\sqrt{\chi_\nu^2} = 1.44$. This implies that the a priori standard deviation of the difference was too optimistic. An a posteriori standard deviation of 1.23 mm is found, indicated in fig. 6.17 by the outer error bars.

6.6.3   Discussion

For a specific atmospheric situation in March 1996, precipitable water vapor obtained from SAR interferometry is validated by GPS time delay analysis combined with Meteosat 6.7 μm WV channel observations. The interferometric phase observations are converted to relative signal delay observations and consecutively processed to precipitable water vapor. For a point location, Meteosat brightness temperature time series are converted to precipitable water vapor using a parameterization obtained from GPS wet signal delay observations. Applying this parameterization spatially for a profile of brightness temperatures in two WV channel images, acquired at nearly the same time as the two SAR images, enables a direct comparison between the two sources.

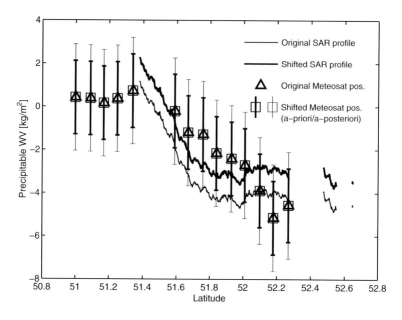

Fig. 6.17. Goodness-of-fit analysis, based on best χ_ν^2 value. The error bars express the scaled 2σ (95%) confidence interval based on the combined variance of the Meteosat and InSAR profile. The original (arbitrary) values of the InSAR profile are shown as the thin line, the best-fit values by the bold line. The original Meteosat positions are indicated by the triangles and the slightly shifted best-fit values by the squares. The best χ_ν^2 value is obtained by scaling the a priori variances (bold inner error bars) with factor 1.44, resulting in the thin outer error bars.

The results show that the phase gradient observed in the SAR interferogram is fully accounted for by a subsidence inversion that moved over the interferogram area during the two SAR acquisitions. The subsidence inversion resulted in temporal precipitable water vapor variations over a range of approximately 6 kg m^{-2}, as observed by GPS and Meteosat. Accounting for the relative character of the InSAR observations, and the positioning uncertainty of the Meteosat images, both datasets describe the same phenomenon in a minimal reduced chi-squared sense with 1.23 kg m^{-2} standard deviation. As the range of the signal encompasses approximately 6 kg m^{-2}, signal to noise ratio values are sufficient to identify the same signal in both data sources. This supports the statement that radar interferometry can be used to derive the spatial variations in precipitable water vapor.

Although the parameterization used in eq. (6.6.4) appears to be sufficient to explain the observations, it needs to be stressed that the use of this method is feasible only for a limited area around the GPS receiver. Moreover, in situations with severe cloud cover, the conversion from brightness temperatures to precipitable water vapor is not likely to succeed, due to insufficient penetration caused by absorption. Future applications of the technique need to be focused on consecutive acquisitions of SAR images to obtain "cascade" series of interferograms. This approach can result in resolving the ambiguity between the atmospheric states during the acquisitions.

6.7 Spatial moisture distribution during CLARA 96

In this study, published by Hanssen and Weckwerth (1999), a specific interferogram is analyzed, acquired at 23 and 24 April 1996 over the southwestern part of the Netherlands, see fig. 6.18. The two SAR acquisitions coincide with the "Clouds and Radiation" (CLARA) experiment, which is aimed at a better understanding of the interaction between clouds and radiation by improving the measurement of cloud properties (van Lammeren et al., 1997; Feijt, 2000). Since the development of clouds is highly influenced by the water vapor distribution, the SAR data can be used to estimate the precipitable water content of clouds relative to the ambient environment. The high resolution of the data reveals regions of enhanced moisture leading to cloud formation and rain.

Similar space-geodetic techniques, such as GPS, are influenced in nearly the same way as SAR. Therefore, these data are used to describe the tropospheric situation during the experiment and identify specific features in the interferogram, in combination with an extensive array of supporting instrumentation. A statistical evaluation of the dataset is presented using two-dimensional structure functions, and conclusions on the potential and limits of radar interferometry as a meteorological technique are discussed.

6.7.1 Methodology

At 23 April 1996, 10:38:07 UTC (12:38:07 local time), the ERS-1 SAR acquired frame 2565 during orbit 24960, covering an area of 100×100 km. Exactly 24 hours later, the twin instrument onboard ERS-2, orbit 5287, acquired the same part of the earth's surface, from a nearly identical position as ERS-1 one day before. Both satellite positions differ only 38 m in the oblique look direction of the radar, and 78 m perpendicular to the look direction.

Using a reference elevation model for the test area a differential interferogram is obtained. In this image, phase differences are only due to atmospheric delay and perhaps long wavelength trends due to orbit inaccuracies. The latter can be approximated using the variance in the orbital state vectors and tie-points at the surface. In this procedure, all long wavelength (> 200 km) atmospheric information is effectively eliminated. The residual differential interferogram can be regarded as an "atmosphere-only" image.

Phase differences in the differential interferogram need to be interpreted with some precaution. First, it is important to realize that the interferogram only contains relative phase information. There is no absolute calibration point. As a result, the value of one single pixel is useless. Second, the variation of the phase is due to the spatial variability during *two* acquisition times with a different state of the atmosphere. This ambiguity is a limiting factor for data interpretation if only two SAR images are used, as in this study. Using a number of SAR images, interferograms

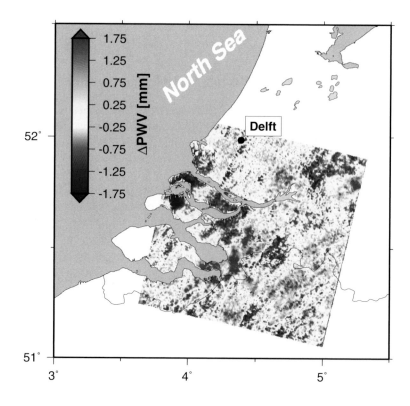

Fig. 6.18. Differential SAR interferogram of 23 and 24 Apr 1996, 10:38 UTC using a Mercator projection and WGS84 coordinates. All topographic information has been removed using a reference elevation model. The interferometric phase has been converted to zenith delay differences with zero average and subsequently to differential precipitable water vapor (ΔPWV) using a mean surface temperature of 15.1°C (Davis et al., 1985). The colorbar is linear between -1.5 and 1.5 mm. Bluish colors correspond with signal on 23 Apr, while red-yellow regions are related to 24 Apr. The black dot indicates the position of Delft, where most CLARA in situ observations were performed.

can be made between different combinations, which enables one to overcome this ambiguity. For this study, the sign of the phase values is used to attribute the phase effect to the SAR image of 23 April or 24 April. This approach can be used for strong disturbances—for small phase variation it is not possible to attribute it to one specific acquisition using only two images.

Using eqs. (6.1.3) and (6.1.4) slant and zenith delay differences are derived from the phase differences. Assuming smooth lateral pressure and temperature variation over scales less than 50 km, the observed delay variation is mainly due to variations in the wet part of the refractivity. Using the method proposed by Bevis et al. (1992) precipitable water variations, labeled differential precipitable water (ΔPWV), are derived from the delay interferogram and surface temperature measurements. After geocoding the ΔPWV map from radar coordinates to the WGS84 ellipsoid, the result is shown in fig. 6.18.

To validate the meteorological interpretation of the ΔPWV map, additional datasets from the CLARA experiment are used. GPS zenith delay observations, radiosonde profiles, lidar, infrared and microwave radiometer data were available at the Delft station, indicated by the black dot in fig. 6.18. Additional Meteosat and NOAA-AVHRR data were used for comparison as well.

For the analysis of the spatial variability of the delay or water vapor signal, the two-dimensional structure function, D_δ, defined as

$$D_\delta(dx, dy) = E\{[\delta(x + dx, y + dy) - \delta(x, y)]^2\}, \qquad (6.7.1)$$

is used. Here, δ is the zenith delay signal and dx and dy are the distances between two arbitrary points in km. The structure function is the expectation value of the squared difference between two points at a certain distance R and azimuth α in the image. It exists for all random functions with stationary increments. In section 4.1.4, the structure function was discussed in more detail.

6.7.2   Results

Here, a first interpretation of the interferogram is made, followed by the general meteorological interpretation of both days using the existing CLARA ground truth data. Then, a comparison with the GPS data and a statistical evaluation is performed.

Interferogram interpretation

Figure 6.18 shows the differential PWV map for 23 and 24 April 1996. The colorbar is linear between -1.5 and 1.5 mm in order to use the full color resolution and suppress the effect of a few localized points with very high values. The yellow-red range of the colorbar corresponds with delay signal on 24 April, blue shades correspond with 23 April. The variation can be interpreted as due to water vapor and clearly indicates bandedness along the wind direction on 24 April: SW–WSW. In the lower-left part of the ΔPWV map, it can be observed that moisture converges into linear bands, which expand to the large anomaly at $51.4°$N, $4.3°$E. The value and shape of this anomaly indicates a deeper column of moist air, which is expected for cumulus–cumulonimbus clouds and for regions penetrated by precipitation. In those sub-cloud regions, the relatively dry air will become saturated by evaporating liquid particles. Although it is in general not possible to determine whether precipitation occurs from the interferometric data only, localized showers often give a strong signal. The shape and magnitude of the anomalies suggest very localized concentrations of water vapor, as expected for convective weather with developing cumulus clouds.

The blue regions, e.g., at $51.8°$N, $4.0°$E and $51.5°$N, $4.7°$E, corresponding to 23 April also indicate strong concentrations of water vapor. The existence of clouds in the area is very likely, although probably with less variation than at 24 April.

In fig. 6.19A, the weather radar reflectivity of both days, 10:30 UTC, is shown, converted to rain rate in mm/h. To facilitate comparison with the interferogram,

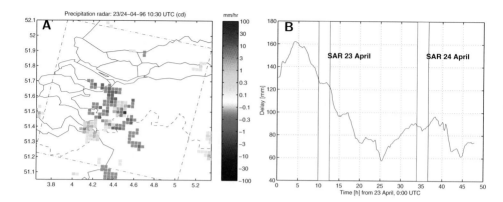

Fig. 6.19. (A) Weather radar reflectivity of 23 and 24 Apr 1996, 10:30 UTC, converted to rain rate in mm/h. The data of the two days are superposed in one image using negative values for 23 Apr and positive values for 24 Apr. Comparison with fig. 6.18 shows that rain areas correspond with strong values for ΔPWV in the interferogram. This, however, does not hold the other way around, e.g., at 51.75°N, 4.0°E a strong anomaly during 23 Apr is not connected to rainfall. **(B)** GPS zenith wet delay variation from 23 Apr, 0:00 UTC to 24 Apr, 23:51 UTC. Two intervals of 2.45 hours, centered around the SAR acquisitions are indicated by the vertical lines.

precipitation during 23 April is plotted using negative (blue) colors, and precipitation during 24 April using positive (yellow) colors. For the grid points where precipitation occurred during both days, the average value is shown. The correlation between the precipitation regions and large ΔPWV values is remarkable, although the absence of precipitation at 51.8°N, 4.0°E supports the hypothesis that the ΔPWV signal can be caused by towering cumulus alone.

General interpretation based on additional data

According to the weather maps of 23 April, a cold front has just passed over the interferogram area during the SAR acquisition. There is no obvious frontal passage signature in neither the satellite imagery nor the Delft soundings nor the Delft surface measurements of temperature, humidity or pressure, thus the cold front was quite weak. Behind it, there are isolated showers in a neutral atmosphere, as derived from weather radar and soundings. Cloud base is 2 km, as determined from the IR radiometer cloud base temperature of 5°C, which corresponds to 2 km on the sounding. This is confirmed with the ESTEC lidar. The radiosonde data reveal strong winds, 10–15 m/s, from SSW. Cloud types are a combination of stratocumulus and cumulus. Near the time of the frontal passage at Delft, there were more clouds and they were more continuous (determined from liquid retrieval of microwave radiometer; also IR radiometer shows continuous cloud cover), and there was a maximum in water vapor. Then cloud cover diminished as the water vapor amounts decreased.

Table 6.3. Basic statistics for the zenith wet delay derived from the two GPS days (24 hour interval), the difference between GPS at day 23 and 24, and the SAR interferogram (instantaneous) [mm].

	Mean	RMS	Range
GPS 23 Apr 96	117.4	30.4	89.1
GPS 24 Apr 96	78.8	10.7	39.1
GPS 23 − 24 Apr 96	38.7	31.0	87.9
SAR	0.0	5.3	54.8

During the acquisition of the second SAR image, at 24 April, there were more isolated showers, but not as many as during the first day. The radiosonde observations indicate a more stable atmosphere with colder, drier air at the surface. The cloud base is 2500 m (determined from the ESTEC lidar), which is consistent with the IR radiometer cloud base temperature of $-10°C$ and sounding temperature at this height of about $-6°C$. The cloud cover is thin and broken, and consists of mostly cumulus, as indicated by the IR radiometer, the ragged look of the liquid water retrieval from the microwave radiometer, and by the order of magnitude less in liquid water content from the microwave radiometer on 24 April compared with 23 April. Winds are weaker during this day, 8–10 m/s, from WSW. Cloud streets, mainly developing cumulus, are visible over the interferogram area in the Meteosat double visual image at 10:30 UTC.

Comparison between SAR and GPS data

GPS observations were performed at station Delft during the two days of the interferogram. In fig. 6.19B, the derived zenith wet delay is shown using a 6-min sampling interval. Hydrostatic delay components are removed using surface pressure data. Two pairs of vertical lines indicate an interval of 2.45 hours around the SAR acquisitions. This interval was determined from an average wind speed of 10 m/s and the interferogram size of 100 km.

From the signal variation within these two intervals, it is expected that the signal variation in the interferogram is less than 15 mm. However, the zenith wet delay variation in the interferogram has a range of 54 mm. Table 6.3 lists some statistical values for SAR and GPS.

From fig. 6.19B and table 6.3 it is obvious that a comparison between 24 hours of GPS data and the interferometric data is not possible, as the daily variation in wet delay is much stronger than the variation over a short interval. This behavior is also expected from power-law considerations. On the other hand, only evaluating the 2.45 hours observations centered around the SAR acquisition time is not representative as well, as it seems to underestimate the amount of variation.

The rms of the zenith wet delay variation in one SAR image is assumed to be equal to the variation in the interferogram divided by $\sqrt{2}$, or 3.7 mm.

Regarding the spatial resolution of the GPS zenith delay observations, two remarks need to be made. First, for a cloud base at 2 km and a 20° elevation cut off, the diameter of the GPS cone at that altitude is 11 km. If an array of GPS receivers would be installed with a 11 km posting, independent observations would be guaranteed at that altitude. From Shannons sampling theorem it follows that water vapor signal with a wavelength of 22 km or less will be aliased into the longer wavelengths. For the SAR-derived spatial delay field, using a spatial averaging to 160 m, the observations can be considered independent. On the other hand, for wavelengths larger than, say 50 km, gradient errors due to satellite orbit inaccuracies and hydrostatic effects limit the unambiguous interpretation of the signal. Such considerations can be important when an array of GPS receivers is available.

Secondly, when only one GPS receiver is available, the temporal behavior of the signal might give a first indication of the delay variability. To perform the conversion from temporal to spatial information, Taylors hypothesis of a "frozen" boundary layer is assumed in which the anomalies do not develop but are only transported in their original shape by the wind (Taylor, 1938). With this rather stringent assumption, the wind speed is needed for the conversion. The observed wind speed during 23 April is approximately 10 m/s. With the 11 km diameter of the GPS observation cone at 2 km altitude, the sampling interval should be 18 minutes to interpret the observations as independent. The 6–10 min sample interval used during the CLARA experiment is therefore clearly too short to interpret the short-wavelength signal. For a comparison with the SAR observations, the sampling interval of 20 min should be used, which results in a shortest wavelength of 40 min, or 24 km.

Scale dependent statistics

For the description of boundary-layer (200 m–2 km) turbulence characteristics as well as analyzing the geodetic implications of positioning using space-geodetic methods, the spatial behavior of the delay variation is very important. For the relative radar interferogram the first moment or expectation value of the delay is zero. In fig. 6.20, the histogram of the delay data from the interferogram is displayed. The second moment or dispersion of single observations can be derived but is less interesting as the dispersion of increments. The latter can be described by analyzing the power or the magnitude spectrum or by the structure function and reveals information on the decrease in power for smaller spatial increments. Figure 6.21A is the rotationally averaged amplitude spectrum of a continuous part of the interferogram. It is derived by computing the 2D FFT and averaging all values with the same radial distance to the origin. The diagonal lines indicate the typical $-5/3$ power law decay for higher frequencies. Note that the $-5/3$ slope observed in a rotationally averaged spectrum corresponds with a $-8/3$ slope in a one-dimensional spectrum, consistent with our results in section 4.7 (Turcotte, 1997). For wavelengths between 0.5 and 2.5 km the variation clearly follows this decay. A small deviation at 2.5 km (0.4 cycles/km) could indicate the effective height of the boundary layer in which 3D turbulence characteristics play a dominant role. For longer wavelengths, the decay is less strong (approaching $-2/3$ for the rotationally averaged or $-5/3$ for a one-dimensional signal) as expected for effective 2D turbulent characteristics.

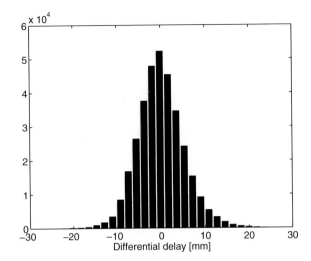

Fig. 6.20. Histogram of the differential delay in the interferogram showing Gaussian characteristics. Assuming the same delay characteristics at both days, the interferogram standard deviation of 5.3 mm corresponds with 3.7 mm during one SAR acquisition.

The two-dimensional structure function $D_\delta(dx, dy)$ reveals information on the correlation of single pixels with other pixels in the image. In fig. 6.21B, the square root of the structure function is shown. The non-isotropy of the field can be clearly derived, in this case mainly caused by the two strong delay signatures in separate acquisitions. As the image is derived from averaging all combinations of pixels, there are less realizations available for the edges and corners of the image. Therefore, areas outside 70 km from the origin need to be interpreted with care. For geodetic applications, it is important to increase the variance of phase difference measurements with increasing distance.

6.7.3 Discussion

Radar interferometric data can be used to obtain a relatively accurate and high-resolution interpretation of the state of the boundary layer. Of all the observational datasets available for the CLARA project, none of them give a two-dimensional spatial overview. Weather radar gives a spatial view only if precipitation occurs, while Meteosat and NOAA-AVHRR lack resolution to observe the fine details in water vapor. All other instrumentation is situated at a single point and gives temporal or vertical information. Therefore, the radar interferograms add considerable complementary value in terms of high resolution and quantitative information.

A disadvantage of many new sensors (e.g., water vapor DIALs) is that they do not work in cloud. Therefore, radar interferometry has an advantage with respect to that as well. In the scientific community there is strong interest in water vapor measurements within clouds (Weckwerth et al., 1999). There may be supersaturation or subsaturation within clouds, which affect droplet size distribution and particle type and therefore the radiation. With existing instrumentation, this is very difficult to determine.

Nevertheless, major drawbacks are currently in the unambiguous interpretation of

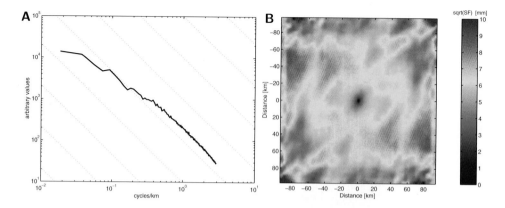

Fig. 6.21. (A) One-dimensional amplitude spectrum of the interferogram showing the decrease in signal amplitude for decreasing wavelengths. The spectrum is a rotational average of the 2D amplitude spectrum. The diagonal lines indicate the $-5/3$ power law decay, corresponding with $-8/3$ for a 1D spectrum. **(B)** Two-dimensional structure function, derived from the original interferogram converted to delay observations. The square-root of the structure function values are displayed to simplify interpretation of the delay differences. For every combination (dx, dy), the standard deviation of the difference between two pixels is shown. The 2D structure function is symmetric around the origin.

two superposed atmospheric situations. With only two SAR images to form one interferogram, this will be a limiting factor. Suitable SAR platforms with more frequent revisit times are expected to solve this problem. Moreover, design considerations might result in, e.g., larger swath widths or reduced resolution to speed up data processing.

6.8   Conclusions

Satellite radar interferometry can be applied to study vertically integrated atmospheric refractivity variations with a spatial resolution of 20 m and an accuracy of ~2 mm, irrespective of cloud cover or solar illumination, over most land areas. Therefore, Interferometric Radar Meteorology (IRM) is a new technique for meteorological applications and atmospheric studies. The data are derived from the difference between the radar signal delay variations within the imaged area during two acquisitions with a temporal separation of one or more days. Hence, they reflect the superposition of the refractivity distribution during these two acquisitions. On short spatial scales integrated refractivity variations appear to be dominantly caused by spatial heterogeneities in the water vapor distribution, only slightly modulated by influences of liquid water and temperature. Hydrostatic delay gradients can be commonly ignored, as well as ionospheric influences for C-band radar.

The configuration of contemporary SAR satellites is currently far from optimal for operational meteorology. This is mainly due to the generally infrequent repeat-acquisition interval. A related problem is the superposition of two atmospheric states. It has been shown, however, that the interferometric combination of a series of SAR acquisitions can be used to identify the atmospheric influence per single acquisition (Ferretti et al., 2000). Power consumption, high data rates, and significant signal processing effort form another challenge for real-time operational applications.

Nonetheless, IRM can be readily used for studying atmospheric dynamics such as mesoscale shallow convection (MSC), and provides unsurpassed insights, particularly in mapping the small-scale water vapor distribution. The main advantage of the technique is the combination of its imaging possibilities with quantitative delay estimates. ESA's tandem mission (August 1995–April 1996) acquired more than 110.000 tandem interferograms covering nearly the total global land surface (Duchossois et al., 1996). Since surface deformation is practically excluded in these data, and topography can be considered known for many areas (especially after the data release from the SRTM mission), almost all of these interferograms can be used for atmospheric studies.

This chapter shows a number of case studies that validate the radar interferometric results based on the comparison with weather radar, spaceborne radiometer, GPS, radiosonde, and synoptic (acquired at the same time around the world) ground-based observations. For a specific situation, a cross-validation combining GPS, radiometer, and InSAR proves possible, resulting in quantitative measures of water vapor content. It proves possible to study causal connections between mesoscale wind patterns and moisture distribution in the lower boundary layer by combining the radar backscatter intensity over water and the radar signal delay over land areas. Analyses are reported on the relation between wind and moisture associated with boundary layer rolls and with a cold-frontal rain band.

There are a number of limitations for the application of IRM. First, the technique is only applicable over land or ice areas. Second, whereas the time interval between the acquisitions should be as short as possible to reduce possible temporal decorrelation, it has to be long enough to allow for the necessary atmospheric decorrelation. In other words, there needs to be some change in the atmospheric refractivity distribution. The latter condition can be somewhat relaxed if acquisitions are available very frequently. Third, topography needs to be known as a DEM with a vertical accuracy of ~10 m and a spatial resolution as fine as the SAR resolution. Using the results of the SRTM mission, this is currently feasible for latitudes between 58°S and 60°N. If the area is characterized by large elevation differences, atmospheric signal due to vertical stratification needs to be accounted for. Fourth, satellite orbits need to be of sufficient quality to remove residual phase gradients as good as possible, and to prevent DEM scaling errors. Very reliable orbit control allows for the retrieval of hydrostatic gradients. An ideal future system would benefit from a wide swath, e.g. by applying a ScanSAR mode, and could operate with a reduced spatial resolution. Such a system would have unique complementary characteristics with conventional meteorological techniques.

Chapter 7

Conclusions and recommendations

The general problem statement of this study was formulated in chapter 1 as: "How can the interpretation and analysis of repeat-pass spaceborne radar interferometric data be improved in a systematic way via a model-based quantification of the error sources?" Here we present the main contributions of this research and recommendations for further studies and practical application of radar interferometry as a geodetic technique.

key words: *Conclusions, Quality assessment, Error analysis and propagation, Atmospheric signal*

This study discussed the technique of repeat-pass spaceborne radar interferometry from a geodetic point of view. The main conclusion of this research is that the interpretation and analysis of radar interferometric data can be improved by formulating the problem using a standard Gauss-Markoff model and by including the error sources as variances and covariances in the stochastic part of this model. Especially covariance modeling is an important extension of current quality assessment strategies. Second, this study presented a novel technique for meteorology, referred to as Interferometric Radar Meteorology (IRM), which provides fine-resolution maps of water vapor distribution.

The conclusions in relation to the four specific research questions listed in chapter 1 are discussed below, followed by the recommendations for further research.

7.1 Contributions of this research

7.1.1 Model formulation

Specific research question 1. "How can we parameterize the formation and the quality of a radar interferogram and its derived interferometric products?"

The general problem statement in chapter 1 was motivated by the need for a geodetic formulation of the observation equations and of the variance-covariance matrix of the observations. A Gauss-Markoff model (see chapter 3) can be used as a framework to combine both the functional relations between observations and parameters and the stochastic characteristics of the observations. This formulation allows for a direct evaluation of the degree of redundancy, quantitative statements on precision and reliability, and standard methodology for data adjustment and filtering.

In the generic functional part of the model, every interferometric phase observation can be related to at least five parameters: topographic height, surface deformation, atmospheric signal during the two acquisitions, and the integer phase ambiguity, hence n observations are related to $5n$ parameters. Based on this generic (underdetermined) model, the specific model for topographic mapping, deformation mapping, or atmospheric mapping can be derived, taking into account the geometric or temporal characteristics of the acquisitions as well as the availability of supportive data such as a digital elevation model or meteorological information.

By considering atmospheric signal as additive noise on the phase observations, it is possible to exclude it from the functional model, and include it in the stochastic model. As such, atmospheric signal results in a dramatic increase in the variance of the observations combined with the introduction of covariances as a function of distance.

7.1.2   Error sources and propagation

Specific research question 2. "Which error sources limit the quality of the interferogram and how does their influence propagate in the interferometric processing?"

Error sources that limit the quality of the interferogram have been discussed in chapter 4. Single-point observation statistics show that the probability density function of the interferometric phase can be expressed as a function of the coherence and the number of effective looks. For distributed scattering mechanisms, coherence can be estimated directly from the data, although this hampers the identification of single coherent point scatterers. Moreover, it is shown that the coherence estimators are biased and have an increased variance for low coherence values and a low number of effective looks. Hence, there is a strict trade-off between coherence estimator accuracy, i.e. phase variance estimation, and spatial resolution. In fact, point scatterers may have a phase variance which is sufficiently low to allow for reliable phase estimates.

A system theoretical estimate of coherence can be obtained for geometry-induced effects such as geometric (baseline) decorrelation and Doppler centroid decorrelation. Coherence effects induced by interferometric processing procedures, such as coregistration, interpolation, complex multiplication, and filtering can be treated the same way. The effects of thermal noise can be derived from the design parameters of the SAR system and estimates for the normalized radar cross section.

Apart from a highly unlikely micro-scale modeling of the scattering characteristics for a specific area, it is impossible to predict the amount of temporal decorrelation for each resolution cell quantitatively. Many indications, such as land use, vegetation, anthropogenic activity, climatic conditions (such as snow), and possible surface deformation can give an indication of the amount of decorrelation to be expected, in combination with the radar wavelength. Nevertheless, these parameters are usually not quantitative and depend on operator experience and a priori information. The propagation of the various sources of decorrelation error to the phase variance is summarized in fig. 4.39, p. 155.

An important characteristic of the interferometric phase is the 2π-ambiguity of its PDF. If we intend to refrain from non-verifiable assumptions on the maximum phase gradient, the location of the phase PDF is unknown. By introducing pseudo-observations, the range of the PDF can be limited, and by using a sufficient number of aligned SAR images with varying spatial and temporal baselines, an optimal solution can be obtained. These topics are outside the scope of this study, but will be of major importance for future research. For many interferometric data analyzed here, heuristic assumptions on maximum phase gradient were used, allowing for "traditional" phase unwrapping algorithms to be applied.

Correlation between the phase observations of resolution cells in the image is introduce by orbit errors and variable atmospheric delay. The effect of orbit (baseline) errors is shown to have long-wavelength characteristics for the reference phase and a scaling error when deriving topographic height differences. It is shown that a correction of the baseline by counting the residual fringes may result in an increased error in the perpendicular baseline. The amount of residual fringes based on the a priori estimation of the radial and across-track orbit errors are approximated.

Errors in the interferometric phase due to the atmospheric refractivity distribution can be distinguished into two contributions: turbulent mixing and vertical stratification. Turbulent mixing induced refractivity distribution results in a power-law (fractal) distribution of the spatial delay variability. Such spatial variability can be conveniently described using structure functions or covariance functions. An atmospheric delay model is introduced based on Kolmogorov turbulence theory. Experimental data over many interferograms support this model.

Differences in vertical stratification between the two SAR acquisitions significantly influences the interferometric phase. An empirical relation between the size of the phase error due to vertical stratification has been derived for different height intervals and temporal baselines, using a set of radiosonde observations. Correction of the error due to vertical stratification is only possible using vertical profile measurements. Surface observations combined with a tropospheric model are in general unreliable, whereas integrated refractivity observations as obtained using GPS are insufficient for correction.

Propagation of errors to the final interferometric products, i.e., a deformation or elevation map, is relatively straightforward and is discussed in section 4.9. An important contribution of this research, with respect to previous studies on InSAR accuracy, is the parameterization of the errors in terms of both covariances and variances. For example, the height difference between two pixels in an elevation map is generally more accurate if the pixels are close together. This is valuable information for the end user of the map, which now becomes available as a complementary product.

7.1.3   Modeling of atmospheric error

Specific research question 3. "To what extent does the atmospheric re-
fractivity distribution affect the interferograms. Is it possible to deter-
mine this influence using additional measurements, and how should it be
modeled stochastically?"

The influence of the atmospheric refractivity distribution can significantly affect
repeat-pass radar interferograms. A study performed within the framework of this
research shows that each interferogram of a series of 26 tandem acquisitions over the
Netherlands is affected by atmospheric signal (Hanssen, 1998). The observed spatial
scales reach from hundreds of meters to 100–200 km. For scales reaching the full
image size, it is, however, difficult to discern them from errors in the satellite orbits.

Although rms-measures of the magnitude of the effects do not describe the spatial
roughness of the signal and are a function of the spatial extent of the evaluation area,
they yield a comprehensible approximation of the atmospheric signal. Observed
rms-values, taken over an area of approximately 40×40 km, reach from 2–16 mm.
Assuming a Gaussian distribution, this implies that 95% of the observed phase
values varies over a minimal range of 8 mm up to a maximal range of 64 mm.
Extreme ranges of more than 120 mm delay variation are observed in the case of
thunderstorms during the two SAR acquisitions.

It is shown that (section 4.7) although the state of the atmospheric refractivity distribution can be very different depending on the weather situation, the shape of
the power spectrum shows three distinct scaling regimes, which are similar independent of the state of the atmosphere. Nevertheless, the absolute values of the power
spectra differ two orders of magnitude. A model for atmospheric delay variation is
presented, which incorporates these scaling characteristics, and which can be initialized for various weather or climate situations. Using the model, it is possible to
quantitatively describe the covariance function of an interferogram using only one
parameter. This parameter can be estimated from a small patch of the interferogram, with either even terrain or without deformation—the bootstrap method. This
first order estimate may then be used to describe covariances in the whole interferogram. An alternative approach is the estimation of the initialization parameter using
GPS time series—GPS initialization. In this way, a single GPS receiver in the interferogram area would be sufficient to quantify the covariances in the interferometric
data.

7.1.4   Interferometric radar meteorology

Specific research question 4. "Is it possible to analyze the atmospheric
error signal as a meteorological source of information?"

Interferometric Radar Meteorology (IRM) is a new technique for meteorology and
atmospheric studies in general. It can be applied to study vertically integrated atmospheric refractivity variations with a spatial resolution of 20 m and an accuracy
of ~2 mm, irrespective of cloud cover or solar illumination, over most land and

ice areas. On short spatial scales integrated refractivity variations are dominantly caused by spatial heterogeneities in the water vapor distribution. IRM can be readily used for studying atmospheric dynamics such as mesoscale shallow convection, by mapping the small-scale water vapor distribution. The main advantage of the technique is the combination of its imaging possibilities with quantitative delay estimates. Readily attainable integrated precipitable water amounts over large areas may make pinpoint weather forecasting a possibility. The delay maps could eventually be used by bench forecasters or serve as an additional constraint in variational data assimilation models.

A number of case studies validate the radar interferometric results based on the comparison with various meteorological data and GPS data. The combination of the instantaneous observations of radar backscatter intensity over water and radar signal delay over land areas enables the study of causal connections between mesoscale wind patterns and moisture distribution in the lower boundary layer. Analyses are reported on the relation between wind and moisture associated with boundary layer rolls and with a cold-frontal rain band.

Contemporary satellites, however, have an orbital repeat period that is far from ideal for operational applications, resulting in loss of coherent phase information over many areas. Ideally, SAR data should be acquired with multiple short temporal spacings in order to maintain high coherence and permit a large number of interferogram pairs to be generated. Multiple observations give not only better precision in time, but keep coherence and hence data quality high. Proper use of airborne and spaceborne SAR systems with short repeat periods could lead to much greater accuracy in meteorological understanding and forecasting.

7.2 Recommendations

The mathematical model introduced in this study is only a first step towards a unified approach for the analysis of radar interferometric data. A first implementation of this theory in software algorithms is necessary to validate the concepts.

The implementation of the theory discussed in this study in operational algorithms can be performed in two stages. In the *quality assessment* stage, the accuracy of (i) a single pixel parameter and (ii) the difference between two pixels can be determined, assuming distributed scattering. For example, if the final product is an elevation map it will be possible to choose an arbitrary pixel and obtain both the height and variance of the height for that pixel. More important, it will be possible to choose two pixels in the image and obtain the height difference and the variance of the difference, using $\sigma^2_{\Delta h} = \sigma^2_{h_1} + \sigma^2_{h_2} - 2\sigma_{h_1,h_2}$. The same reasoning holds for deformation mapping or atmospheric mapping. The quality assessment stage is relatively easy to implement by combining the coherence image, the interferometric product derived from the interferogram, the atmospheric initialization parameter P_0, and the a priori estimates for the radial and across-track orbit errors.

The second stage involves the implementation of the variance-covariance matrix. Using a full-scene single-look interferogram, this will be a matrix with millions of elements. Matrix operations such as inversions will be vary laborious, if possible at all. Nevertheless, the well-organized structure of the matrix (BTTB) suggests that standard inversion procedures might not be the most economic. For example, it could be investigated whether it is possible to derive the weight matrix Q_φ^{-1} directly based on theoretical grounds. These issues need to be considered, combined with problems on numerical stability and parallel processing.

The PDF of the interferometric phase has a 2π-ambiguity. As data adjustment using a Gauss-Markoff model is conventionally limited to observations with an unambiguous PDF, this poses a limitation in using the model for wrapped data. The use of more than two interferograms for a specific goal (e.g., topography or deformation) results in a different sensitivity for height or deformation, depending on the spatial and temporal baseline, respectively. Integration of this information in the model will result in a more generic approach in the data adjustment.

Although the different sensitivities for topography and deformation result in different scaling of the PDF, it theoretically still extends to infinity. Therefore, pseudo-observations could be introduced, which decrease the likelihood of solutions outside the physical reality. It needs to be investigated how these pseudo-observations need to be structured and incorporated into the model.

The use of many, or all available, SAR acquisitions for a certain area for analyzing a specific problem will become more important. The combination of time-series of interferograms (stacking) with the spatial deformation pattern within an interferogram has already lead to new applications of the technique, previously unattainable. In this light, the use of the *permanent scatterers* technique is a major contribution, which deserves more attention (Ferretti et al., 2000).

Finally, regarding deformation measurements there appears to be a gap between the derived slant-range deformation pattern and the determination of geophysical parameters. Often, forward models are applied to minimize the difference between the observed and the modeled deformation pattern. It needs to be investigated whether the proposed geodetic model discussed in this study can be converted to directly derive the geophysical parameters of interest. The advantage of such an approach is that accuracy and reliability can be directly derived from observations and model, and that model adjustment is possible based on the characteristics of the observations.

Addenda

Appendix A

Comparison neutral delay GPS and InSAR

Neutral (tropospheric) delay can be an important source of error when using radar interferometry for deformation or topographic studies. Since the signals of the Global Positioning System are affected by the troposphere in a similar way as the radar signals, it is investigated whether GPS data can be used for the correction or quantification of the tropospheric error in radar interferometric data. Two experiments discuss the use of a spatially distributed set of receivers and of one stationary receiver, using time series analysis.

key words: *GPS, Troposphere, Signal delay, GPS network, GPS time series*

Atmospheric delay in radio signal propagation is known to be a major source of error in high-precision positioning applications such GPS and InSAR (see, e.g. Bevis et al. (1992) for GPS and section 4.7 and chapter 6 for InSAR measurements). In GPS data processing, the dispersive ionospheric part of the delay is largely eliminated by using a linear combination of the two GPS frequencies. The non-dispersive delay in the neutral part of the atmosphere cannot be eliminated this way, and is often estimated as an extra unknown in the GPS data processing (Brunner and Welsch, 1993). In this procedure it is often assumed that there are no horizontal variations and that the observations can be related to zenith delay using a mapping function (Herring, 1992; Niell, 1996). If the tropospheric delay is the main parameter of interest, it is often decomposed in a hydrostatic ("dry") component, which can be estimated with a 1 mm accuracy using surface pressure measurements, and a "wet" component which is highly variable, both spatially as well as temporally (Davis et al., 1985). Nowadays, the wet delay can be derived from the GPS observations at a fixed station with an accuracy of about 6–10 mm in the zenith direction (Bevis et al., 1992; van der Hoeven et al., 1998).

Since the troposphere is not dispersive for frequencies less than 30 GHz, the GPS signal delays are affected the same way as, e.g., X-band, C-band, or L-band SAR observations. Therefore, if GPS stations are available in the interferogram area, it might be possible to use the GPS-derived zenith delays as a priori information in the interpretation of the interferogram. There are two possibilities to combine GPS and InSAR observations. In the first, deterministic, approach the relative delay differences between the stations at the time of the image acquisitions are determined, interpolated, and subtracted from the interferogram data. The second, stochastic,

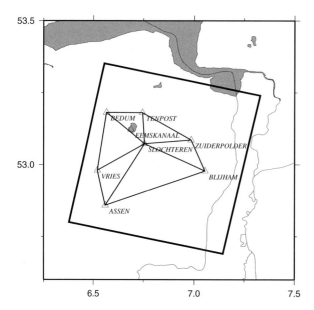

Fig. A.1. Location of the SAR image and the GPS receivers during the GISARE-GPS experiment (northern part of the Netherlands, Mar 1996). The lines connecting 7 stations are the analyzed GPS baselines.

approach does not aim at full correction, but uses time series observed at a stationary receiver to estimate the amount and variability of delay variation. This way, a quantitative stochastic measure of the delay behavior can be obtained. Experiments which investigate both possibilities are discussed in the sequel. Section A.1 explores the possibilities of a spatial network of receivers, whereas section A.2 covers the single stationary receiver.

A.1   Spatial networks

The experimental data for this study were collected during the Groningen Interferometric SAR Experiment (GISARE), performed during 14–18 March 1996 in the northeast of the Netherlands. In the framework of this experiment tandem SAR data were acquired during five months by the satellites ERS-1 and ERS-2. During two of these acquisitions, simultaneous GPS measurements were performed using a temporary network of 8 GPS receivers. By comparing the GPS zenith signal delays with the interferometric data, an analysis can be performed of the correspondence and differences, regarding spatial and temporal resolution, accuracy, complementarity and feasibility (Hanssen, 1996; Stolk, 1997; Stolk, Hanssen, van der Marel and Ambrosius, 1997).

The difference in the interferometric phase between two points m and k, in terms of tropospheric delay S_{t_i} at acquisition t_i can be written as

$$\Delta\varphi_{\mathrm{sar},mk} = \frac{4\pi}{\lambda}\left([S_{t_1} - S_{t_2}]_k - [S_{t_1} - S_{t_2}]_m\right) + \epsilon_{\mathrm{sar}} \qquad (\mathrm{A.1.1})$$

where λ_{sar} is the SAR wavelength. Using a simple mapping function, $M(\theta_{inc}) = 1/\cos(\theta_{inc})$, with local incidence angle θ_{inc}, the tropospheric delays can be converted to zenith delays:

$$\Delta\varphi_{sar,mk} = \frac{4\pi}{\lambda_{sar}} M(\theta_{inc}) \left([S_{t_1,z} - S_{t_2,z}]_k - [S_{t_1,z} - S_{t_2,z}]_m\right) + \epsilon_{sar,mk}. \quad (A.1.2)$$

The interferometric phase values represent the temporal difference in tropospheric delay, but only in a relative sense. By comparing different locations m and k in the interferogram, the spatial distribution of these differences can be studied.

The relative tropospheric delays can be derived from the GPS phase observations as well. Clock errors are eliminated using double-difference observations. Ionospheric delays are eliminated by using the ionospheric-free linear combination (L3). The effective wavelength of L3 is denoted by λ_{L3}. Assuming known satellite ephemerides and solving for the integer phase ambiguities, the double-difference residuals can be obtained:

$$\Delta\varphi_{GPS,mk,t_1}^{qp} = \frac{2\pi}{\lambda_{L3}} \left(S_{mk,t_1}^{qp} + d_{mk,t_1}^{qp}\right) + \epsilon_{mk,t_1}^{qp}, \quad (A.1.3)$$

where S_{mk,t_1}^{qp} is the double-difference tropospheric delay between stations m and k and satellites p and q at $t = t_1$. Multipath is denoted by d_{mk,t_1}^{qp} and measurement noise by ϵ_{mk,t_1}^{qp}. Separating the double differences to the two stations and applying a mapping function to derive zenith delays results in

$$\Delta\varphi_{GPS,mk,t_1}^{qp} = C_1 (S_k^z - S_m^z)_{t_1} - C_2 S_{m,t_1}^z + d_{mk,t_1}^{qp} + \epsilon_{mk,t_1}^{qp}, \quad (A.1.4)$$

with

$$C_1 = \frac{2\pi}{\lambda_{L3}} [M(\alpha_k^p) - M(\alpha_k^q)],$$

$$C_2 = \frac{2\pi}{\lambda_{L3}} ([M(\alpha_m^p) - M(\alpha_m^q)] - [M(\alpha_k^p) - M(\alpha_k^q)]).$$

$$= \frac{2\pi}{\lambda_{L3}} ([M(\alpha_m^p) - M(\alpha_k^p)] - [M(\alpha_m^q) - M(\alpha_k^q)]).$$

Mapping function M is a function of the elevation angle α to the satellites. In eq. (A.1.4) the tropospheric delay occurs in two terms: (i) the relative spatial tropospheric zenith delay difference between station k and m, and (ii) in the absolute value of the tropospheric zenith delay for station m. Both terms can in principle be derived from the double-difference residuals of a measurement epoch. However, the coefficient of the absolute term C_2 is dependent of the length of the baseline between m and k; relatively short baselines imply nearly identical elevation angles for both stations and both satellites. Therefore, $C_2 \approx 0$ for these baselines, and no information on the absolute delay can be obtained. A similar problem can occur for coefficient C_1 in case both satellites are observed at the same elevation angle. In these cases (which are only a temporary problem) the relative tropospheric delay difference cannot be determined. In general, however, it is possible to derive estimates for relative tropospheric delay differences from double-difference, ionosphere-free residuals, independent of the length of the baseline.

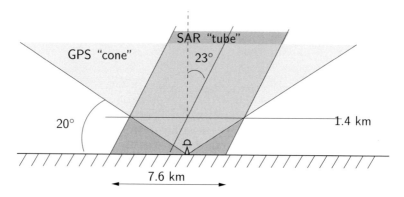

Fig. A.2. Cross-section of the difference in geometry for SAR and GPS, as used for comparing zenith delays. GPS zenith signal delays are obtained using all observations higher than a critical elevation angle of 20°. The SAR delay is averaged within a "tube" under a zenith angle of 23°. At 1.4 km height, the overlap between both configurations is used for comparison.

Comparing the GPS relative tropospheric delay differences at the SAR acquisition times t_1 and t_2, and assuming that multipath effects will cancel in the differencing of a 24 hr interval, we find

$$(S_k^z - S_m^z)_{t_1} - (S_k^z - S_m^z)_{t_2} = (S_{t_1}^z - S_{t_2}^z)_k - (S_{t_1}^z - S_{t_2}^z)_m. \qquad (A.1.5)$$

Comparing the right-hand side of eq. (A.1.5) with the right-hand side of eq. (A.1.2) and assuming zero-mean Gaussian noise we find

$$E\{\Delta\varphi_{\text{sar},mk}\} = E\{\Delta\varphi_{\text{GPS},mk,t_1}^{qp}\} - E\{\Delta\varphi_{\text{GPS},mk,t_2}^{qp}\}, \qquad (A.1.6)$$

hence, both the GPS and SAR observations can be compared.

A.1.1 Experiment set-up

During the GISARE-GPS experiment, between 14 and 18 March 1996, simultaneous GPS observations where performed on eight NAM locations in the northeast part of the Netherlands. Using a survey epoch of 5 days, it is possible to determine the position of the receivers with an accuracy higher than 1.2 mm. Eight GPS receivers (4 *Ashtech MDL Z-12*, 4 *Turbo-Rogue's SNR-12RM*) with Dorne-Margolin antennas were used simultaneously during this period. At 16 and 17 March the area was imaged by ERS-1 and ERS-2, respectively, see figs. A.1 and A.3. Since the perpendicular baseline was 26 m, the sensitivity for topographic height is very small, see 2.4.1, excluding this influence on the interferometric phase. Since the data were acquired within 24 hours surface deformation can be excluded as well, making atmospheric signal the most prominent phase signal.

A.1.2   Methodology

To retrieve the double-difference ionosphere-free residuals from the GPS phase observations, it is necessary to determine the positions of the satellites and receivers,
and to solve for the integer ambiguities. Satellite positions where made available by
the International GPS Service for Geodynamics (IGS). Using the Bernese GPS software all receiver stations are tied to a network of independent baselines, to solve for
the station coordinates and the integer ambiguities (Rothacher and Mervart, 1996).
For every baseline, the double-difference ionosphere-free residuals are determined.
Since the baselines are relatively short, it is not possible to determine the absolute
tropospheric delay. It is, however, possible to estimate the relative delay differences.
A relative tropospheric delay is estimated every 30 seconds, using a cut-off elevation
angle of 20°. For every baseline, this results in a time series spanning one day. To
eliminate noise in these time series, the data are low-pass filtered. From these filtered data, the value of the relative tropospheric delay is recorded at the time of the
SAR acquisitions.

The interferometric data are multilooked by 2×10 pixels, resulting in a spatial
posting of approximately 40×40 m. After phase unwrapping, the phase difference
between adjacent pixels can be interpreted as a spatial difference in tropospheric
delay during the two acquisitions. Since orbit errors can result in a residual phase
ramp this is accounted for in the comparison with the GPS data.

The part of the atmosphere that contributes to the observed delays is different for
GPS and InSAR. The atmospheric contribution for a GPS receiver is a reversed
cone, with the receiver in the focal point and the imaginary rays to several GPS
satellites located within or at the edges of the cone, see fig. A.2. The aperture angle
of the cone is defined by the cut-off elevation angle, below which observations are
discarded. All phase observations between receiver and the satellites, averaged over
a certain time interval, contribute to the derived zenith delay observation. For SAR
interferometry, the delay observation at a single pixel can be regarded as one single
ray, ranging from that pixel to the satellite, at the time of the image acquisitions.
These observations are spatially relative, and are formed by differencing the delays
at two, nearly uncorrelated, acquisition times. To ensure analysis of comparable
quantities, both data sources need to be tuned.

Regarding both sources of data, it is assumed that especially the local variations in
the wet delay, dominantly water vapor, contribute to the relative differences. The
analysis of radiosonde data acquired at noon shows that more than half of the wet
delay originates from the lower 1.4 km of the troposphere. The intersection of the
cone and the 1.4 km level forms a circle with a diameter of 7.6 km, and it is assumed
that atmosphere within that circle is the main contribution to the GPS wet zenith
delay observation. To tune the high-resolution interferometric data to the same area,
the interferogram is averaged over all data within that circle, considering a lateral
shift of the circle due to the SAR incidence angle. This geometry is visualized in
fig. A.2. As a result, the SAR data regard a larger fraction of the atmosphere below
1.4 km, whereas the GPS data consider a larger atmospheric volume above this level.

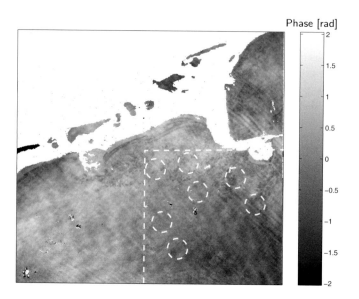

Fig. A.3. Interferogram of 16 and 17 Mar 1996, with interferometric phase in radians. One radian corresponds with 5 mm delay. Both short (diagonal) and long wavelength phase differences are visible. The GPS receivers were distributed over the south-eastern quadrant of the interferogram. The white circles indicate the area around every receiver to be compared with the GPS zenith delay estimates.

A.1.3 Results

Figure A.3 shows the full-scene interferogram of the SAR acquisitions of 16 and 17 March 1996. The area covered by the eight GPS receivers covers only the southeastern quadrant of this image. Atmospheric signal is limited to a weak long-wavelength anomaly in the southeast (the dark part) and short-wavelength waves over the larger part of the image. The latter are probably caused by gravity waves—waves at the boundary between two masses of air with different densities—or due to wind shear. Local variations of less than 1 cm are observed, while the wavelength of the gravity waves is ~250–300 m.

The averaged areas around each GPS receiver, used for the quantitative comparison with GPS are indicated in fig. A.3, and consist of ~25000 pixels. The difference between two of these circles can be compared with the GPS observations related to that specific baseline.

Additional meteorological information is obtained from Meteosat and NOAA-AVHRR (Advanced Very High Resolution Radiometer) satellite imagery, synoptic surface observations, radiosonde launches, and weather radar, see Hanssen (1998). Both days had full (8/8) cloud cover during the SAR acquisitions. Stratus and stratocumulus clouds were observed at 300–400 m, while during the second acquisition also higher cloud cover was observed. The AVHRR images show varying wave patterns over

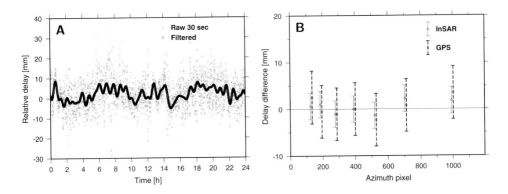

Fig. A.4. (A) Row 30 s data and low-pass filtered relative tropospheric delay for baseline Assen-Blijham, 16 Mar 1996, modified from Stolk (1997). **(B)** Relative tropospheric differences for GPS and InSAR, as a function of the GPS location in azimuth direction. (GPS positions are slightly offset to the left for comparison). Slochteren (fourth position) is used as fixed reference station. The error bars express a 95% confidence interval.

the test area. The wind direction varied from east at the surface to south aloft. Temperatures during the SAR acquisitions were a few degrees above the freezing level.

Of the eight installed GPS receivers, one failed after some hours of operation. Of the seven remaining stations, six independent baselines are defined. Relative tropospheric delay differences can be derived for every baseline, using the double-difference ionospheric free residuals for 16 and 17 March, every 30 s. Figure A.4A, modified from Stolk (1997), shows an example of the relative troposphere estimations for day 76, baseline Assen-Blijham. A considerable amount of noise is apparent in these data, making a direct comparison of the raw 30 s GPS data with the interferogram impossible. The result of low-pass filtering the data, with a cut-off frequency of 2×10^{-4} Hz, is shown overlaid on fig. A.4A. Unfortunately, the temporal variation in the tropospheric delay of day 76 for the 33 km baseline Assen-Blijham is very limited, less than 2 cm.

Using the low-pass filter for all baselines for both days enables the extraction of the relative tropospheric wet delay values during the two SAR acquisitions. These values for both days are then subtracted to mimic the difference between the two days in the interferogram. Using a reference station, where the values for both methods are assigned equal (null) values, the absolute difference in tropospheric zenith delay observations between both acquisition times is determined at every GPS station. This is shown in fig. A.4B. On the horizontal axes the location of the receivers in azimuth direction is plotted. The error bars represent a 95% ($\pm 2\sigma$) confidence interval. Although it is obvious that the atmospheric signal during this experiment is limited, a first quantitative evaluation between the InSAR and the GPS results shows agreement between both methods: the observed rms of the difference is ~3 mm. Recognizing the possibility of a phase trend in the interferogram, and solving for such a first-order trend reduces the rms of the difference to ~2 mm. In fig. A.6, the comparison between the six GPS baselines with respect to station Slochteren and the SAR data is shown. The location of the points, along the diagonal, suggests a correlation between both data sources, although the variances of the data are relatively large.

The possibility to "correct" the interferometric data using the (interpolated) GPS data is investigated and shown in fig. A.5. Figure A.5A shows the interferogram, which might include a linear phase trend due to orbit errors. In the corresponding histogram of the phase values, the interferogram including trend is shown in gray, while the interferogram corrected for a bilinear phase trend is shown in black. The GPS data, see fig. A.5B, are interpolated using an isotropic harmonic spline, with local maxima/minima occurring only at the data points (Wessel and Smith, 1998). It is obvious that the extrapolation of the data outside the GPS station framework can be erroneous, although it could as well be an indication for a true trend in the atmospheric delay. There is no way to solve this problem without any additional data, although a significant trend in the delay is not likely, considering the weather conditions. The histogram corresponding to the GPS data reflects the trend in the data. Eliminating the trend in the InSAR data by fitting it to the GPS data yields a "corrected" result. In this result, the interferogram adopts the GPS trend and should be corrected for atmospheric signal with, say, 10–20 km wavelengths, based on the distance between the receivers. In fig. A.5C, we subtract the trend, which would disturb the visual comparison and cannot be validated, and show only the residual signal for comparison. Evaluation of the histograms for the detrended data, in fig.A.5A and C, shows no significant improvement of the data distribution.

A.1.4   Discussion

Due to the fact that the amount of spatial variability in the atmospheric delay was extremely limited during this specific experiment, the methodological validation of an atmospheric correction of interferograms using GPS networks is difficult. Moreover, the L3 relative delays in the GPS baselines are too noisy for a direct comparison with the interferogram, resulting in error margins that exceed the atmospheric delay variability in this situation, even though there is correlation between both datasets, see fig. A.6. Therefore, it is recommended to use L1 and/or L2 processing instead of the ionospheric-free (L3) linear combination.

Low-pass filtering the data (effectively averaging over periods of ~80 minutes) reduces the noise, but is rather arbitrary and crude. Therefore, the results obtained from this experiment do not allow for general statements on the feasibility of correction of radar interferograms using a GPS network. Further experiments, during variable weather conditions, are necessary to achieve better understanding of the possibilities. Nevertheless, based on this experiment we can comment on some important issues.

The possibility for correction of the interferograms is dependent on the spatial distribution of the GPS receivers. Since (continuous) GPS networks are usually designed for other purposes, sampling criteria for SAR purposes are not optimal, and the re-

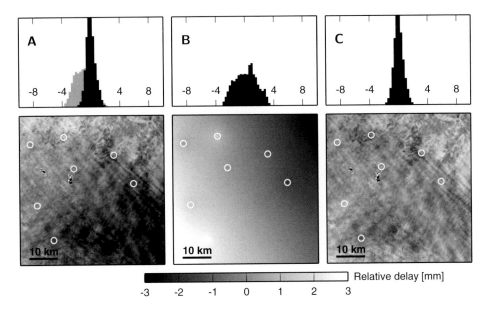

Fig. A.5. Comparison of interferometric delays with GPS delays. The top row shows the histograms of the delay distribution in millimeters, the bottom row the images. **(A)** SAR interferogram, converted to zenith delay in millimeters, including possible orbital phase ramp. The white circles indicate the position of the GPS receivers. The histogram plot shows the distribution including the trend in gray and the distribution without trend in black. **(B)** Interpolated GPS delay signal, using a harmonic spline solution, with local maxima/minima occurring only at the data points. **(C)** Difference between GPS-corrected and original interferogram. A bilinear trend was removed from this image for clarity.

duction of the atmospheric noise using interpolation between the receivers is mainly achieved over areas directly surrounding the GPS receivers. On the other hand, considering a densification of a GPS network there are limitations related to the zenith delay reduction of the data. Using a 20-degree elevation angle, the zenith delay estimate will reflect atmospheric variation within a wide range, and cannot be considered as a point measurement, see fig.A.2. For atmospheric signal at 1, 2, or 3 km altitude, this elevation angle results in circles with a diameter of 5, 11, and 16 km respectively. Therefore, the zenith delay will reflect some kind of weighted average of all atmospheric variation within that area, and it will not be possible to remove atmospheric artifacts with smaller scales from the interferogram. This results in an upper limit to the effective density of the GPS stations, in effect a smoothing of the signal whenever the GPS receivers are too close. This is discussed in more detail in appendix A.2.2. Therefore, future studies need to address the problem of slant-delay measurements instead of zenith-delay measurements.

For deformation studies, there is another, more fundamental, reason that limits the

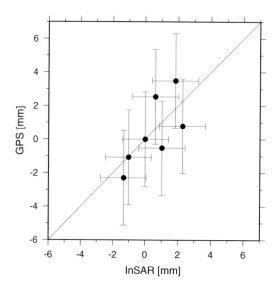

Fig. A.6. Scatterplot showing the GPS delay differences against the InSAR-derived delay differences with respect to station Slochteren. The error bars span $\pm 1\sigma$.

value of GPS networks for atmospheric corrections in SAR interferograms. With a continuous GPS network it is possible to measure the long wavelength displacement field in a region which may be deforming. Unfortunately, deformations on small spatial scales cannot be observed, due to the limitations in the receiver density. Filtering the long wavelength atmospheric signal from a SAR interferogram still leaves all atmospheric (and deformation) signal on these small spatial scales in the image. In fact, the technique even introduces ambiguous short wavelength signal due to the limitations in the interpolation. As a result, for short wavelengths the interpretation of a GPS-filtered interferogram can be even more difficult than without the filtering, since it is a superposition of atmospheric and interpolation errors. On the other hand, for long wavelengths, the deformation signal was already known from GPS, which makes SAR superfluous.

A.2 Single-station time-series

A single GPS station has been selected to investigate the situation when no dense network of GPS receivers is available for the observation period. Unfortunately, this is common practice for most of the situations where research with SAR images is involved. Moreover, although a dense (permanent) network of GPS receivers would enable a direct comparison of the spatial delay variability, there are considerable differences as well. Limitations in this approach are (i) the weighted averaging to zenith delays and (ii) the introduction of interpolation errors due to the inhomogeneous distribution of GPS receivers. If such techniques are applied "blindly" and used for extracting an atmospheric phase screen from the SAR interferograms, the geophysical interpretation of the results may be more ambiguous than without such a correction, as demonstrated in the previous section.

This experiment is not an attempt to correct the interferograms, but rather to obtain reasonable estimates of the delay variations and spatial scales to be expected. Such quantitative estimates could significantly improve the interpretation of interferometric SAR data, using only a single GPS receiver as additional source of information (van der Hoeven and Hanssen, 1999).

Five case-studies are performed, using ERS-1 and ERS-2 SAR data acquired with a one-day interval, hereby excluding crustal deformation signal. Swaths of 200 × 100 km are processed differentially, removing the topographic signal using a reference elevation model (TDN/MD, 1997), see section 2.4.2. After applying a simple cosine mapping function, the residual signal consists of zenith delay variations in each pair of data acquisitions (Hanssen et al., 1999). GPS data are obtained at the Kootwijk Observatory for Satellite Geodesy (KOSG), a permanent GPS receiving station. The GPS observations are processed using GIPSY-OASIS while solving for a free network. Data from 13 widely spread IGS stations are used to solve for a number of parameters including satellite clocks, station positions, gradients, and tropospheric delays. The receiver clocks were estimated as white noise processes, while the wet tropospheric delays were estimated as random-walk processes, using the Niell mapping function to convert slant delays from a minimum elevation of 10o to zenith delays (Niell, 1996). The a priori hydrostatic delay is calculated using the Saastamoinen model. To reduce the noise in the zenith delay estimates, the data obtained at a sampling rate of 30 s are averaged into 6 min intervals. The fluctuations in the zenith delays consist of the combined hydrostatic and wet delay—the same contributions as measured by InSAR.

A.2.1 Methodology of comparison

The absolute zenith tropospheric delays derived from GPS constitute time-series at a fixed position. In contrary, InSAR generates a spatial image of the relative zenith delays, differenced between two fixed acquisition times. A conversion needs to be performed to connect the relative-spatial and absolute-temporal observations in order to validate the two datasets.

Two assumptions have been applied to match the two techniques. First, Taylors approximation is applied, treating the local atmosphere as a *frozen* atmosphere moving over the area without changing its refractivity distribution during a preset time interval (Taylor, 1938; Treuhaft and Lanyi, 1987). This refractivity distribution is displaced by the mean wind speed and direction, obtained from surface or radiosonde observations. Second, wind speed and wind direction are assumed to be approximately equal during both observation days. The latter assumption will certainly fail for longer time intervals, but can be more likely for a one-day interval, a common choice for, e.g., DEM generation with InSAR.

Applying these assumptions the GPS time-series can be converted to spatial zenith delay profiles, corresponding with a ground trace in the wind direction and stretched by the wind velocity. These profiles are computed for the two SAR observations from about 4 hours before until about 2 hours after the SAR acquisition (21.41 UTC). The non-symmetric epoch was chosen to avoid delay errors caused by discontinuities

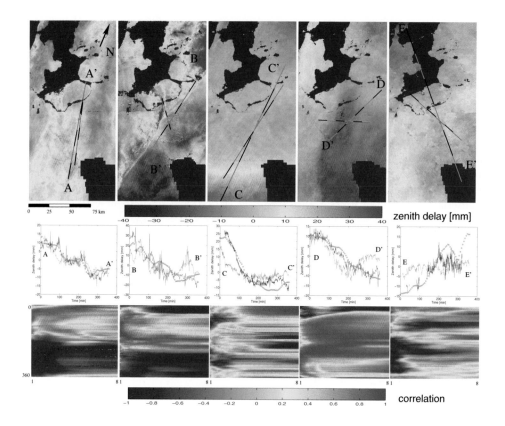

Fig. A.7. The top row of figures shows the differential SAR interferograms together with the synoptic (black-red) and best-fit profiles (black-blue) based on GPS delay time series. The line segments indicate 1-hour time intervals. The red star indicates the position of the GPS station KOSG. The second row of figures shows the GPS time series (black dots) together with the synoptic profile (red) and best-fit profile (blue). In the lower row of figures the correlations are shown between GPS and InSAR for all combinations of wind speed and direction between 1–8 m/s, respectively 0–360° (van der Hoeven and Hanssen, 1999).

in the GPS satellite orbit estimation at the day break, a common problem in contemporary GPS processing. Subsequently the two profiles are differenced, resulting in a differential delay profile, which is comparable to a cross-section in the interferogram. The arbitrary bias between the two sources of data is removed by subtracting their means. This approach is referred to as the *synoptic* approach.

To verify the feasibility of using synoptic wind observations, an alternative approach is performed by rotating and stretching the ground trace over the InSAR-image while recording the correlation between the two profiles. The GPS observations made at the time of the SAR observations are used as axis of rotation. Maximum correlation

Table A.1. Comparison between GPS time series and SAR interferogram profiles.

			Aug 95	Oct 95	Mar 96	Apr 96	Jun 96
fig.			A.7A	A.7B	A.7C	A.7D	A.7E
correlation	best fit		0.95	0.90	0.95	0.95	0.81
	synoptic		0.84	0.56	0.81	0.57	0.44
v_w	best fit	[m/s]	5.2	2.1	8.0	3.8	8.0
	synoptic	[m/s]	4.1	4.1	6.2	3.1	3.6
θ_w	best fit	[deg]	353	148	003	208	143
	synoptic	[deg]	350	195	015	075	115
rms	best fit	[mm]	2.2	6.6	3.9	2.3	6.4
	synoptic	[mm]	3.5	10.3	9.1	6.6	8.8
range	$\mathrm{SAR_{syn}}$	[mm]	24	40	24	24	15
	$\mathrm{SAR_{b.f.}}$	[mm]	31	59	39	29	20
	GPS	[mm]	19	35	31	21	30

v_w, wind speed; θ_w, wind direction. Interferogram profiles are retrieved (i) based on a best-fit criterion between GPS and InSAR and (ii) based on the wind speed and direction from synoptic surface observations. Total range of the delay variation and rms-values are given for signal-to-noise comparison.

corresponds with the best-fit between GPS and InSAR. The length of the best-fit ground trace divided by the total GPS observation time is a measure for the wind speed. The angle of rotation of the best-fit ground trace gives the effective wind direction relative to the local north. The derived wind speed and direction can then be compared with the synoptic observations for validation.

A.2.2 Results

The results of the tropospheric delays estimated by GPS and InSAR are shown in fig. A.7 and table A.1. The first row of fig. A.7 shows the five differential interferograms together with the estimated *best-fit* profile in black-blue, and the profile based on the *synoptic* observations in black-red. The color changes in the profiles represent 1 hour time intervals. The location of the GPS receiver is indicated by the red star. All interferograms have been converted to relative zenith delay differences, expressed in mm. One colorbar is used, to enable the comparison of the magnitude in variation in every case.

The profiles below the interferograms correspond with the GPS observations (black dots), the scaled SAR profile which shows a best fit with the GPS time series (blue), and the SAR profile which is obtained using the surface wind speed and direction (red). In the lower row of plots, correlation coefficients are depicted which resulted in the best-fit profile. The vertical axis corresponds with the direction of the profile, the horizontal axis scales the profile with the wind speed in m/s.

Synoptic data from a meteo-station 30 km north of KOSG were used. Average wind speeds were calculated using the data from 21.00 and 22.00 UTC for both days.

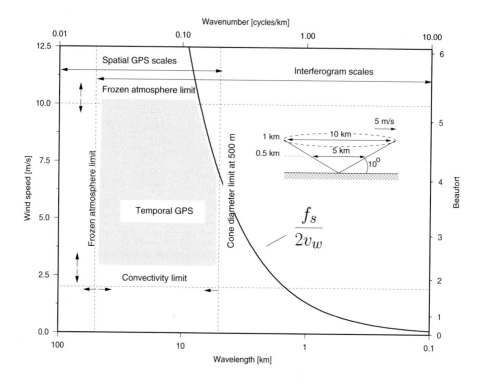

Fig. A.8. Sketch indicating the sensitivity ranges of GPS and InSAR delay observations. The shaded region indicates the wavelengths of atmospheric signal which can be detected by GPS time series, under the assumptions discussed in the text. Wavelengths to be detected by a spatial GPS network and by SAR interferometry are indicated by the arrows.

When the average wind speed was below 2.5 m/s it was assumed that no significant delay changes were present and so this synoptic data was not used in the averaging.

Table A.1 lists, for every interferogram, the correlation coefficient between the profiles and the interferometric data. The wind speed and wind direction can be derived based on the amount of rotation and stretching of the GPS profile. This can be compared with the observed synoptic wind speeds. The comparison between GPS and SAR yields an rms of difference, which can be compared to the total range of delay variation.

A.2.3 Discussion

The five analyzed interferograms represent very different weather situations, ranging from large-scale humidity variation, a narrow and a wide cold front, to a relatively undisturbed, well-mixed boundary layer. Nevertheless, relatively high correlations are found between the GPS and the SAR data. Table A.1 shows that the best-fit correlations between GPS and InSAR for the five analyzed interferograms are 0.8 or better, while the estimated wind directions are accurate to within 30deg and wind speeds differ maximally 4.0 m/s. In case the synoptic data are used to situate the SAR profile, correlations are 0.75 or better, except for the May 1996 interferogram, where a correlation of −0.01 is found.

RMS values between GPS and the synoptic SAR profiles are generally worse than those between GPS and the best-fit SAR profiles, which is also indicated by the correlation values. This can be due to (i) a difference in wind speed and wind direction between the two days, and (ii) the difference in wind speed and direction at surface level and aloft. Furthermore, the assumption of a frozen atmosphere might not be valid in all cases or limited in time.

The sensitivity of GPS observations and SAR interferograms for horizontal refractivity variations at different scales is dependent on the spatial sampling characteristics, the data processing strategy, and the measurement accuracy. Using the frozen atmosphere model with constant wind speed and direction, GPS time series can be regarded as spatial observations along a straight line, with a sampling rate determined by the wind speed. Figure A.8 is a conceptual sketch of the sensitivity of GPS and InSAR for atmospheric variation at different wavelengths.

For InSAR, the sensitivity ranges vary theoretically from the resolution cell size to the size of the interferogram. Since measurement noise affects the phase measurements at single resolution cells considerably, an averaging to 100–200 m is used to suppress this noise. For large wavelengths, orbit inaccuracies may cause nearly linear trends in the interferogram, see section 4.6. Both effects limit the effective part of the spectrum to wavelengths between 50 km and 100 m.

A GPS network approach is independent of wind speed, assuming the zenith delays can be obtained instantaneous. Effectively, the lower part of the spectrum is not limited, as it is dependent of the amount of receivers and the spatial extent of the network. If the system would be able to measure true zenith delays, the short wavelength part of the spectrum would simply be determined by the spacing between the receivers. Unfortunately, accurate zenith delay estimates are currently only obtained using a spatial and temporal averaging procedure over many satellites above an a priori defined elevation cut-off angle, see the inset at fig. A.8. This procedure effectively acts as a low-pass filter on the spatial wavelengths, with a cut-off wavenumber determined by the minimum elevation angle and the scale height of the wet troposphere, below which most of the horizontal refractivity variations occur. This scale is indicated in the figure as the cone diameter limit.

The use of GPS time series, as described above, implies a dependency of the wind speed. The horizontal lines in fig. A.8 indicate situations in which the wind speed is either too low or too high for the assumption of a frozen atmosphere to be valid. The left-most vertical line shows that for large spatial scales it will take very long for the atmosphere to drift over the GPS receiver—too long to regard it as *frozen*. The conversion from temporal GPS observations with an effective sampling frequency of

$f_s = 1/360$ Hz to a spatial sampling, as a function of wind speed v_w, yields a spatial Nyquist wavenumber of

$$f_N = \frac{f_s}{2v_w}, \tag{A.2.1}$$

indicated by the curved line in fig. A.8. It is shown that for wind speeds higher than ~ 7 m/s, the data are undersampled, which might result in aliasing effects. On the other hand, for lower wind speeds the data are effectively oversampled and can be regarded as bandlimited. Low-pass filtering should be applied to suppress short wavelengths higher than the bandwidth determined by the cone diameter. The effective range of wavelengths and wind speeds for which GPS time series can be applied is indicated by the shaded region in fig. A.8.

A.2.4   Conclusions

When comparing water vapor time-series from a single GPS station with water vapor profiles derived from SAR interferograms it is necessary to have comparable wind speeds and wind directions on both observation days. For the mentioned cases, where the wind speed doesn't differ more than 30 degrees between both days, and the wind speed is equal within 10%, correlations are found in the range of 0.91-0.98. During the situations with larger differences, 30 degrees and 50% respectively, the correlations decreased to 0.81-0.85. This technique only works for a limited strip of the interferogram, i.e., the area where the measured time series covers the SAR image. Nevertheless, the study shows a clear and strong correspondence between wet signal delay as observed by GPS and InSAR. As such, it aids the interpretation of GPS studies, recognizing the spatial variability of the wet delay and the validity of assumptions on homogeneous or gradient atmospheric models. Currently, the GPS data are used in experiments to parameterize the stochastic model of atmospheric signal for SAR interferometry.

Structure function and power spectrum

This appendix elaborates on the relation between the structure function of the refractivity and the structure function of the (integrated) signal delay, as used in chapter 4. In the second part, the analogy between the structure function and the power spectrum is discussed.

key words: Structure function, Power spectrum, Refractivity, Signal delay

B.1   Structure function for refractivity and signal delay

It can be shown that the structure function of the delay $D_S(\rho)$, see eq. (4.7.6), can be related to the structure function of the refractivity $D_N(R)$, eq. (4.7.1) (Tatarski, 1961; Treuhaft and Lanyi, 1987; Coulman and Vernin, 1991):

$$D_S(\rho) = \frac{1}{\sin^2 \theta_{inc}} \iint_{0 \to h} [D_N(R_2) - D_N(R_1)] dz_1 dz_2, \tag{B.1.1}$$

with $\rho = R_2 - R_1$ and

$$R_1 = \frac{z_1 - z_2}{\sin \theta_{inc}}, \quad \text{and} \tag{B.1.2}$$

$$R_2 = \sqrt{\rho^2 + R_1^2 + 2\rho R_1 \cos \theta_{inc}}$$

$$= \sqrt{\rho^2 + (\frac{z_1 - z_2}{\sin \theta_{inc}})^2 + 2\rho \frac{z_1 - z_2}{\sin \theta_{inc}} \cos \theta_{inc}} \tag{B.1.3}$$

$$= \sqrt{\rho^2 + (\frac{z_1 - z_2}{\sin \theta_{inc}})^2 + 2\rho(z_1 - z_2) \cot \theta_{inc}}.$$

The geometric configuration used in this equation is sketched in fig. B.1.

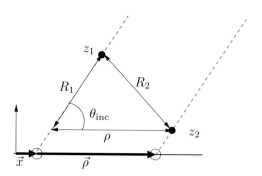

Fig. B.1. Geometric configuration for the integration along two paths, z_1 and z_2, expressed in eq. (B.1.1).

Derivation. Starting from the definition of the structure function in eq. (4.7.6) we substitute eq. (4.7.5):

$$D_S(\rho) = E\{[\frac{1}{\sin\theta_{\text{inc}}}\int_0^h N(\vec{x}+\vec{\rho}+\vec{r}(\theta_{\text{inc}}, z_2))dz_2$$
$$-\frac{1}{\sin\theta_{\text{inc}}}\int_0^h N(\vec{x}+\vec{r}(\theta_{\text{inc}}, z_1))dz_1]^2\}. \tag{B.1.4}$$

First we decompose the quadratic term. As the expectation of a sum is equal to the sum of the expectations, we can exchange integration and expectation (or ensemble average):

$$D_S(\rho) = \frac{1}{\sin^2\theta_{\text{inc}}}\Big[\int_0^h E\{N^2(\vec{\rho}+\vec{r}(\theta_{\text{inc}}, z))\}dz + \int_0^h E\{N^2(\vec{r}(\theta_{\text{inc}}, z))\}dz$$
$$-2\iint_{0\to h} E\{N(\vec{\rho}+\vec{r}(\theta_{\text{inc}}, z_1))\, N(\vec{r}(\theta_{\text{inc}}, z_2))\}dz_1 dz_2\Big]. \tag{B.1.5}$$

We removed the (horizontal) $\vec{x}$ term, assuming that $E\{N\}$ is independent of the horizontal position. Writing the integral outside of the summation, we can simplify eq. (B.1.5) to

$$D_S(\rho) = \frac{1}{\sin^2\theta_{\text{inc}}}\Big[2\int_0^h E\{N^2(\vec{r}(\theta_{\text{inc}}, z_1))\}dz_1$$
$$-2\iint_{0\to h} E\{N(\vec{\rho}+\vec{r}(\theta_{\text{inc}}, z_1))\, N(\vec{r}(\theta_{\text{inc}}, z_2))\}dz_1 dz_2\Big]$$
$$= \frac{1}{\sin^2\theta_{\text{inc}}}\iint_{0\to h}\Big[2E\{N^2(\vec{r}(\theta_{\text{inc}}, z_1))\}$$
$$-2E\{N(\vec{\rho}+\vec{r}(\theta_{\text{inc}}, z_1))\, N(\vec{r}(\theta_{\text{inc}}, z_2))\}\Big]dz_1 dz_2. \tag{B.1.6}$$

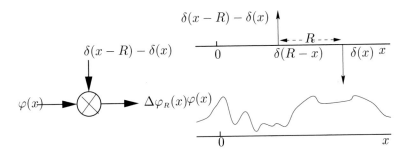

Fig. B.2. Dirac impulse convolution with a random signal $\varphi(x)$.

For second-order stationary signals we can write the structure function of N, eq. (4.7.1), as

$$D_N(R) = 2E\{N^2(\vec{r})\} - 2E\{N(\vec{r}+\vec{R})N(\vec{r})\}, \tag{B.1.7}$$

which yields an expression of the structure function of the delay as a function of the structure function of the refractivity N:

$$D_S(\rho) = \frac{1}{\sin^2 \theta_{\text{inc}}} \underset{0 \to h}{\iint} [D_N(R_2) - D_N(R_1)]dz_1 dz_2, \tag{B.1.8}$$

as in eqs. (B.1.1) and (4.7.7).
End of derivation

B.2   Structure function and power-law spectrum

The structure function $D_\varphi(R)$ of a function $\varphi(x)$ is defined as

$$D_\varphi(R) = E\{[\varphi(x+R) - \varphi(x)]^2\} = E\{[\Delta\varphi_R(x)]^2\}, \tag{B.2.1}$$

where a new signal $\Delta\varphi_R(x)$ is defined as

$$\Delta\varphi_R(x) = \varphi(x+R) - \varphi(x). \tag{B.2.2}$$

To relate the structure function $D_\varphi(R)$ to the power-law spectral form of eq. (4.7.12), we can rewrite $\Delta\varphi_R(x)$ in eq. (B.2.2) in a more general form using the convolution notation, see fig. B.2 (Bracewell, 1986):

$$\Delta\varphi_R(x) = \varphi(x) * [\delta(x - R) - \delta(x)], \tag{B.2.3}$$

where the δ-functions act as a spatial filter. The power spectrum of $\Delta\varphi_R(x)$ can be written as the product of two power spectra:

$$P_{\Delta\varphi}(f) = |G_R(f)|^2 P_\varphi(f), \tag{B.2.4}$$

where $P_\varphi(f)$ is the power spectrum of $\varphi(x)$, and $|G_R(f)|^2$ the power spectrum of $[\delta(x - R) - \delta(x)]$, with

$$[\delta(x - R) - \delta(x)] \xrightarrow{\mathscr{F}} G_R(f). \tag{B.2.5}$$

Using $\delta(x) \xrightarrow{\mathscr{F}} 1$, and using the shift theorem (Bracewell, 1986), we find

$$\delta(x - R) \xrightarrow{\mathscr{F}} e^{-j\omega R} 1, \tag{B.2.6}$$

and using the linearity conditions of Fourier transforms we obtain

$$[\delta(x - R) - \delta(x)] \xrightarrow{\mathscr{F}} e^{-j\omega R} - 1. \tag{B.2.7}$$

Using $e^{-j\omega R} = \cos\omega R - j\sin\omega R$ we can retrieve the power spectrum $|G_R(f)|^2$ by complex multiplication

$$\begin{aligned}
|G_R(f)|^2 &= (\cos\omega R - 1 - j\sin\omega R)(\cos\omega R - 1 + j\sin\omega R) \\
&= 2 - 2\cos\omega R \\
&= 4\frac{1}{2}(1 - \cos(2\omega\frac{1}{2}R)) \\
&= 4\sin^2(\frac{1}{2}\omega R) \\
&= 4\sin^2(\pi f R).
\end{aligned} \tag{B.2.8}$$

Hence, the power spectrum of $\Delta\varphi_R(x)$, see eq. (B.2.4) is given by

$$P_{\Delta\varphi}(f) = 4\sin^2(\pi f R)P_\varphi(f), \tag{B.2.9}$$

The variance of a random process $\Delta\varphi_R$ is equal to the integral of its power spectrum over all frequencies, using eqs. (B.2.8) and (B.2.4):

$$E\{\Delta\varphi_R^2\} = \int_0^\infty P_{\Delta\varphi}(f)df = \int_0^\infty 4P_\varphi(f)\sin^2(\pi f R)df. \tag{B.2.10}$$

If we now insert the special power spectral form of (4.7.12), we can write (B.2.10) as

$$E\{\Delta\varphi_R^2\} = \int_0^\infty 4P_0(f/f_0)^\nu \sin^2(\pi f R)df. \tag{B.2.11}$$

Substituting $u = \pi f R$ this can be written as

$$E\{\Delta\varphi_R^2\} = \int_0^\infty 4P_0 u^\nu (\pi R)^{-\nu} f_0^{-\nu}\pi^{-1}R^{-1} \sin^2(u)du, \tag{B.2.12}$$

or, as in Agnew (1992)

$$E\{\Delta\varphi_x^2\} = \frac{4P_0}{f_0^\nu} \frac{R^{-(\nu+1)}}{\pi^{\nu+1}} \int_0^\infty u^\nu \sin^2 u \, du. \qquad (B.2.13)$$

We use the following definite integral (Gradshteyn et al., 1994)

$$\int_0^\infty x^{\mu-1} \sin^2 ax \, dx = -\frac{\Gamma(\mu)\cos\frac{\mu\pi}{2}}{2^{\mu+1}a^\mu} \quad \text{for} \quad [a > 0, -2E\{\mathrm{Re}\{\mu\}\} < 0], \qquad (B.2.14)$$

with $\nu = \mu - 1$ and $a = 1$. This results in

$$\begin{aligned}
E\{\Delta\varphi_R^2\} &= \frac{4P_0}{f_0^\nu} \frac{R^{-(\nu+1)}}{\pi^{\nu+1}} \frac{\Gamma(\nu+1)\cos(\nu\pi/2+\pi/2)}{2^{\nu+2}} \\
&= \frac{-4\Gamma(\nu+1)\sin(\nu\pi/2)}{2^{\nu+2}\pi^{\nu+1}} \frac{P_0}{f_0^\nu} R^{-(\nu+1)}.
\end{aligned} \qquad (B.2.15)$$

From the recurrence relation, see, e.g., (Arfken, 1985) or (Monin et al., 1975)

$$\Gamma(\nu)\Gamma(-\nu) = -\frac{\pi}{\nu \sin \pi\nu}, \qquad (B.2.16)$$

we derive

$$\begin{aligned}
\Gamma(\nu)\Gamma(1-\nu) &= \frac{\pi}{\sin \pi\nu} \quad \text{or} \\
\frac{\Gamma(1+\nu)\sin \pi\nu}{\pi} &= \frac{\nu}{\Gamma(1-\nu)},
\end{aligned} \qquad (B.2.17)$$

and using $\sin(2a) = 2\sin a \cos a$ we find

$$\frac{\Gamma(1+\nu)\sin(\nu\pi/2)}{\pi} = \frac{\nu}{2\cos(\nu\pi/2)\Gamma(1-\nu)}. \qquad (B.2.18)$$

Including (B.2.18) in (B.2.15) yields

$$E\{\Delta\varphi_R^2\} = \frac{-2^2\nu}{2^{\nu+2}\pi^\nu 2\cos(\nu\pi/2)\Gamma(1-\nu)} \frac{P_0}{f_0^\nu} R^{-(\nu+1)}. \qquad (B.2.19)$$

Using $\Gamma(p+1) = p\Gamma(p)$ and by rearranging terms we finally find

$$\begin{aligned}
E\{\Delta\varphi_R^2\} &= \frac{-1}{2^{\nu+1}\pi^\nu \Gamma(-\nu)\cos(\nu\pi/2)} \frac{P_0}{f_0^\nu} R^{-(\nu+1)} \\
&= C_\nu \frac{P_0}{f_0^\nu} R^{-(\nu+1)},
\end{aligned} \qquad (B.2.20)$$

or

$$D_\varphi(R) = C_\nu \frac{P_0}{f_0^\nu} R^{-(\nu+1)}, \qquad (B.2.21)$$

as in eq. (4.7.13).
End of derivation.

Bibliography

Abramowitz, M. and Stegun, I. A. (1970), *Handbook of Mathematical Functions*, Dover books on advanced mathematics, Dover, New York, 9th edn.

Adam, N., Eineder, M., Breit, H. and Suchandt, S. (1999), Shuttle Radar Topography Mission: DLR's Interferometric SAR Processor for the Generation of a Global Digital Elevation Model, in: *Second International Workshop on ERS SAR Interferometry, 'FRINGE99', Liège, Belgium, 10–12 Nov 1999*, ESA.

Afraimovich, E. L., Terekhov, A. I., Udodov, M. Y. and Fridman, S. V. (1992), Refraction distortions of transionospheric radio signals caused by changes in a regular ionosphere and by travelling ionospheric disturbances, *Journal of Atmospheric and Terrestrial Physics*, **54**(7/8):1013–1020.

Agnew, D. C. (1992), The time-domain behaviour of power-law noises, *Geophysical Research Letters*, **19**(4):333–336.

Ahmed, S., Warren, H. R., Symonds, D. and Cox, R. P. (1990), The Radarsat system, *IEEE Transactions on Geoscience and Remote Sensing*, **28**:598–602.

Alaska SAR Facility (1997), *ERS-1 and ERS-2 SAR Images*, ASF, (Accessed 22 March, 1998), <http://www.asf.alaska.edu/>.

Alpers, W. and Brümmer, B. (1994), Atmospheric boundary layer rolls observed by the synthetic aperture radar aboard the ERS-1 satellite, *Journal of Geophysical Research*, **99**(C6):12613–12621.

Amelung, F., Galloway, D. L., Bell, J. W., Zebker, H. A. and Laczniak, R. J. (1999), Sensing the ups and downs of Las Vegas: InSAR reveals structural control of land subsidence and aquifer-system deformation, *Geology*, **27**(6):483–486.

Amelung, F., Jónssen, S., Zebker, H. and Segall, P. (2000a), Widespread uplift and trap door faulting on Galápagos volcanoes observed with radar interferometry, *Nature*, **407**(6807):993–996.

Amelung, F., Oppenheimer, C., Segall, P. and Zebker, H. (2000b), Ground deformation near Gada 'Ale Volcano, Afar, observed by radar interferometry, *Geophysical Research Letters*, **27**(19):3093–3097.

Ammar, G. S. and Gragg, W. B. (1988), Superfast solution of real positive definite Toeplitz Systems, *SIAM J. Matrix Anal. Appl,*, **9**(1):61–76.

Arfken, G. (1985), *Mathematical Methods for Physicists*, Academic Press, Boston, 3rd edn.

Armstrong, J. W. and Sramek, R. A. (1982), Observations of tropospheric phase scintillations at 5 GHz on vertical paths, *Radio Science*, **17**(6):1579–1586.

Askne, J. and Hagberg, J. O. (1993), Potential of Interferometric SAR for Classification of Land Surfaces, in: *International Geoscience and Remote Sensing Symposium, Tokyo, Japan, 18–21 August 1993*, pp. 985–987.

Askne, J. and Nordius, H. (1987), Estimation of tropospheric delay for microwaves from surface weather data, *Radio Science*, **22**:379–386.

Atkinson, B. W. and Zhang, J. W. (1996), Mesoscale shallow convection in the atmosphere, *Reviews of Geophysics*, **34**(4):403–431.

Attema, E. P. W. (1991), The Active Microwave Instrument On-Board the ERS-1 Satellite, *Proceedings of the IEEE*, **79**(6):791–799.

Avallone, A., Zollo, A., Briole, P., Delacourt, C. and Beauducel, F. (1999), Subsidence of Campi Flegrei (Italy) detected by SAR interferometry, *Geophysical Research Letters*, **26**(15):2303–2306.

Baby, H. B., Golé, P. and Lavergnat, J. (1988), A model for the tropospheric excess path length of radio waves from surface meteorological measurements, *Radio Science*, **23**(6):1023–1038.

Baer, G., Sandwell, D., Williams, S., Bock, Y. and Shamir, G. (1999), Coseismic deformation associated with the November 1995, M$_w$=7.1 Nuweiba earthquake, Gulf of Elat (Aqaba), detected by synthetic aperture radar interferometry, *Journal of Geophysical Research*, **104**(B11):25221–25232.

Balakrishnan, A. V. (1995), *Introduction to random processes in engineering*, John Wiley, New York.

Baltsavias, E. (1993), Fotogrammetrie III, Lecture notes, Delft University of Technology, Faculty of Geodetic Engineering.

Bamler, R. (1992), A comparison of range-Doppler and wave-number domain SAR focusing algorithms, *IEEE Transactions on Geoscience and Remote Sensing*, **30**(4):706–713.

Bamler, R., Adam, N., Davidson, G. W. and Just, D. (1998), Noise-Induced Slope Distortion in 2-D Phase Unwrapping by Linear Estimators with Application to SAR Interferometry, *IEEE Transactions on Geoscience and Remote Sensing*, **36**(3):913–921.

Bamler, R., Davidson, G. W. and Adam, N. (1996a), On the Nature of Noise in 2-D Phase Unwrapping, in: Franceschetti et al. (1996).

Bamler, R., Eineder, M. and Breit, H. (1996b), The X-SAR Single-Pass Interferometer on SRTM: Expected Perfomance and Processing Concept, in: *EUSAR'96, Königswinter, Germany*, pp. 181–184.

Bamler, R. and Hartl, P. (1998), Synthetic aperture radar interferometry, *Inverse Problems*, **14**:R1–R54.

Bamler, R. and Just, D. (1993), Phase Statistics and Decorrelation in SAR Interferograms, in: *International Geoscience and Remote Sensing Symposium, Tokyo, Japan, 18–21 August 1993*, pp. 980–984.

Bamler, R. and Schättler, B. (1993), SAR Data Acquisition and Image Formation, in: Schreier, G., ed., *SAR Geocoding: data and systems*, pp. 53–102, Wichmann Verlag, Karlsruhe.

Barber, B. C. (1993a), The non-isotropic two-dimensional random walk, *Waves in Random Media*, **3**:243–256.

Barber, B. C. (1993b), The phase statistics of a multichannel radar interferometer, *Waves in Random Media*, **3**:257–266.

Barbieri, M., Lichtenegger, J. and Calabresi, G. (1999), The Izmit Earthquake: A Quick Post-Seismic Analysis with Satellite Observations, *ESA Bulletin*, **100**:107–110.

Barka, A. (1999), The 17 August 1999 Izmit Earthquake, *Science*, **285**:1858–1859.

Bean, B. R. and Dutton, E. J. (1968), *Radio Meteorology*, Dover, New York.

Bendat, J. S. and Piersol, A. G. (1986), *Random Data: Analysis and Measurement Procedures*, Wiley-Interscience, New York, 2nd edn.

Bennett, J. R. and Cumming, I. G. (1979), A Digital Processor for the Production of Seasat Synthetic Aperture Radar Imagery, Tech. Rep. ESA SP-154, ESA.

Bevington, P. R. and Robinson, D. K. (1992), *Data reduction and analysis for the physical sciences*, McGraw-Hill, Boston, 2nd edn.

Bevis, M., Businger, S., Herring, T. A., Anthes, R. A., Rocken, C. and Ware, R. H. (1994), GPS meteorology: mapping zenith wet delays onto precipitable water, *Journal of Applied Meteorology*, **33**(3).

Bevis, M., Businger, S., Herring, T. A., Rocken, C., Anthes, R. A. and Ware, R. H. (1992), GPS Meteorology: Remote sensing of atmospheric water vapor using the Global Positioning System, *Journal of Geophysical Research*, **97**:15,787–15,801.

Bevis, M., Chiswell, S., Businger, S., Herring, T. A. and Bock, Y. (1996), Estimating wet delays using numerical weather analysis and predictions, *Radio Science*, **31**(3):477–487.

Bitmead, R. R. and Anderson, B. D. O. (1980), Asymptotically Fast Solution of Toeplitz and Related Systems of Linear Equations, *Linear Algebra and Its Applications*, **34**:103–116.

Bochner, S. (1959), *Lectures on Fourier Integrals; with a supplement on monotonic functions, Stieltjes integrals, and harmonic analysis;*, Annals of mathematics studies 42, Princeton University Press, Princeton, transl. from German by M. Tenenbaum and H. Pollard.

Born, M. and Wolf, E. (1980), *Electromagnetic Theory of Propagation Interference and Diffraction of Light*, Pergamon Press, New York.

Born, M., Wolf, E. and Bhatia, A. B. (1959), *Principle of optics; electromagnetic theory of propagation, interference and diffraction of light*, Pergamon Press, New York.

Box, G. E. P., Jenkins, G. M. and Reinsel, G. C. (1994), *Time series analysis: forecasting and control*, Prentice-Hall, Englewood Cliffs, 3rd edn.

Bracewell, R. N. (1986), *The Fourier transform and its applications*, McGraw-Hill, New York, 3rd edn.

van Bree, R. J. P., Gens, R., Groot, J. S., van Halsema, D., Hanssen, R. F., van den Hout, P., Klees, R., de Min ans E J O Schrama, E. J. and Usai, S. (2000), *Deformation measurements with SAR*, vol. USP-2, 99-16, Beleidscommissie Remote Sensing, Delft.

Brent, R. P. (1991), Parallel Algorithms for Toeplitz Systems, in: Golub, G. G. and Dooren, P. V., eds., *Numerical Linear Algebra, Digital Signal Processing and Parallel Algorithms*, vol. 70 of *NATO ASI Series, Series F: Computer and Systems Sciences*, pp. 75–92, Springer-Verlag, Berlin.

Briole, P., Massonnet, D. and Delacourt, C. (1997), Post-eruptive deformation associated with the 1986-87 and 1989 lava flows of Etna detected by radar interferometry, *Geophysical Research Letters*, **24**:37–40.

Brown, L. G. (1992), A Survey of Image Registration Techniques, *ACM Computing Surveys*, **24**(4):325–376.

Brown, R. A. (1980), Longitudinal instabilities and secondary flows in the planetary boundary layer: A review, *Reviews in Geophysics and Space Physics*, **18**:683–697.

Brunner, F. K. and Welsch, W. M. (1993), Effect of the Troposphere on GPS Measurements, *GPS World*, pp. 42–51.

Buderi, R. (1996), *The invention that changed the world*, Simon & Schuste, New York.

Bürgmann, R., Schmidt, D., Nadeau, R. M., d'Alessio, M., Fielding, E. J., Manaker, D., McEvilly, T. V. and Murray, M. H. (2000), Earthquake Potential Along the Northern Hayward Fault, California, *Science*, **289**:1178–1182.

Calais, E. and Minster, J. B. (1995), Detection of Ionospheric Perturbations Following an Earthquake with the Global Positioning System: Implications for Nuclear Tests Discrimination, in: Console, R. and Nikolaev, A., eds., *Earthquakes Induced by Underground Nuclear Explosions*, vol. 4 of *NATO ASI Series, Partnership Subseries, 2. Environment*, pp. 235–253, Springer-Verlag, Berlin.

Calais, E. and Minster, J. B. (1996), GPS detection of ionospheric perturbations following a Space Shuttle ascent, *Geophysical Research Letters*, **23**(15):1897–1900.

Calais, E. and Minster, J. B. (1998), GPS, earthquakes, the ionosphere, and the Space Shuttle, *Physics of the Earth and Planetary Interiors*, **105**:167–181.

Carnec, C. and Fabriol, H. (1999), Monitoring and modeling land subsidence at the Cerro Prieto geothermal field, Baja California, Mexico, using SAR interferometry, *Geophysical Research Letters*, **26**(9):1211–1215.

Carnec, C., Massonnet, D. and King, C. (1996), Two examples of the use of SAR interferometry on displacement fields of small spatial extent, *Geophysical Research Letters*, **23**(24):3579–3582.

Carrasco, D., Alonso, J. and Broquetas, A. (1995), Accuracy assessment of SAR interferometry using the ERS-1, in: *International Geoscience and Remote Sensing Symposium, Florence, Italy, 10–14 July 1995*, pp. 781–783.

Cattabeni, M., Monti-Guarnieri, A. and Rocca, F. (1994), Estimation and Improvement of Coherence in SAR Interferograms, in: *International Geoscience and Remote Sensing Symposium, Pasadena, CA, USA, 8–12 August 1994*, pp. 720–722.

Chen, C. W. and Zebker, H. A. (2000a), Network approaches to two-dimensional phase unwrapping: intractability and two new algorithms, *Journal of the Optical Society of America A.*, **17**(3):401–414.

Chen, C. W. and Zebker, H. A. (2000b), Two-dimensional phase unwrapping using statistical models for cost functions in nonlinear optimization, *Journal of the Optical Society of America A.*, **in press**.

Cheng, K. and Huang, Y.-N. (1992), Ionospheric disturbances observed during the period of Mount Pinatubo eruptions in June 1991, *Journal of Geophysical Research*, **97**(A11):16,995–17,004.

Chilès, J.-P. and Delfiner, P. (1999), *Geostatistics: modeling spatial uncertainty*, Wiley Series in Probability and Statistics, John Wiley & Sons, New York.

Ching, N. H., Rosenfeld, D. and Braun, M. (1992), Two-dimensional phase unwrapping using a minimum spanning tree algorithm, *IEEE Transactions on Image Processing*, 1:355–365.

Clarke, P. J., Paradissis, D., Briole, P., England, P. C., Parsons, B. E., Billiris, H., Veis, G., and Ruegg, J.-C. (1996), Geodetic investigation of the 13 May 1995 Kozani-Grevena (Greece) earthquake, *Geophysical Research Letters*, **24**:707–710.

Coltelli, M., Fornaro, G., Franceschetti, G., Lanari, R., Migliaccio, M., ao R Moreira, J., Papatanassiou, K. P., Puglisi, G., Riccio, D. and Schwäbisch, M. (1996), SIR-C/X-SAR multifrequency multipass interferometry: A new tool for geological interpretation, *Journal of Geophysical Research*, **101**(E10):23127–23148.

Costantini, M. (1996), A Phase Unwrapping Method Based on Network Programming, in: *'FRINGE 96' workshop on ERS SAR Interferometry, Zürich, Switzerland, 30 Sep–2 October 1996*.

Costantini, M. (1998), A Novel Phase Unwrapping Method Based on Network Programming, *IEEE Transactions on Geoscience and Remote Sensing*, **36**(3):813–821.

Coulman, C. E. and Vernin, J. (1991), Significance of anisotropy and the outer scale of turbulence for optical and radio seeing, *Applied Optics*, **30**(1):118–126.

Coulson, S. N. (1995), *SAR Interferometry with ERS-1*.

Crook, N. A. (1996), Sensitivity of moist convection forced by boundary layer processes to low-level thermodynamic fields, *Monthly Weather Review*, **124**:1767–1785.

Curlander, J. C. and McDonough, R. N. (1991), *Synthetic aperture radar: systems and signal processing*, John Wiley & Sons, Inc, New York.

Dainty, J. C., ed. (1975), *Laser Speckle and Related Phenomena*, vol. 9 of *Topics in Applied Physics*, Springer-Verlag, Heidelberg.

Dammert, P. B. G. (1997), Accuracy of INSAR Measurements in Forested Areas, in: *'FRINGE 96' workshop on ERS SAR Interferometry, Zürich, Switzerland, 30 Sep–2 October 1996*, ESA SP-406, Vol II.

Daniels, M. J. and Cressie, N. (1999), A hierarchical approach to covariance function estimation for time series, submitted to Journal of Time Series Analysis.

Davenport, Jr, W. B. and Root, W. L. (1987), *An Introduction to the Theory of Random Signals and Noise*, IEEE Press, New York.

Davis, A., Marshak, A., Wiscombe, W. and Cahalan, R. (1994), Multifractal characterization of nonstationarity and intermittency in geophysical fields: Observed, retrieved, or simulated, *Journal of Geophysical Research*, **99**(D4):8055–8072.

Davis, J. L., Herring, T. A., Shapiro, I. I., Rogers, A. E. E. and Elgered, G. (1985), Geodesy by radio interferometry: Effects of atmospheric modelling errors on estimates of baseline length, *Radio Science*, **20**(6):1593–1607.

Delacourt, C., Briole, P. and Achache, J. (1998), Tropospheric corrections of SAR interferograms with strong topography. Application to Etna, *Geophysical Research Letters*, **25**(15):2849–2852.

Delacourt, C., Briole, P., Achache, J., Fruneau, B. and Carnec, C. (1997), Correction of the tropospheric delay in SAR interferometry and application to 1991-93 eruption of Etna volcano, Italy, in: *AGU Fall meeting, December 8-12, San Francisco, USA,*.

Delaunay, B. N. (1934), Sur la sphère vide, *Izvestia Akademia Nauk SSSR, VII Seria, Otdelenie Matematicheskii i Estestvennyka Nauk*, **7**:793–800.

Dravskikh, A. F. and Finkelstein, A. M. (1979), Tropospheric limitations in phase and frequency coordinate measurements in astronomy, *Astrophysics and Space Science*, **60**:251–265.

Duchossois, G., Kohlhammer, G. and Martin, P. (1996), Completion of the ERS Tandem Mission, *Earth Observation Quarterly*, **55**.

Elachi, C. (1988), *Spaceborne radar remote sensing: applications and techniques*, Institute of Electrical and Electronics Engineers, New York.

Elachi, C., Bicknell, T., Jordan, R. L. and Wu, C. (1982), Spaceborne Synthetic-Aperture Imaging Radar: Applications, Techniques and Technology, *Proceedings of the IEEE*, **70**(10):1174–1209.

Elgered, G. (1982), Tropospheric wet path-delay measurements, *IEEE Trans. on Antennas and Propagation*, **AP-30**:502–505.

Emanuel, K., Raymond, D., Betts, A., Bosart, L., Bretherton, C., Droegemeier, K., Farrell, B., Fritsch, J. M., Houze, R., LeMone, M., Lilly, D., Rotunno, R., Shapiro, M., Smith, R. and Thorpe, A. (1995), Report of the first prospectus development team of the U. S. weather research program to NOAA and the NSF, *Bull. Amer. Meteor. Soc.*, **76**:1194–1208.

Emardson, T. R. and Derks, H. J. P. (2000), On the relation between the wet delay and the integrated precipitable water vapour in the European atmosphere, *Meteorological Applications*, **7**(1):61–68.

European Space Agency (1999), Active Microwave Instrument, (Accessed 13 Aug, 1999) <http://earth1.esrin.esa.it/ipgami>.

Evans, D. L., Elachi, C., Stofan, E. R., Holt, B., Way, J. B., Kobrick, M., Öttl, H., Pampaloni, P., Vogt, M., Wall, S., van Zyl, J. and Schier, M. (1993), The Shuttle Imaging Radar-C and X-SAR Mission, *EOS*, **74**(13).

Evans, J. V. and Hagfors, T. (1968), *Radar Astronomy*, McGraw-Hill, New York.

Fabriol, H. and Glowacka, E. (1997), Seismicity and fluid reinjection at CerroPrieto geothermal field: preliminary results, in: *Twenty-Second Workshop on Geothermal Reservoir Engineering*, pp. 11–17, Stanford University, Jan 27-29, 1997.

Feigl, K. L., Sergent, A. and Jacq, D. (1995), Estimation of an earthquake focal mechanism from a satellite radar interferogram: Application to the December 4, 1992 Landers aftershock, *Geophysical Research Letters*, **22**(9):1037–1040.

Feijt, A. J. (2000), *Quantitative Cloud Analysis using Meteorological Satellites*, Ph.D. thesis, Wageningen University.

Ferretti, A., Monti-Guarnieri, A., Prati, C. and Rocca, F. (1998), Multi-image DEM Reconstruction, in: *International Geoscience and Remote Sensing Symposium, Seattle, Washington, USA, 6–10 July 1998*, pp. 1367–1369.

Ferretti, A., Prati, C. and Rocca, F. (1999a), Multibaseline InSAR DEM Reconstruction: The Wavelet Approach, *IEEE Transactions on Geoscience and Remote Sensing*, **37**(2):705–715.

Ferretti, A., Prati, C. and Rocca, F. (1999b), Non-Uniform Motion Monitoring Using the Permanent Scatterers Technique, in: *Second International Workshop on ERS SAR Interferometry, 'FRINGE99', Liège, Belgium, 10–12 Nov 1999*, pp. 1–6, ESA.

Ferretti, A., Prati, C. and Rocca, F. (2000), Nonlinear Subsidence Rate Estimation using Permanent Scatterers in Differential SAR Interferometry, *IEEE Transactions on Geoscience and Remote Sensing*, **38**(5):2202–2212.

Ferretti, A., Prati, C., Rocca, F. and Monti Guarnieri, A. (1997), Multibaseline SAR Interferometry for Automatic DEM Reconstruction, in: *Third ERS Symposium—Space at the Service of our Environment, Florence, Italy, 17–21 March 1997*, ESA SP-414, pp. 1809–1820.

Feynman, R. P., Leighton, R. B. and Sands, M. (1963), *The Feynman Lectures on Physics: Commemorative Issue*, vol. 1, Addison-Wesley company, Inc., Reading, Massachusetts.

Fialko, Y. and Simons, M. (2000), Deformation and seismicity in the Coso geothermal area, Inyo County, California: Observations and modeling using satellite radar interferometry, *Journal of Geophysical Research*, **105**(B9):21781–21794.

Fielding, E., Wright, T., Parsons, B., England, P., Rosen, P., Hensley, S. and Bilham, R. (1999), Topography of Northwest Turkey from SAR Interferometry: Applications to the 1999 Izmit earthquake Geomorphology and Co-seismic Strain, *EOS Transactions, AGU*, **80**(46):F663.

Fitch, J. P. (1988), *Synthetic aperture radar*, Springer-Verlag, New York.

Flynn, T. J. (1997), Two-dimensional phase unwrapping with minimum weighted discontinuity, *Journal of the Optical Society of America A.*, **14**:2692–2701.

Fornaro, G., Franceschetti, G. and Lanari, R. (1996a), Interferometric SAR Phase Unwrapping Using Green's Formulation, *IEEE Transactions on Geoscience and Remote Sensing*, **34**(3):720–727.

Fornaro, G., Franceschetti, G., Lanari, R. and Sansosti, E. (1996b), Robust phase unwrapping Techniques: A Comparison, *Journal of the Optical Society of America A.*, **13**(12):2355–2366.

Foster, M. R. and Guinzy, N. J. (1967), The coefficient of coherence: its estimation and use in geophysical data processing, *Geophysics*, **32**(4):602–616.

Franceschetti, G., Oliver, C. J., Shiue, J. C. and Tajbakhsh, S., eds. (1996), *Microwave Sensing and Synthetic Aperture Radar*, SPIE, Bellingham.

Fruneau, B., Rudant, J.-P., Obert, D. and Raymond, D. (1998), Small displacements detected by SAR interferometry on the city of Paris, (France), in: *Second Int. Workshop on "Retrieval of Bio- and Geophysical Parameters from SAR Data for Land Applications" 21-23 Oct*, Noordwijk, The Netherlands, ESTEC.

Fujii, N., T Nakano, T. O. and Yamaoka, K. (1994), Detection of Ground Deformations by the Multi–pass Differential SAR–interferometry. Examples of the Active Volcanic Area, in: *paper presented at the 1st workshop on SAR interferometry, Tokyo, Japan, December*

1994, NASDA.

Fujiwara, S., Yarai, H., Ozawa, S., Tobita, M., Murakami, M., Nakagawa, H., Nitta, K., Rosen, P. A. and Werner, C. L. (1998), Surface displacement of the March 26, 1997 Kagoshima-kenhokuseibu earthquake in Japan from synthetic aperture radar interferometry, *Geophysical Research Letters*, **25**(24):4541–4544.

Fuliwara, S., Rosen, P., Tobita, M. and Murakami, M. (1998), Crustal deformation measurments using repeat-pass JERS-1 synthetic aperture radar interferometry near the Izu Peninsula, Japan, *Journal of Geophysical Research*, **103**(B2):2411–2426.

Gabriel, A. K. and Goldstein, R. M. (1988), Crossed orbit interferometry: theory and experimental results from SIR-B, *Int.J. Remote Sensing*, **9**(5):857–872.

Gabriel, A. K., Goldstein, R. M. and Zebker, H. A. (1989), Mapping small elevation changes over large areas: differential radar interferometry, *Journal of Geophysical Research*, **94**(B7):9183–9191.

Galloway, D. L., Hudnut, K. W., Ingebritsen, S. E., Phillips, S. P., Peltzer, G., Rogez, F. and Rosen, P. A. (1998), Detection of acquifer system compaction and land subsidence using interferometric synthetic aperture radar, Antelope Valley, Mojave Desert, California, *Water Resources Research*, **34**(10):2573–2585.

Gatelli, F., Monti Guarnieri, A., Parizzi, F., Pasquali, P., Prati, C. and Rocca, F. (1994), The wavenumber shift in SAR Interferometry, *IEEE Transactions on Geoscience and Remote Sensing*, **32**(4):855–865.

Gauss, C. F. (1809), *Theoria Motus Corporum Coelestium*, Perthes und Besser, Hamburg.

Geudtner, D. (1995), *Die Interferometrische Verarbeitung von SAR-Daten des ERS-1*, Ph.D. thesis, Universität Stuttgart, Draft.

Geudtner, D., Schwäbisch, M. and Winter, R. (1994), SAR-Interferometry with ERS-1 data, in: *PIERS'94 ESA/ESTEC Noordwijk*.

Geudtner, D., Winter, R. and Vachon, P. W. (1996), Flood mointoring using ERS-1 SAR Interferometry coherence maps, Presented at IGARSS'96.

Ghiglia, D. C. and Pritt, M. D. (1998), *Two-dimensional phase unwrapping: theory, algorithms, and software*, John Wiley & Sons, Inc, New York.

Ghiglia, D. C. and Romero, L. A. (1994), Robust two-dimensional weighted and unweighted phase unwrapping that uses fast transforms and iterative methods, *Journal of the Optical Society of America A.*, **11**(1):107–117.

Ghiglia, D. C. and Romero, L. A. (1996), Minimum L^p-norm two-dimensional phase unwrapping, *Journal of the Optical Society of America A.*, **13**:1999–2013.

Goldstein, R. (1995), Atmospheric limitations to repeat-track radar interferometry, *Geophysical Research Letters*, **22**(18):2517–2520.

Goldstein, R. M. (1964), Venus characteristics by earth-based radar, *Astronomical Journal*, **69**(1):12–18.

Goldstein, R. M. (1969), Preliminary Venus Radar Results, *Radio Science*, **96D**:1623–1625.

Goldstein, R. M., Caro, E. R. and Wu, C. (1985), Method and apparatus for contour mapping using synthetic aperture radar, *United States Patent, No. 4551724*.

Goldstein, R. M. and Carpenter, R. L. (1963), Rotation of Venus: Period Estimated from Radar Measurements, *Science*, **139**:910–911.

Goldstein, R. M., Engelhardt, H., Kamp, B. and Frolich, R. M. (1993), Satellite radar interferometry for monitoring ice sheet motion: Application to an Antarctic ice stream, *Science*, **262**:1525–1530.

Goldstein, R. M. and Gillmore, W. F. (1963), Radar observations of Mars, *Science*, **141**:1171–1172.

Goldstein, R. M. and Werner, C. L. (1998), Radar interferogram filtering for geophysical applications, *Geophysical Research Letters*, **25**(21):4035–4038.

Goldstein, R. M. and Zebker, H. A. (1987), Interferometric radar measurement of ocean surface currents, *Nature*, **328**(20):707–709.

Goldstein, R. M., Zebker, H. A. and Werner, C. L. (1988), Satellite radar interferometry: Two-dimensional phase unwrapping, *Radio Science*, **23**(4):713–720.

Golub, G. H. and Van Loan, C. F. (1996), *Matrix Computations*, Johns Hopkins University Press, Baltimore, 3rd edn.

Goodman, J. W. (1975), Statistical Properties of Laser Speckle Patterns, in: Dainty (1975), chap. 2, pp. 9–75.

Goodman, N. R. (1963), Statistical analysis based on a certain multivariate complex Gaussian distribution (an introduction), *Annals of Mathematical Statistics*, **34**(1):152–177.

Goovaerts, P. (1997), *Geostatistics fo Natural Resources Evaluation*, Applied Geostatistics Series, Oxford University Press, New York.

Gradshteyn, I. S., Ryzhik, I. M. and Jeffrey, A. (1994), *Table of integrals, series, and products*, Academic Press, Boston, 5th edn., Original Title: Tablitsy integralov, summ, ryadov i proizvediniy, 1965.

Graham, L. C. (1974), Synthetic Interferometer Radar for Topographic Mapping, *Proceedings of the IEEE*, **62**(6):763–768.

Gray, A. L. and Farris-Manning, P. J. (1993), Repeat-Pass Interferometry with Airborne Synthetic Aperture Radar, *IEEE Transactions on Geoscience and Remote Sensing*, **31**(1):180–191.

Gray, A. L., Mattar, K. E. and Sofko, G. (2000), Influence of Ionospheric Electron Density Fluctuations on Satellite Radar Interferometry, *Geophysical Research Letters*, **27**(10):1451–1454.

Gray, L., Mattar, K. and Short, N. (1999), Speckle Tracking for 2-Dimensional Ice Motion Studies in Polar Regions: Influence of the Ionosphere, in: *Second International Workshop on ERS SAR Interferometry, 'FRINGE99', Liège, Belgium, 10–12 Nov 1999*, pp. 1–8, ESA.

Hagberg, J. O., Ulander, L. M. H. and Askne, J. (1995), Repeat-Pass SAR Interferometry over Forested Terrain, *IEEE Transactions on Geoscience and Remote Sensing*, **33**(2):331–340.

Hall, M. P. M., Barclay, L. W. and Hewitt, M. T., eds. (1996), *Propagation of Radiowaves*, The Institution of Electrical Engineers, London.

Haltiner, G. J. and Martin, F. L. (1957), *Dynamical and Physical Meteorology*, McGraw-Hill, New York.

Hannan, E. J. and Thomson, P. J. (1971), The estimation of coherence and group delay, *Biometrika*, **58**(3-481):469.

Hanssen, R. (1994), *The Application of Radon transformation for improved analysis of sparsely sampled data: a feasibility study*, Faculty of Geodetic Engineering, Rep.93.1, Delft University of Technology, Delft.

Hanssen, R. (1996), GISARE: GPS metingen voor de schatting van atmosferische parameters, *GPS Nieuwsbrief*, **11**(1):17–21, In Dutch.

Hanssen, R. (1998), *Atmospheric heterogeneities in ERS tandem SAR interferometry*, Delft University Press, Delft, the Netherlands.

Hanssen, R., Amelung, F. and Zebker, H. (1998a), Geodetic interpretation of land subsidence measurements at the Cerro Prieto geothermal field monitored by radar interferometry, *EOS Transactions, AGU*, **79**(45):F37.

Hanssen, R. and Bamler, R. (1999), Evaluation of Interpolation Kernels for SAR Interferometry, *IEEE Transactions on Geoscience and Remote Sensing*, **37**(1):318–321.

Hanssen, R. and Feijt, A. (1996), A first quantitative evaluation of atmospheric effects on SAR interferometry, in: *'FRINGE 96' workshop on ERS SAR Interferometry, Zürich*,

Switzerland, 30 Sep–2 October 1996, pp. 277–282, ESA SP-406.

Hanssen, R. and Usai, S. (1997), Interferometric phase analysis for monitoring slow deformation processes, in: *Third ERS Symposium—Space at the Service of our Environment, Florence, Italy, 17–21 March 1997*, ESA SP-414, pp. 487–491.

Hanssen, R., Vermeersen, B., Scharroo, R., Kampes, B., Usai, S., Gens, R. and Klees, R. (2000a), Deformatiepatroon van de aardbeving van 17 augustus 1999 in Turkije gemeten met satelliet radar interferometrie, *Remote Sensing Nieuwsbrief*, **90**:42–44, In Dutch.

Hanssen, R., Weinreich, I., Lehner, S. and Stoffelen, A. (2000b), Tropospheric wind and humidity derived from spaceborne radar intensity and phase observations, *Geophysical Research Letters*, **27**(12):1699–1702.

Hanssen, R., Zebker, H., Klees, R. and Barlag, S. (1998b), On the use of meteorological observations in SAR interferometry, in: *International Geoscience and Remote Sensing Symposium, Seattle, Washington, USA, 6–10 July 1998*, pp. 1644–1646.

Hanssen, R. F., Feijt, A. J. and Klees, R. (2001), Comparison of water vapor column derived from radar interferometry and Meteosat WV channel analysis, *Journal of Atmospheric and Oceanic Technology*, in press.

Hanssen, R. F. and Weckwerth, T. M. (1999), Evaluation of spatial moisture distribution during CLARA 96 using spaceborne radar interferometry, in: *Remote Sensing of cloud parameters: retrieval and validation. 21-22 Oct 1999, Delft, The Netherlands*, pp. 15–20.

Hanssen, R. F., Weckwerth, T. M., Zebker, H. A. and Klees, R. (1999), High-Resolution Water Vapor Mapping from Interferometric Radar Measurements, *Science*, **283**:1295–1297.

Hartl, P. (1991), Application of Interferometric SAR-Data of the ERS-1 Mission for High Resolution Topographic Terrain Mapping, *GIS*, **2**:8–14.

Hartl, P., Reich, M., Thiel, K.-H. and Xia, Y. (1993), SAR interferometry applying ERS-1: some preliminary test results, in: *First ERS-1 Symposium—Space at the Service of our Environment, Cannes, France, 4–6 November 1992*, ESA SP-359, pp. 219–222.

Hartl, P. and Thiel, K.-H. (1993), Bestimmung von topographischen Feinstrukturen mit interferometrischem ERS-1-SAR, *Zeitschrift für Photogrammetrie und Fernerkundung*, **3**:108–114.

Hartl, P., Thiel, K.-H. and Wu, X. (1994a), Information extraction from ERS-1 SAR data by means of INSAR and D-INSAR techniques in Antarctic research, in: *Second ERS-1 Symposium—Space at the Service of our Environment, Hamburg, Germany, 11–14 October 1993*, ESA SP-361, pp. 697–701.

Hartl, P., Thiel, K. H., Wu, X., Doake, C. and Sievers, J. (1994b), Application of SAR Interferometry with ERS-1 in the Antarctic, *Earth Observation Quarterly*, **43**:1–4.

Hartl, P. and Xia, Y. (1993), Besonderheiten der Datenverarbeitung bei der SAR-Interferometrie, *Zeitschrift für Photogrammetrie und Fernerkundung*, **6**:214–222.

Henderson, F. M. and Lewis, A. J., eds. (1998), *Principles and applications of Imaging Radar*, vol. 2 of *Manual of Remote Sensing*, John Wiley & Sons, Inc., New York, 3rd edn.

Hernandez, B., Cotton, F., Campillo, M. and Massonnet, D. (1997), A Comparison between short term (co-seismic) and long term (one year) slip for the Landers earthquake: Measurements from strong motion and SAR interferometry, *Geophysical Research Letters*, **24**(13):1579–1582.

Herring, T. A. (1992), Modeling atmospheric delays in the analysis of space geodetic data, in: de Munck, J. C. and Spoelstra, T. A. T., eds., *Symposium on Refraction of transatmospheric signals in geodesy*, pp. 157–164, Delft, Netherlands Geodetic Commission.

Ho, C. M., Mannucci, A. J., Lindqwister, U. J., Pi, X., Tsurutani, B. T., Sparks, L., Iijima, B. A., Wilson, B. D., Harris, I. and Reyes, M. J. (1998), Global ionospheric TEC variations during January 10, 1997 storm, *Geophysical Research Letters*, **25**(14):2589–2592.

Hoen, E. W. and Zebker, H. A. (2000), Penetration Depths Inferred from Interferometric Volume Decorrelation Observed over the Greenland Ice Sheet, *IEEE Transactions on Geoscience and Remote Sensing*, **38**(6):2571–2583.

van der Hoeven, A. G. A., Ambrosius, B. A. C., van der Marel, H., Derks, H., Klein Baltink, H., van Lammeren, A. and Kösters, A. J. M. (1998), Analysis and Comparison of Integrated Water Vapor Estimation from GPS, in: *Proc. 11th Intern. Techn. Meeting Sat. Div. Instit. of Navigation*, pp. 749–755.

van der Hoeven, A. G. A. and Hanssen, R. F. (1999), Cross-Validation of Tropospheric Water Vapor Measurements by GPS and SAR interferometry, in: *General Assembly European Geophysical Society, The Hague, The Netherlands, 19-23 April 1999*.

Hogg, D. C., Guiraud, F. O. and Dekker, M. T. (1981), Measurement of Excess Radio Transmission Length on Earth-space Paths, *Astronomy and Astrophysics*, **95**:304–307.

Homer, J. and Longstaff, I. D. (1995), Improving Height Resolution of Interferometric SAR, in: *International Geoscience and Remote Sensing Symposium, Florence, Italy, 10-14 July 1995*, pp. 210–212.

Hovanessian, S. A. (1980), *Introduction to Synthetic Array and Imaging Radars*, Artech House, Inc., Dedham, MA.

Hubert-Ferrari, A., Barka, A., Jacques, E., Nalbant, S. S., Meyer, B., Armijo, R., Tapponnier, P. and King, G. C. P. (2000), Seismic hazard in the Marmara Sea region following the 17 August 1999 Izmit earthquake, *Nature*, **404**:269–273.

Hulsmeyer, C. (1904), Herzian-Wave Projecting and Receiving Apparatus Adopted to Indicate or Give Warning of the Presence of a Metallic Body such as a Ship or a Train, *British Patent No. 13,170*.

Hunt, B. R. (1979), Matrix formulation of the reconstruction of phase values from phase differences, *Journal of the Optical Society of America A.*, **69**(3):393–399.

Huschke, R. E. (1959), *Glossary of meteorology*, American Meteorological Society, Boston.

Huygens, C. (1690), *Traité de la lumière, où sont expliquées les causes de ce qui luy arrive dans la reflexion, et dans la refraction: et particulierement dans l'étrange refraction du cristal d'Islande: avec un discours de la cause de la pesanteur*, Van der Aa, Leiden.

Ishimaru, A. (1978), *Wave Propagation and Scattering in Random Media*, vol. 2, Academic Press, New York.

Jakowski, N., Bettac, H. D. and Jungstand, A. (1992), Ionospheric corrections for Radar Altimetry and Geodetic Positioning Techniques, in: *Proc. of the Symp. Refraction of Transatmospheric Signals in Geodesy, the Hague, 19–22 may 1992*, pp. 151–154.

Jónsson, S., Adam, N. and Björnsson, H. (1998), Effects of Subglacial Geothermal Activity Observed by Satellite Radar Interferometry, *Geophysical Research Letters*, **25**(7):1059–1062.

Jónsson, S., Zebker, H., Cervelli, P., Segall, P., Garbeil, H., Mouginis-Mark, P. and Rowland, S. (1999), A Shallow-Dipping Dike fed the 1995 Flank Eruption at Fernandina Volcano, Galápagos, Observed by Satellite Radar Interferometry, *Geophysical Research Letters*, **26**(8):1077–1080.

Jordan, R. L., Caro, E. R., Kim, Y., Kobrick, M., Shen, Y., Stuhr, F. V. and Werner, M. U. (1996), Shuttle radar topography mapper (SRTM), in: Franceschetti et al. (1996), pp. 412–422.

Joughin, I., Gray, L., Bindschadler, R., Price, S., Morse, D., Hulbe, C., Mattar, K. and Werner, C. (1999), Tributaries of West Antarctic Ice Streams Revealed by RADARSAT Interferometry, *Science*, **286**(5438):283–286.

Joughin, I. R. (1995), *Estimation of Ice-Sheet Topography and Motion Using Interferometric*

Synthetic Aperture Radar, Ph.D. thesis, University of Washington.

Joughin, I. R., Fahnestock, M., Ekholm, S. and Kwok, R. (1997), Balance Velocities of the Greenland ice sheet, *Geophysical Research Letters*, **24**(23):3045–3048.

Joughin, I. R., Kwok, R. and Fahnestock, M. A. (1998), Interferometric Estimation of Three-Dimensional Ice-Flow Using Ascending and Descending Passes, *IEEE Transactions on Geoscience and Remote Sensing*, **36**(1):25–37.

Joughin, I. R., Winebrenner, D., Fahnestock, M., Kwok, R. and Krabill, W. (1996), Measurement of ice-sheet topography using satellite-radar interferometry, *Journal of Glaciology*, **42**(140):10–22.

Joughin, I. R. and Winebrenner, D. P. (1994), Effective Number of Looks for a Multilook Interferometric Phase Distribution, in: *International Geoscience and Remote Sensing Symposium, Pasadena, CA, USA, 8–12 August 1994*, pp. 2276–2278.

Journel, A. G. and Huijbregts, C. J. (1978), *Mining Geostatistics*, Academic Press, London.

Just, D. and Bamler, R. (1994), Phase statistics of interferograms with applications to synthetic aperture radar, *Applied Optics*, **33**(20):4361–4368.

Kailath, T., Vieira, A. and Morf, M. (1978), Inverses of Toeplitz Operators, innovations, and orthogonal polynomials, *SIAM Review*, **20**(1):106–119.

Kawai, S. and Shimada, M. (1994), Detections of Earth Surface Deformation Change by means of INSAR technique, in: *paper presented at the 1st workshop on SAR interferometry, Tokyo, Japan, December 1994*, NASDA.

Kellogg, O. D. (1929), *Foundations of Potential Theory*, Frederick Ungar, New York.

Keys, R. G. (1981), Cubic Convolution Interpolation for Digital Image Processing, *IEEE Transactions on Acoustics, Speech, and Signal Processing*, **ASSP-29**(6):1153–1160.

King, G. C. P., Stein, R. S. and Lin, J. (1994), Static stress changes and the triggering of earthquakes, *Bulletin of the Seismological Society of America*, **84**:935–953.

Klinger, Y., Michel, R. and Avouac, J.-P. (2000), Co-seismic deformation during the Mw 7.3 Aqaba earthquake (1995) from ERS-SAR interferometry, *Geophysical Research Letters*, **27**(22):3651–3655.

Koch, K.-R. (1999), *Parameter Estimation and Hypothesis Testing in Linear Models*, Springer-Verlag, New York, 2nd edn.

Kolmogorov, A. N. (1941), Dissipation of energy in locally isotropic turbulence, *Doklady Akad. Nauk SSSR*, **32**(16), German translation in "Sammelband zur Statistischen Theorie der Turbulenz", Akademie Verlag, Berlin, 1958, p77.

van der Kooij, M. (1999a), Personal communication.

van der Kooij, M. (1999b), Production of DEMs from ERS tandem Data, in: *Second International Workshop on ERS SAR Interferometry, 'FRINGE99', Liège, Belgium, 10–12 Nov 1999*, ESA.

van der Kooij, M. W. A., van Halsema, D., Groenewoud, W., Mets, G. J., Overgaauw, B. and Visser, P. N. A. M. (1995), *SAR Land Subsidence Monitoring*, vol. NRSP-2 95-13, Beleidscommissie Remote Sensing, Delft.

Krul, L. (1982), Principles of radar measurement, in: *Radar calibration, Proceedings of EARSeL Workshop, Alpbach, Austria, 6–10 December 1982*, pp. 11–20.

Kuettner, J. P. (1971), Cloud bands in the earth's atmosphere, *Tellus*, **23**:404–425.

Kursinski, E. R. (1997), *The GPS radio occultation concept: theoretical performance and initial results*, Ph.D. thesis, California Institute of Technology.

Kursinski, E. R., Haij, G. A., Schofield, J. T. and Linfield, R. P. (1997), Observing Earth's atmosphere with radio occultation measurements using the Global Positioning System, *Journal of Geophysical Research*, **102**(D19):23,429–23,465.

Kwok, R. and Fahnestock, M. A. (1996), Ice Sheet Motion and Topography from Radar Interferometry, *IEEE Transactions on Geoscience and Remote Sensing*, **34**(1):189–200.

van Lammeren, A. C. A. P., Russchenberg, H. W. J., Apituley, A. and ten Brink, H. (1997), CLARA: a data set to study sensor synergy, in: *Proceedings Workshop on Synergy of Active Instruments in the Earth Radiation Mission, November 12-14, Geesthacht, Germany*, ESA EWP 1968 or GKSS 98/eE10.

Lanari, R., Fornaro, G., Riccio, D., Migliaccio, M., Papathanassiou, K. P., ao R Moreira, J., Schwäbisch, M., Dutra, L., Puglisi, G., Franceschetti, G. and Coltelli, M. (1996), Generation of Digital Elevation Models by Using SIR-C/X-SAR Multifrequency Two-Pass Interferometry: The Etna Case Study, *IEEE Transactions on Geoscience and Remote Sensing*, **34**(5):1097–1114.

Lanari, R., Lundgren, P. and Sansosti, E. (1998), Dynamic deformation of Etna volcano observed by satellite radar interferometry, *Geophysical Research Letters*, **25**:1541–1544.

Lay, O. P. (1997), The temporal power spectrum of atmospheric fluctuations due to water vapor, *Astronomy and Astrophysics Supplement Series*, **122**:535–545.

Lee, J. S., Hoppel, K. W., Mango, S. A. and Miller, A. R. (1994), Intensity and phase statistics of multilook polarimetric and interferometric SAR imagery, *IEEE Transactions on Geoscience and Remote Sensing*, **30**:1017.

Lehner, S., Horstmann, J., Koch, W. and Rosenthal, W. (1998), Mesoscale wind measurements using recalibrated ERS SAR images, *Journal of Geophysical Research*, **103**(C4):7847–7856.

LeMone, M. A. (1973), The structure and dynamics of horizontal roll vortices in the planetary boundary layer, *J. Atmos. Sci.*, **30**:1077–1091.

LeMone, M. A. and Pennell, W. T. (1976), The relationship of trade wind cumulus distribution to subcloudlayer fluxes and structure, *Monthly Weather Review*, **104**:524–539.

Li, F. and Goldstein, R. (1987), Studies of Multi-baseline Spaceborne Interferometric Synthetic Aperture Radar, in: *International Geoscience and Remote Sensing Symposium, Ann Arbor, 18–21 May 1987*, pp. 1545–1550.

Li, F. K. and Goldstein, R. M. (1990), Studies of Multibaseline Spaceborne Interferometric Synthetic Aperture Radars, *IEEE Transactions on Geoscience and Remote Sensing*, **28**(1):88–97.

Lin, Q., Vesecky, J. F. and Zebker, H. A. (1992), New approaches in Interferometric SAR data processing, *IEEE Transactions on Geoscience and Remote Sensing*, **30**(3):560–567.

Lu, Z., Fatland, R., Wyss, M., Li, S., Eichelberger, J., Dean, K. and Freymueller, J. (1997), Deformation of New Trident Volcano measured by ERS 1 SAR interferometry, Katmai National Park, Alaska, *Geophysical Research Letters*, **24**(6):695–698.

Lu, Z. and Freymueller, J. T. (1999), Synthetic aperture radar interferometry coherence analysis over Katmai volcano group, Alaska, *Journal of Geophysical Research*, **103**(B12):29887–29894.

Madsen, S. N. (1986), *Speckle Theory: Modelling, analysis, and applications related to Synthetic Aperture Radar Data*, Ph.D. thesis, Technical University of Denmark.

Madsen, S. N. (1989), Estimating the Doppler Centroid of SAR Data, *IEEE Transactions on Aerospace and Electronic Systems*, **25**(2):134–140.

Mandelbrot, B. B. (1983), *The Fractal Geometry of Nature*, Freeman, New York.

Mandelbrot, B. B. and Ness, J. W. V. (1968), Fractional Brownian motions, fractional noises and applications, *SIAM Review*, **10**(4):422–438.

Margot, J. L., Campbell, D. B., Jurgens, R. F. and Slade, M. A. (1999a), Topography of the Lunar Poles from Radar Interferometry: A Survey of Cold Trap Locations, *Science*, **284**:1658–1660.

Margot, J.-L., Campbell, D. B., Jurgens, R. F. and Slade, M. A. (1999b), The topography of Tycho Crater, *Journal of Geophysical Research*, **104**(E5):11875–11882.

Markoff, A. A. (1912), *Wahrscheinlichkeitsrechnung*, Teubner, Leibzig.

Massonnet, D. and Adragna, F. (1993), A full-scale validation of Radar Interferometry with ERS-1: the Landers earthquake, *Earth Observation Quarterly*, **41**.

Massonnet, D., Adragna, F. and Rossi, M. (1994a), CNES General-Purpose SAR Correlator, *IEEE Transactions on Geoscience and Remote Sensing*, **32**(3):636–643.

Massonnet, D., Briole, P. and Arnaud, A. (1995), Deflation of Mount Etna monitored by spaceborne radar interferometry, *Nature*, **375**:567–570.

Massonnet, D., Feigl, K., Rossi, M. and Adragna, F. (1994b), Radar interferometric mapping of deformation in the year after the Landers earthquake, *Nature*, **369**:227–230.

Massonnet, D. and Feigl, K. L. (1995a), Discrimination of geophysical phenomena in satellite radar interferograms, *Geophysical Research Letters*, **22**(12):1537–1540.

Massonnet, D. and Feigl, K. L. (1995b), Satellite radar interferometric map of the coseismic deformation field of the M = 6.1 Eureka Valley, California earthquake of May 17, 1993, *Geophysical Research Letters*, **22**(12):1541–1544.

Massonnet, D. and Feigl, K. L. (1998), Radar interferometry and its application to changes in the earth's surface, *Reviews of Geophysics*, **36**(4):441–500.

Massonnet, D., Feigl, K. L., Vadon, H. and Rossi, M. (1996a), Coseismic deformation field of the M=6.7 Northridge, California earthquake of January 17, 1994 recorded by two radar satellites using interferometry, *Geophysical Research Letters*, **23**(9):969–972.

Massonnet, D., Holzer, T. and Vadon, H. (1997), Land subsidence caused by the East Mesa geothermal field, California, observed using SAR interferometry, *Geophysical Research Letters*, **24**(8):901–904, Correction in GRL v.25,n.16,p.3213,1998.

Massonnet, D., Rossi, M., Carmona, C., Adagna, F., Peltzer, G., Feigl, K. and Rabaute, T. (1993), The displacement field of the Landers earthquake mapped by radar interferometry, *Nature*, **364**(8):138–142.

Massonnet, D., Thatcher, W. and Vadon, H. (1996b), Detection of postseismic fault zone collapse following the Landers earthquake, *Nature*, **382**:489–497.

Massonnet, D. and Vadon, H. (1995), ERS-1 Internal Clock Drift Measured by Interferometry, *IEEE Transactions on Geoscience and Remote Sensing*, **33**(2):401–408.

Massonnet, D., Vadon, H. and Rossi, M. (1996c), Reduction of the Need for Phase Unwrapping in Radar Interferometry, *IEEE Transactions on Geoscience and Remote Sensing*, **34**(2):489–497.

McCarthy, D. D., ed. (1996), *IERS Conventions (1996)*, IERS Technical Note 21, Observatoire de Paris.

Melsheimer, C., Alpers, W. and Gade, M. (1996), Investigation of multifrequency/multipolarization radar signatures of rain cells over the ocean using SIR-C/X-SAR data, submitted to JGR.

Meyer, B., Armijo, R., Massonnet, D., de Chabalier, J. B., Delacourt, C., Ruegg, J. C., Achache, J., Briole, P. and Panastassiou, D. (1996), The 1995 Grevena (Northern Greece) earthquake: fault model constrained with tectonic observations and SAR interferometry, *Geophysical Research Letters*, **23**:2677–2680.

Michel, R., Avouac, J.-P. and Taboury, J. (1999), Measuring ground displacement from SAR amplitude images: application to the Landers earthquake, *Geophysical Research Letters*, **26**(7):875–878.

Mogi, K. (1958), Relations Between Eruptions of Various Volcanoes and the Deformation of the Ground Surface Around Them, *Bulletin of the Earthquake Research Institute, University of Tokyo*, **36**:99–134.

Mohr, J. J. (1997), *Repeat Track SAR Interferometry. An investigation of its Utility for Studies of Glacier Dynamics*, Ph.D. thesis, Technical University of Denmark, Copenhagen.

Mohr, J. J., Reeh, N. and Madsen, S. N. (1998), Three-dimensional glacial flow and surface elevation measured with radar interferometry, *Nature*, **291**:273–276.

Monin, A. S., Yaglom, A. M. and Lumley, J. L. (1975), *Statistical Fluid Mechanics: Mechanics of Turbulence*, vol. 2, MIT Press, Cambridge, Original title: Statisicheskaya gidromekhanika – Mekhanika Turbulenosti.

Monti Guarnieri, A. and Prati, C. (1997), SAR Interferometry: A "Quick and Dirty" Coherence Estimator for Data Browsing, *IEEE Transactions on Geoscience and Remote Sensing*, **35**(3):660–669.

Mossop, A., Murray, M., Owen, S. and Segall, P. (1997), Subsidence at the Geysers geothermal field: results and simple models, in: *Twenty-Second Workshop on Geothermal Reservoir Engineering*, pp. 377–382, Stanford University, Jan 27-29, 1997.

Mouginis-Mark, P. J. (1995a), Analysis of volcanic hazards using Radar interferometry, *Earth Observation Quarterly*, **47**:6–10.

Mouginis-Mark, P. J. (1995b), Preliminary Observations of Volcanoes with the SIR-C Radar, *IEEE Transactions on Geoscience and Remote Sensing*, **33**(4):934–941.

Mourad, P. D. and Walter, B. A. (1996), Viewing a cold air outbreak using satellite-based synthetic aperture radar and advanced very high resolution radiometer imagery, *Journal of Geophysical Research*, **101**(C7):16391–16400.

Mueller, C. K., Wilson, J. W. and Crook, N. A. (1993), The utility of sounding and mesonet data to nowcast thunderstorm initiation, *Weather and Forecasting*, **8**:132–146.

Murakami, M., Tobita, M., Fujiwara, S. and Saito, T. (1996), Coseismic crustal deformations of 1994 Northridge, California, earthquake detected by interferometric JERS-1 Synthetic aperture radar, *Journal of Geophysical Research*, **101**(B4):8605–8614.

NAM (2000), *Bodemdaling door Aardgaswinning: Groningen veld en randvelden in Groningen, Noord-Drenthe en het Oosten van Friesland. Status Rapport 2000 en Prognose tot het jaar 2050.*, NAM rapport nr. 2000-02-00410, Nederlandse Aardolie Maatschappij B.V, Assen, in Dutch.

Newland, D. E. (1993), *Random vibrations, spectral and wavelet analysis*, Longman, Harlow, 3rd edn.

Niell, A. E. (1996), Global mapping functions for the atmosphere delay at radio wavelengths, *Journal of Geophysical Research*, **101**(B2):3227–3246.

NorthWest Research Associates, I. (2000), Atmospheric Sciences, (Accessed 15 Jun, 2000) <http://www.nwra.com/>.

Oberhettinger, F. (1970), Hypergeometric Functions, in: Abramowitz and Stegun (1970), chap. 15, pp. 555–566.

Obukov, A. M. (1941), On the distribution of energy in the spectrum of turbulent flow, *Doklady Akad. Nauk SSSR*, **32**(19).

Ohkura, H. (1998), Applications of SAR data to monitoring Earth surface changes and displacements, *Advances in Space Research*, **21**(3):485–492.

Olmsted, C. (1993), *Alaska SAR Facility Scientific SAR User's Guide*.

Otten, M. P. G. (1998), Review of SAR processing techniques, in: *28th EuMC workshop proceedings*, pp. 33–41, EuMW Amsterdam, 9 October 1998.

Ozawa, S., Murakami, M., Fujiwara, S. and Tobita, M. (1997), Synthetic aperture radar interferogram of the 1995 Kobe earthquake and its geodetic inversion, *Geophysical Research Letters*, **24**(18):2327–2330.

Papoulis, A. (1968), *Systems and Tronsforms with Applications in Optics*, McGraw-Hill series in Systems Science, McGraw-Hill, New York.

Papoulis, A. (1991), *Probability, Random variables, and stochastic processes*, McGraw-Hill series in Electrical Engineering, McGraw-Hill, New York.

Park, S. K. and Schowengerdt, R. A. (1983), Image Reconstruction by Parametric Cubic Convolution, *Computer Vision, Graphics and Image Processing*, **23**(3):258–272.

Parsons, D. B. (1992), An Explanation for Intense Frontal Updrafts and Narrow Cold-Frontal Rainbands, *Journal of the Atmospheric Sciences*, **49**(19):1810–1825.

Parsons, T., Toda, S., Stein, R. S., Barka, A. and Dieterich, J. H. (2000), Heightened Odds of Large Earthquakes Near Istanbul: An Interaction-Based Probability Calculation, *Science*, **288**:661–665.

Peitgen, H. O. and Saupe, D., eds. (1988), *The Science of Fractal Images*, Springer-Verlag, New York.

Peltzer, G., Crampé, F. and King, G. (1999), Evidence of the Nonlinear Elasticity of the Crust from Mw7.6 Manyi (Tibet) Earthquake, *Science*, **286**(5438):272–276.

Peltzer, G., Hudnut, K. W. and Feigl, K. L. (1994), Analysis of coseismic surface displacement gradients using radar interferometry: New insights into the Landers earthquake, *Journal of Geophysical Research*, **99**(B11):21971–21981.

Peltzer, G. and Rosen, P. (1995), Surface Displacement of the 17 May 1993 Eureka Valley, California Earthquake Observed by SAR Interferometry, *Science*, **268**:1333–1336.

Peltzer, G., Rosen, P., Rogez, F. and Hudnut, K. (1996), Postseismic rebound in fault step-overs caused by pore fluid flow, *Science*, **273**:1202–1204.

Prati, C. and Rocca, F. (1990), Limits to the resolution of elevation maps from stereo SAR images, *Int. J. Remote Sensing*, **11**(12):2215–2235.

Prati, C. and Rocca, F. (1992), Range Resolution Enhancement with multiple SAR Surveys Combination, in: *International Geoscience and Remote Sensing Symposium, Houston, Texas, USA, May 26–29 1992*, pp. 1576–1578.

Prati, C. and Rocca, F. (1993), Improving Slant-Range Resolution With Multiple SAR Surveys, *IEEE Transactions on Aerospace and Electronic Systems*, **29**(1):135–144.

Prati, C. and Rocca, F. (1994a), DEM generation with ERS-1 interferometry, in: Sansò, F., ed., *Geodetic Theory Today*, pp. 19–26, Berlin, Third Hotine-Marussi Symposium on Mathematical Geodesy, Springer-Verlag.

Prati, C. and Rocca, F. (1994b), Use of the spectral shift in SAR interferometry, in: *Second ERS-1 Symposium—Space at the Service of our Environment, Hamburg, Germany, 11–14 October 1993*, ESA SP-361, pp. 691–696.

Prati, C., Rocca, F. and Monti Guarnieri, A. (1989), Effects of speckle and additive noise on the altimetric resolution of interferometric SAR (ISAR) surveys, in: *Proc. IGARSS'89, Vancouver*, pp. 2469–2472.

Prati, C., Rocca, F. and Monti Guarnieri, A. (1993), SAR interferometry experiments with ERS-1, in: *First ERS-1 Symposium—Space at the Service of our Environment, Cannes, France, 4–6 November 1992*, ESA SP-359, pp. 211–217.

Prati, C., Rocca, F., Monti Guarnieri, A. and Damonti, E. (1990), Seismic Migrataion For SAR Focussing: Interferometrical Applications, *IEEE Transactions on Geoscience and Remote Sensing*, **28**(4):627–640.

Price, E. J. (1999), *Coseismic and Postseismic Deformations Associated With the 1992 Landers, California Earthquake Measured by Synthetic Aperture Radar Interferometry*, Ph.D. thesis, University of California, San Diego.

Price, E. J. and Sandwell, D. T. (1998), Small-scale deformations associated with the 1992 Landers, California, earthquake mapped by synthetic aperture radar interferometry phase gradients, *Journal of Geophysical Research*, **103**(B11):27001–27016.

Priestley, M. B. (1981), *Spectral Analysis and Time Series*, Academic Press, London.

Pritt, M. D. (1996), Phase Unwrapping by Means of Multigrid Techniques for Interferometric SAR, *IEEE Transactions on Geoscience and Remote Sensing*, **34**(3):728–738.

Radarsat (1999), RADARSAT Specifications, (Accessed 2 Nov, 1999), <http://radarsat.space.gc.ca/info/description/specifications.html>.

Raney, R. K. (1998), Radar Fundamentals: Technical Perspective, in: Henderson and Lewis

(1998), chap. 2, pp. 9–130.

Raney, R. K., Runge, H., Bamler, R., Cumming, I. G. and Wong, F. H. (1994), Precision SAR Processing Using Chirp Scaling, *IEEE Transactions on Geoscience and Remote Sensing*, **32**(4):786–799.

Reilinger, R. E., Ergintav, S., Bürgmann, R., McClusky, S., Lenk, O., Barka, A., Gurkan, O., Hearn, L., Feigl, K. L., Cakmak, R., Aktug, B., Ozener, H. and Töksoz, M. N. (2000), Coseismic and Postseismic Fault Slip for the 17 August 1999, M = 7.5, Izmit, Turkey Earthquake, *Science*, **289**(5484):1519–1524.

Resch, G. M. (1984), Water Vapor Radiometry in Geodetic Applications, in: Brunner, F. K., ed., *Geodetic refraction*, pp. 53–84, Springer-Verlag, Berlin.

Richelson, J. T. (1991), The future of Space Reconnaisance, *Scientific American*, **264**(1):38–45.

Richman, D. (1982), Three-dimensional Azimuth-correcting Mapping Radar, *United States Patent, No. 4321601*, Orgininally filed in 1971.

Rignot, E. J., Gogineni, S. P., Krabill, W. B. and Ekholm, S. (1997), North and northeast Greenland ice discharges from satellite radar interferometry, *Science*, **276**:934–937.

Rind, D., Chiou, E. W., Chu, W. P., Larsen, J. C., Oltmans, S., Lerner, J., McCormick, M. P. and McMaser, L. R. (1991), Positive water vapor feedback in climate models confirmed by satellite data, *Nature*, **349**:7500–7503.

Riva, R., Hanssen, R., Vermeersen, L. L. A., Aoudia, K. and Sabadini, R. (2000), Coseismic deformation of the 17 Aug 1999, Izmit, Turkey earthquake from normal mode modeling and satellite radar interferometry (SAR) imaging, in: *EGS XXV General Assembly, Nice, France, 25-29 April 1999, Geophys. Res Abstracts, 2*.

Rodriguez, E. and Martin, J. M. (1992), Theory and design of interferometric synthetic aperture radars, *IEE Proceedings-F*, **139**(2):147–159.

Rogers, A. E. and Ingalls, R. P. (1969), Venus: Mapping the surface reflectivity by Radar Interferometry, *Science*, **165**(3895):797–799.

Rosen, P., Werner, C., Fielding, E., Hensley, S. and Vincent, S. B. P. (1998), Aseismic creep along the San Andreas fault northwest of Parkfield, CA measured by radar interferometry, *Geophysical Research Letters*, **25**(6):825–828.

Rosen, P. A., Hensley, S., Zebker, H. A., Webb, F. H. and Fielding, E. J. (1996), Surface deformation and coherence measurements of Kilauea Volcano, Hawaii, from SIR-C radar interferometry, *Journal of Geophysical Research*, **101**(E10):23109–23125.

Roth, A., Adam, N., Schwäbisch, M., Müschen, B., Böhm, C. and Lang, O. (1997), Observation of the effects of the subglacial volcano eruption underneath the Vatnajökull glacier in Iceland with ERS-SAR data, in: *Proc. of the third ERS symposium, Florence, Italy, 17-20 March 1997*.

Rothacher, M. and Mervart, L., eds. (1996), *Bernese GPS Software version 4.0*, Astronomical Institute University of Berne, Berne.

Rott, H., Schechl, B., Siegel, A. and Grasemann, B. (1999), Monitoring very slow slope movements by means of SAR interferometry: A case study from a mass waste above a reservoir in the Ötztal Alps, Austria, *Geophysical Research Letters*, **26**(11):1629–1632.

Rott, H. and Siegel, A. (1997), Glaciological Studies in the Alps and in Antarctica Using ERS Interferometric SAR, in: *'FRINGE 96' workshop on ERS SAR Interferometry, Zürich, Switzerland, 30 Sep–2 October 1996*, pp. 149–159, ESA SP-406, Vol II.

Rott, H., Stuefer, M., Siegel, A., Skvarca, P. and Eckstaller, A. (1998), Mass fluxes and dynamics of Moreno Glacier, Southern Patagonia Icefield, *Geophysical Research Letters*, **25**(9):1407–1410.

Ruf, C. S. and Beus, S. E. (1997), Retrieval of Tropospheric Water Vapor Scale Height from Horizontal Turbulence Structure, *IEEE Transactions on Geoscience and Remote*

Sensing, **35**(2):203–211.

Rummel, R. (1990), *Physical Geodesy 2, Lecture Notes*, Delft University of Technology, Department of Geodetic Engineering, Delft.

Runge, H. and Bamler, R. (1992), A novel high-precision SAR processing algorithm based on chirp scaling, in: *International Geoscience and Remote Sensing Symposium, Houston, Texas, USA, May 26–29 1992*.

Russchenberg, H. (2000), Personal communication.

Ryle, M. and Vonberg, D. D. (1946), Solar radiation on 175 Mc/s, *Nature*, **158**:339–340.

Saastamoinen, J. (1972), Introduction to Practical Computation of Astronomical Refraction, *Bulletin Geodesique*, **106**:383–397.

Saito, A., Fukao, S. and Miyazaki, S. (1998), High resolution mapping of TEC perturbations with the GSI GPS network over Japan, *Geophysical Research Letters*, **25**(16):3079–3082.

Samson, J. (1996), *Coregistration in SAR interferometry*, Master's thesis, Faculty of Geodetic Engineering, Delft University of Technology.

Sandwell, D. T. and Price, E. J. (1998), Phase gradient approach to stacking interferograms, *Journal of Geophysical Research*, **103**(B12):30183–30204.

Sandwell, D. T. and Sichoix, L. (2000), Topographic Phase Recovery from Stacked ERS Interferometry and a Low Resolution Digital Elevation Model, *Journal of Geophysical Research*, **105**(B12):28211–28222.

Sandwell, D. T., Sichoix, L., Agnew, D., Bock, Y. and Minster, J.-B. (2000), Near real-time radar interferometry of the Mw 7.1 Hector Mine Earthquake, *Geophysical Research Letters*, **27**(19):3101–3104.

Scharroo, R. and Visser, P. (1998), Precise orbit determination and gravity field improvement for the ERS satellites, *Journal of Geophysical Research*, **103**(C4):8113–8127.

Schmetz, J., Menzel, W., Velden, C., Wu, X., van de Berg, L., Nieman, S., Hayden, C., Holmlund, K. and Geijo, C. (1995), Monthly mean Large-scale analyses of upper-tropospheric humidity and wind field divergence derived from three Geostationary Satellites, *Bulletin of the American Meteorological Society*, **76**(9):1578–1584.

Schwäbisch, M. (1995a), *Die SAR-Interferometrie zur Erzeugung digitaler Geländemodelle*, Ph.D. thesis, Stuttgart University.

Schwäbisch, M. (1995b), Die SAR-Interferometrie zur Erzeugung digitaler Geländemodelle, Forschungsbericht 95-25, Deutsche Forschungsanstalt für Luft- un Raumfahrt, Oberpfaffenhofen.

Schwäbisch, M. and Geudtner, D. (1995), Improvement of Phase and Coherence Map Quality Using Azimuth Prefiltering: Examples from ERS-1 and X-SAR, in: *International Geoscience and Remote Sensing Symposium, Florence, Italy, 10–14 July 1995*, pp. 205–207.

Segall, P. (1985), Stress and Subsidence Resulting From Subsurface Fluid Withdrawal in the Epicentral Region of the 1983 Coalinga Earthquake, *Journal of Geophysical Research*, **90**(B8):6801–6816.

Segall, P. and Davis, J. L. (1997), GPS applications for geodynamics and Earthquake studies, *Annu. Rev. Earth Planet. Sci.*, **25**:301–336.

Seymour, M. S. and Cumming, I. G. (1994), Maximum Likelyhood Estimation For SAR Interferometry, in: *International Geoscience and Remote Sensing Symposium, Pasadena, CA, USA, 8–12 August 1994*, pp. 2272–2275.

Shapiro, I. I., Zisk, S. H., Rogers, A. E. E., Slade, M. A. and Thompson, T. W. (1972), Lunar Topography: Global Determination by Radar, *Science*, **178**(4064):939–948.

Sigmundsson, F., Durand, P. and Massonnet, D. (1999), Opening of an eruptive fissure and seaward displacement at Piton de la Fournaise volcano measured by RADARSAT satellite radar interferometry, *Geophysical Research Letters*, **26**(5):533.

Sigmundsson, F., Vadon, H. and Massonnet, D. (1997), Readjustment of the Krafla spreading segment to crustal rifting measured by Satellite Radar Interferometry, *Geophysical Research Letters*, **24**(15):1843–1846.

Simpson, J. E. (1994), *Sea Breeze and Local Wind*, Cambridge University Press, Cambridge.

Skolnik, M. I. (1962), *Introduction to Radar Systems*, McGraw-Hill Kogakusha, Ltd., Tokyo.

Small, D., Werner, C. and Nüesch, D. (1993), Baseline Modelling for ERS-1 SAR Interferometry, in: *International Geoscience and Remote Sensing Symposium, Tokyo, Japan, 18–21 August 1993*, pp. 1204–1206.

Smith, Jr., E. K. and Weintraub, S. (1953), The Constants in the Equation for Atmospheric Refractive Index at Radio Frequencies, *Proceedings of the I.R.E.*, **41**:1035–1037.

Smith, F. G., Little, C. G. and Lovell, A. C. B. (1950), Origin of the fluctuations in the intensity of radio waves from galactic sources, *Nature*, **165**:422–424.

Solheim, F. S., Vivekanandan, J., Ware, R. H. and Rocken, C. (1997), Propagation Delays Induced in GPS Signals by Dry Air, Water Vapor, Hydrometeors and Other Particulates, *draft for J. of Geoph. Res.*

Spoelstra, T. A. T. (1997), Personal communication.

Spoelstra, T. A. T. and Yi-Pei, Y. (1995), Ionospheric scintillation observations with radio interferometry, *Journal of Atmospheric and Terrestrial Physics*, **57**(1):85–97.

SPOT image company (1995), Presentation of SPOT Image Company, (Last modified at 2 Oct 1995), <http://www.orstom.fr/hapex/spot/spotimg.htm>.

Stein, R. S. (1999), The role of stress transfer in earthquake occurence, *Nature*, **402**:605–609.

Stein, R. S., Barka, A. A. and Dieterich, J. H. (1997), Progressive failure on the North Anatolian fault since 1939 by earthquake stress triggering, *Geophysical Journal International*, **128**:594–604.

Stoffelen, A. C. M. (1998), *Scatterometry*, Ph.D. thesis, Utrecht University.

Stolk, R., Hanssen, R., van der Marel, H. and Ambrosius, B. (1997), Troposferische signaalvertraging; een vergelijking tussen GPS en SAR interferometrie, *GPS Nieuwsbrief*, **2**:3–7, In Dutch.

Stolk, R. M. (1997), *GPS and SAR interferometry, a comparative analysis of estimating tropospheric propagation delays: The GISARE GPS Experiment*, Master's thesis, Delft University of Technology, Delft.

Stotskii, A. A. (1973), Concerning the Fluctuation Characteristics of the Earth's Atmosphere, *Radiophysics and quantum electronics*, **16**:620–622.

Stramondo, S., Tesauro, M., Briole, P., Sansosti, E., Salvi, S., Lanari, R., Anzidei, M., Baldi, P., Fornaro, G., Avallone, A., Buongiorno, M. F. and Boschi, G. F. E. (1999), The September 26, 1997 Colfiorito, Italy, earthquakes: modeled coseismic surface displacement from SAR interferometry and GPS, *Geophysical Research Letters*, **26**(7):883–886.

Strang, G. (1986), A Proposal for Toeplitz Matrix Calculations, *Studies in applied mathematics*, **74**:171–176.

Strohmer, T. (1997), Computationally Attractive Reconstruction of Bandlimited Images from Irregular Samples, *IEEE Transactions on Geoscience and Remote Sensing*, **6**(4):540–548.

Stull, R. B. (1995), *Meteorology Today For Scientists and Engineers*, West Publishing, St. Paul, Minneapolis.

Susskind, Rosenfield, J. J., Reuter, D. and Chahine, M. T. (1984), Remote sensing of weather and climate parameters from HIRS2/MSU on TIROS-N, *Journal of Geophysical Research*, **89D**:4677–4697.

Swart, L. M. T. (2000), *Spectral filtering and oversampling for radar interferometry*, Master's thesis, Faculty of Geodetic Engineering, Delft University of Technology.

Tarayre, H. (1996), *Extraction de Modeles Numeriques de Terrain par Interferometrie Radar Satellitaire: Algorithmie et Artefacts Atmospheriques*, Ph.D. thesis, Institut National Polytechnique de Toulouse, Toulouse.

Tarayre, H. and Massonnet, D. (1996), Atmospheric propagation heterogeneities revealed by ERS-1 interferometry, *Geophysical Research Letters*, **23**(9):989–992.

Tatarski, V. I. (1961), *Wave propagation in a turbulent medium*, McGraw-Hill, New York, Translated from Russian.

Taverne, B. G. (1993), *Beginselen en voorbeelden van regelgeving en beleid ten aanzien van de opsporing en winning van aardolie en aardgas*, Delft University Press, Delft, in Dutch.

Taylor, G. I. (1938), The spectrum of turbulence, *Proceedings of the Royal Society London, Series A*, **CLXIV**:476–490.

TDN/MD (1997), TopHoogteMD Digital Elevation Model, Topografische Dienst Nederland and Meetkundige Dienst Rijkswaterstaat.

Tesauro, M., Berardino, P., Lanari, R., Sansosti, E., Fornaro, G. and Franceschetti, G. (2000), Urban subsidence inside the city of Napoli (Italy) observed by satellite radar interferometry, *Geophysical Research Letters*, **27**(13):1961.

Teunissen, P. J. G. (2000), *Adjustment theory; an introduction*, Delft University Press, Delft, 1st edn.

Thatcher, W. and Massonnet, D. (1996), Crustal deformation at Long Valley Caldera. eastern California, 1992-1996 inferred from satellite radar interferometry, *Geophysical Research Letters*, **24**(20):2519–2522.

Thayer, G. D. (1974), An Improved Equation for the Radio Refractive Index of Air, *Radio Science*, **9**:803–807.

Thiel, K.-H., Wu, X. and Hartl, P. (1997), ERS-tandem-interferometric obsertvation of volcanic activities in Iceland, in: *Third ERS Symposium—Space at the Service of our Environment, Florence, Italy, 17–21 March 1997*, ESA SP-414, pp. 475–480.

Thompson, A. R., Moran, J. M. and Swenson, G. W. (1986), *Interferometry and Synthesis in Radio Astronomy*, Wiley-Interscience, New York.

Tough, R. J. A., Blacknell, D. and Quegan, S. (1995), A statistical description of polarimetric and interferometric synthetic aperture radar, *Proceeedings of the Royal Society London A*, **449**:567–589.

Touzi, R. and Lopes, A. (1996), Statistics of the Stokes Parameters and of the Complex Coherence Parameters in One-Look and Multilook Speckle Fields, *IEEE Transactions on Geoscience and Remote Sensing*, **34**(2):519–531.

Touzi, R., Lopes, A., Bruniquel, J. and Vachon, P. W. (1999), Coherence Estimation for SAR Imagery, *IEEE Transactions on Geoscience and Remote Sensing*, **1**(37):135–149.

Touzi, R., Lopes, A. and Vachon, P. W. (1996), Estimation of the coherence function for interferometric SAR applications, in: *EUSAR'96, Königswinter, Germany*, pp. 241–244.

Treuhaft, R. N. and Lanyi, G. E. (1987), The effect of the dynamic wet troposphere on radio interferometric measurements, *Radio Science*, **22**(2):251–265.

Tubbs, B. (1997), Astronomical Optical Interferometry, A Literature Review, (Updated 2000), <http://www.cus.cam.ac.uk/~rnt20/interferometry/ast_opt_int/page1.html>.

Turcotte, D. L. (1997), *Fractals and chaos in geology and geophysics*, Cambridge University Press, Cambridge, 2nd edn.

Ulaby, F. T., Moore, R. K. and Fung, A. K. (1982), *Microwave remote sensing: active and passive. Vol. 2. Radar remote sensing and surface scattering and emission theory*, Addison-Wesley, Reading.

Usai, S. (1997), The use of man-made features for long time scale INSAR, in: *International Geoscience and Remote Sensing Symposium, Singapore, 3–8 Aug 1997*, pp. 1542–1544.

Usai, S., Gaudio, C. D., Borgstrom, S. and Achilli, V. (1999), Monitoring Terrain Deformations at Phlegrean Fields with SAR Interferometry, in: *Second International Workshop
on ERS SAR Interferometry, 'FRINGE99', Liège, Belgium, 10–12 Nov 1999*, pp. 1–5,
ESA.

Usai, S. and Hanssen, R. (1997), Long time scale INSAR by means of high coherence
features, in: *Third ERS Symposium—Space at the Service of our Environment, Florence,
Italy, 17–21 March 1997*, pp. 225–228.

Usai, S. and Klees, R. (1998), On the Feasibility of Long Time Scale INSAR, in: *International Geoscience and Remote Sensing Symposium, Seattle, Washington, USA, 6–10
July 1998*, pp. 2448–2450.

Usai, S. and Klees, R. (1999a), On the interferometric characteristics of anthropogenic
features, in: *International Geoscience and Remote Sensing Symposium, Hamburg, Germany, 28 June–2 July 1999*.

Usai, S. and Klees, R. (1999b), SAR Interferometry On Very Long Time Scale: A Study
of the Interferometric Characteristics Of Man-Made Features, *IEEE Transactions on
Geoscience and Remote Sensing*, **37**(4):2118–2123.

Visser, P. N. A. M., Scharroo, R., Floberghagen, R. and Ambrosius, B. A. C. (1997),
Impact of PRARE on ERS-2 orbit determination, in: *Proceedings of 12th International
Symposium on "Space Flight Dynamics", Darmstadt, Germany, 2-6 June 1997*, pp. 115–
120, SP3-403.

Voss, R. F. (1988), Fractals in nature: From characterization to simulation, in: Peitgen
and Saupe (1988), pp. 21–70.

Wackernagel, H. (1995), *Multivariate Geostatistics: An Introduction with Applications*,
Springer-Verlag, Berlin.

Wadge, G., Scheuchl, B. and Stevens, N. F. (1999), Spaceborne radar measurements of
the eruption of Soufriere Hills Volcano, Montserrat during 1996-99, Submitted to J.
Volcanology and Geothermal Research.

Weckwerth, T. M., Wilson, J. W. and Wakimoto, R. M. (1996), Thermodynamic Variability
within the Convective Boundary Layer Due to Horizontal Convective Rolls, *Monthly
Weather Review*, **124**(5):769–784.

Weckwerth, T. M., Wilson, J. W., Wakimoto, R. M. and Crook, N. A. (1997), Horizontal
Convective Rolls: Determining the environmental conditions supporting their existence
and characteristics, *Monthly Weather Review*, **125**(4):505–526.

Weckwerth, T. M., Wulfmeyer, V., Wakimoto, R. M., Hardesty, R. M., Wilson, J. W.
and Banta, R. M. (1999), NCAR/NOAA Lower-Tropospheric Water Vapor Workshop,
Bulletin of the American Meteorological Society, **80**(11):2339–2357.

Wegmüler, U., Werner, C. L., Nüesch, D. and Borgeaud, M. (1995), Land-Surface Analysis
Using ERS-1 SAR Interferometry, *ESA Bulletin*, **81**:30–37.

Wegmüller, U. and Werner, C. (1997), Retrieval of Vegetation Parameters with SAR Interferometry, *IEEE Transactions on Geoscience and Remote Sensing*, **35**(1):18–24.

Wegmüller, U. and Werner, C. L. (1995), SAR Interferometric Signatures of Forest, *IEEE
Transactions on Geoscience and Remote Sensing*, **33**(5):1153–1161.

Wegmüller, U., Werner, C. L., Nüesch, D. and Borgeaud, M. (1995), Forest Mapping Using
ERS Repeat-Pass SAR Interferometry, *Earth Observation Quarterly*, **49**.

Weldon, R. B. and Holmes, S. J. (1991), Water Vapor Imagery: Interpretation and Applications to Weather Analysis and Forecasting, Tech. Rep. NESDIS 57, NOAA, Washington.

Werner, C. L., Hensley, S., Goldstein, R. M., Rosen, P. A. and Zebker, H. A. (1993),
Techniques and applications of SAR interferometry for ERS-1: Topographic mapping,
change detection and slope measurement, in: *First ERS-1 Symposium—Space at the
Service of our Environment, Cannes, France, 4-6 November 1992*, ESA SP-359, pp.

205–210.

Werner, M. (1998), Shuttle Radar Topography Mission (SRTM). The X-Band SAR Interferometer, in: *28th EuMC workshop proceedings*, EuMW Amsterdam, 9 October 1998.

Wessel, P. and Smith, W. H. F. (1998), New, Improved Version of Generic Mapping Tools Released, *EOS Transactions, AGU*, **79**(47):579.

Weydahl, D. J. (1991), Change detection in SAR images, in: *International Geoscience and Remote Sensing Symposium, Espoo, Finland, 3–6 June 1991*, pp. 1421–1424.

Wicks, Jr, C., Thatcher, W. and Dzurisin, D. (1998), Migration of Fluids Beneath Yellowstone Caldera Inferred from Satellite Radar Interferometry, *Science*, **282**:458–462.

Wiley, C. A. (1954), Pulsed Doppler Radar Methods and Apparatus, *United States Patent, No. 3196436*.

Williams, C. A. and Wadge, G. (1998), The effects of topography on magma chamber deformation models: Application to Mt. Etna and radar interferometry, *Geophysical Research Letters*, **25**(10):1549–1552.

Williams, S., Bock, Y. and Fang, P. (1998), Integrated satellite interferometry: Tropospheric noise, GPS estimates and implications for interferometric synthetic aperture radar products, *Journal of Geophysical Research*, **103**(B11):27,051–27,067.

Wu, C., Barkan, B., Huneycutt, B., Leans, C. and Pang, S. (1981), An introductoin to the interim digital SAR processor and the characteristics of the associated Seasat SAR imagery, *JPL Publications, 81-26*, p. 123.

Wu, X. (1996), *Anwendung der Radarinterferometrie zur Bestimmung der Topographie und der Geschwindigkeitsfelder der Eisoberflächen antarktischer Gebiete*, Ph.D. thesis, Universität Stuttgart.

Yocky, D. A. and Johnson, B. F. (1998), Repeat-Pass Dual-Antenna Synthetic Aperture Radar Interferometric Change-Detection Post-Processing, *Photogrammetric Engineering and Remote Sensing*, pp. 425–429.

Zebker, H. (1996), Imaging Radar and Applications, Class notes EE355.

Zebker, H., Rosen, P., Hensley, S. and Mouginis-Mark, P. (1996), Analysis of active lava flows on Kilauea volcano, Hawaii, using SIR-C radar correlation measurements, *Geology*, **24**:495–498.

Zebker, H. A., Farr, T. G., Salazar, R. P. and Dixon, T. H. (1994a), Mapping the World's Topography Using Radar Interferometry: The TOPSAT Mission, *Proceedings of the IEEE*, **82**(12):1774–1786.

Zebker, H. A. and Goldstein, R. M. (1986), Topographic Mapping From Interferometric Synthetic Aperture Radar Observations, *Journal of Geophysical Research*, **91**(B5):4993–4999.

Zebker, H. A. and Lu, Y. (1998), Phase unwrapping algorithms for radar interferometry: residue-cut least-squares, and synthesis algorithms, *Journal of the Optical Society of America A.*, **15**(3):586–598.

Zebker, H. A., Madsen, S. N., Martin, J., Wheeler, K. B., Miller, T., Lou, Y., Alberti, G., Vetrella, S. and Cucci, A. (1992), The TOPSAR Interferometric Radar Topographic Mapping Instrument, *IEEE Transactions on Geoscience and Remote Sensing*, **30**(5):933–940.

Zebker, H. A., Rosen, P. A., Goldstein, R. M., Gabriel, A. and Werner, C. L. (1994b), On the derivation of coseismic displacement fields using differential radar interferometry: The Landers earthquake, *Journal of Geophysical Research*, **99**(B10):19617–19634.

Zebker, H. A., Rosen, P. A. and Hensley, S. (1997), Atmospheric effects in interferometric synthetic aperture radar surface deformation and topographic maps, *Journal of Geophysical Research*, **102**(B4):7547–7563.

Zebker, H. A. and Villasenor, J. (1992), Decorrelation in interferometric radar echoes, *IEEE*

Transactions on Geoscience and Remote Sensing, **30**(5):950–959.

Zebker, H. A., Werner, C. L., Rosen, P. A. and Hensley, S. (1994c), Accuracy of Topographic Maps Derived from ERS-1 Interferometric Data, *IEEE Transactions on Geoscience and Remote Sensing,* **32**(4):823–836.

Zisk, S. H. (1972a), Lunar Topography: First Radar-interferometer Measurements of the Alphonsus-Arzachel Region, *Science,* **178**(4064):977–980.

Zisk, S. H. (1972b), A New Earth-based Radar Technique for the Measurement of Lunar Topography, *Moon,* **4**:296–306.

About the Author

Ramon Hanssen (1968) studied aerospace engineering (1987-1989) and geodetic engineering (1989-1993) at Delft University of Technology (M.Sc. 1993). His graduation work focused on the use of the Radon transformation for the analysis of sparsely sampled data, with applications in satellite altimetry and imaging tomography. In 1994, he worked as a researcher on potential field (gravity and aeromagnetics) data inversion at the International Institute for Aerospace Surveys and Earth Sciences (ITC). In 1995, he started his PhD research at the Delft Institute for Earth-Oriented Space research (DEOS) of Delft University of Technology on the geodetic analysis of repeat-pass spaceborne radar interferometry, with emphasis on error propagation. During this research, he worked as a visiting scholar at Stuttgart University (1996), at the German Aerospace Center (DLR, 1997) and at Stanford University (1997-1998) on a Fulbright research fellowship. In 1997, he received an ESA grant to study the effect of atmospheric heterogeneities for ERS-1/2 radar interferometry, a project performed in close collaboration with the Royal Netherlands Meteorological Institute (KNMI). At present, he is employed as an assistant-professor in the field of geostatistics at the Faculty of Civil Engineering and Geosciences at DUT. His current research interests are the geodetic analysis of two-dimensional survey data, the influence of atmospheric delay on space-geodetic techniques, and the mathematical modeling and physical interpretation of deformation processes.

Index